Some modern methods of organic synthesis

Cambridge Chemistry Texts

GENERAL EDITORS

E. A. V. Ebsworth, Ph.D.
Professor of Inorganic Chemistry
University of Edinburgh

D. T. Elmore, Ph.D.
Reader in Biochemistry
Queen's University of Belfast

P. J. Padley, Ph.D.
Lecturer in Physical Chemistry
University College of Swansea

K. Schofield, D.Sc.
Reader in Organic Chemistry
University of Exeter

Some modern methods of organic synthesis

W. CARRUTHERS

Chemistry Department
University of Exeter

CAMBRIDGE
at the University Press, 1971

58221

Published by the Syndics of the Cambridge University Press
Bentley House, 200 Euston Road, London N.W.1
American Branch: 32 East 57th Street, New York, N.Y. 10022

© Cambridge University Press 1971

Library of Congress Catalogue Card Number: 71-149427

ISBN: 0 521 08145 9   Clothbound
      0 521 09643 X   Paperback

Printed in Great Britain
by William Clowes & Sons Limited
London, Colchester and Beccles

Contents

Preface   *page* ix

1   Formation of carbon–carbon single bonds   1
    1.1.   *Alkylation; importance of enolate anions*   1
    1.2.   *Alkylation, alkenylation and arylation of relatively acidic
           methylene groups*   4
    1.3.   *γ-Alkylation of 1,3-dicarbonyl compounds*   13
    1.4.   *Alkylation of β-keto-sulphoxides and -sulphones*   15
    1.5.   *Alkylation of ketones*   17
    1.6.   *The enamine reaction*   29
    1.7.   *Reactions of bisthio carbanions*   36
    1.8.   *The dihydro-1,3-oxazine synthesis of aldehydes and ketones*   38
    1.9.   *1,4-Addition of organometallic compounds to αβ-unsatu-
           rated ketones. Lithium dialkyl- and diaryl- cuprates*   41
    1.10. *Coupling of organonickel and organocopper complexes*   45
    1.11. *Synthesis of 1,5-dienes from allyl compounds*   51
    1.12. *Synthetic applications of carbenes and carbenoids*   53
    1.13. *Some photocyclisation reactions*   63

2   Formation of carbon–carbon double bonds   71
    2.1.   *β-Elimination reactions*   71
    2.2.   *Pyrolytic* syn-*eliminations*   75
    2.3.   *The Wittig and related reactions*   81
    2.4.   *Stereoselective synthesis of tri- and tetra-substituted
           ethylenes*   95
    2.5.   *Fragmentation reactions*   106
    2.6.   *Oxidative decarboxylation of carboxylic acids*   108
    2.7.   *Decomposition of toluene-p-sulphonylhydrazones*   110
    2.8.   *Stereospecific synthesis from 1,2-diols*   111
    2.9.   *Claisen rearrangement of allyl vinyl ethers*   112

3   The Diels–Alder reaction   115
    3.1.   *General*   115
    3.2.   *The dienophile*   116
    3.3.   *The diene*   127

3.4. *The retro Diels–Alder reaction* 144
3.5. *Catalysis by Lewis acids* 146
3.6. *Stereochemistry of the Diels–Alder reaction* 147
3.7. *Mechanism of the Diels–Alder reaction* 160
3.8. *Photosensitised Diels–Alder reactions* 165
3.9. *The homo Diels–Alder reaction* 167
3.10. *The 'ene' synthesis* 168

4 Reactions at unactivated C–H bonds 172
4.1. *The Hofmann–Loeffler–Freytag reaction* 173
4.2. *Cyclisation reactions of nitrenes* 178
4.3. *The Barton reaction and related processes* 181
      photolysis of organic nitrites; photolysis of organic hypohalites
4.4. *Reaction of monohydric alcohols with lead tetra-acetate* 197
4.5. *Miscellaneous reactions* 201
      olefinic alcohols from hydroperoxides; cyclobutanols by photolysis of ketones; intramolecular abstraction of hydrogen by carbon radicals

5 Synthetic applications of organoboranes 207
5.1. *Reduction with diborane and dialkylboranes* 207
5.2. *Hydroboration* 211
5.3. *Reactions of organoboranes* 222
      protonolysis; oxidation; isomerisation and cyclisation; coupling; carbonylation; 1,4-addition to αβ-unsaturated aldehydes and ketones; reaction with α-bromoketones and α-bromo-esters; reaction with diazo compounds

6 Oxidation 239
6.1. *Oxidation of hydrocarbons with chromic acid* 241
6.2. *Oxidation of alcohols* 244
6.3. *Oxidation of olefins* 259
6.4. *Oxidations with ruthenium tetroxide and nickel peroxide* 281
6.5. *Baeyer–Villiger oxidation of ketones* 287
6.6. *Photosensitised oxidation of olefins* 291

7 Reduction 299
7.1. *Catalytic hydrogenation* 299
      the catalyst; selectivity of reduction; reduction of functional groups; stereochemistry and mechanism; homogeneous hydrogenation

7.2. *Reduction by dissolving metals* 326
   reduction with metal and an acid; reduction of carbonyl compounds with metal and an alcohol; reduction with metal in liquid ammonia
7.3. *Reduction by hydride-transfer reagents* 352
   aluminium alkoxides; lithium aluminium hydride and sodium borohydride; lithium hydrido-alkoxyaluminates; lithium aluminium hydride–aluminium chloride reagents 367
7.4. *Other methods—Wolff–Kishner reduction, desulphurisation of thio-ketals and -acetals, reductions with di-imide*

References 372

Index 387

Preface

This book is addressed principally to advanced undergraduates and to graduates at the beginning of their research careers, and aims to bring to their notice some of the reactions used in modern organic syntheses. Clearly, the whole field of synthesis could not be covered in a book of this size, even in a cursory manner, and a selection has had to be made. This has been governed largely by consideration of the usefulness of the reactions, their versatility and, in some cases, their selectivity.

A large part of the book is concerned with reactions which lead to the formation of carbon–carbon single and double bonds. Some of the reactions discussed, such as the alkylation of ketones and the Diels–Alder reaction, are well established reactions whose scope and usefulness has increased with advancing knowledge. Others, such as those involving phosphorus ylids, organoboranes and new organometallic reagents derived from copper, nickel, and aluminium, have only recently been introduced and add powerfully to the resources available to the synthetic chemist. Other reactions discussed provide methods for the functionalisation of unactivated methyl and methylene groups through intramolecular attack by free radicals at unactivated carbon–hydrogen bonds. The final chapters of the book are concerned with the modification of functional groups by oxidation and reduction, and emphasise the scope and limitations of modern methods, particularly with regard to their selectivity.

Discussion of the various topics is not exhaustive. My object has been to bring out the salient features of each reaction rather than to provide a comprehensive account. In general, reaction mechanisms are not discussed except in so far as is necessary for an understanding of the course or stereochemistry of a reaction. In line with the general policy in the series references have been kept to a minimum. Relevant reviews are noted but, for the most part, references to the original literature are given only for points of outstanding interest and for very recent work. Particular reference is made here to the excellent book by H. O. House, *Modern Synthetic Reactions* which has been my guide at several points and on which I have tried to build, I fear all too inadequately.

I am indebted to my friend and colleague, Dr K. Schofield, for much helpful comment and careful advice which has greatly assisted me in writing the book.

W. CARRUTHERS

26 October 1970

1 Formation of carbon–carbon single bonds

In spite of the fundamental importance in organic synthesis of the formation of carbon–carbon single bonds there are comparatively few general methods available for effecting this process, and fewer still which proceed in good yield under mild conditions. Many of the most useful procedures involve carbanions, themselves derived from organometallic compounds, or from compounds containing 'activated' methyl or methylene groups. They include reactions which proceed by attack of the carbanion on a carbonyl or conjugated carbonyl group, as in the Grignard reaction, the aldol and Claisen ester condensations and the Michael reaction, and other reactions, with which this chapter will be largely concerned, which involve nucleophilic displacement at a saturated carbon atom, as in the alkylation of ketones and the coupling reactions of some organometallic compounds.

1.1. Alkylation: importance of enolate anions. It is well known that certain unsaturated groups attached to a saturated carbon atom render hydrogen atoms attached to that carbon relatively acidic, so that the compound can be converted into an anion on treatment with an appropriate base. Table 1.1, taken from House (1965), shows the pK_a values for some compounds of this type and for some common solvents and reagents.

The acidity of the C–H bonds in these compounds is due to a combination of the inductive electron-withdrawing effect of the unsaturated groups and resonance stabilisation of the anion formed by removal of a proton (1.1). Not all groups are equally effective in 'activating' a neighbouring CH_2 or CH_3; nitro is the most powerful of the common groups and thereafter the series follows the approximate order $—NO_2 >$ $—COR > —SO_2R > —CO_2R > —CN > —C_6H_5$. Two activating groups reinforce each other, as can be seen by comparing diethyl malonate ($pK_a 13$) with ethyl acetate ($pK_a \sim 24$). Acidity is also increased slightly by electron-withdrawing substituents, and decreased by alkyl groups,

TABLE 1.1   *Approximate acidities of active methylene compounds and other common reagents*

Compound	pK_a	Compound	pK_a
CH₃CO₂H̲	5	C₆H₅CO.CH̲₃	19
CH̲₂(CN)CO₂C₂H₅	9	CH₃CO.CH̲₃	20
CH̲₂(CO.CH₃)₂	9	CH̲₃SO₂CH̲₃	~23
CH₃NO₂	10	CH̲₃CO₂C₂H₅	~24
CH₃CO.CH̲₂CO₂C₂H₅	11	CH₃CO⊖	~24
CH̲₂(CO₂C₂H₅)₂	13	CH₃CN	~25
CH₃OH̲	16	C₆H₅NH₂	~30
C₂H₅OH̲	18	(C₆H₅)₃CH̲	~40
(CH₃)₃COH̲	19	CH̲₃SO.CH̲₃	~40

(Acidic hydrogen atoms are underlined)

H.O. House, *Modern synthetic reactions*, copyright 1965, W.A. Benjamin, Inc., Menlo Park, California.

so that diethyl methylmalonate, for example, has a slightly less acidic C–H group than diethyl malonate itself.

By far the most important activating groups in synthesis are the carbonyl and carboxylic ester groups. Removal of a proton from the

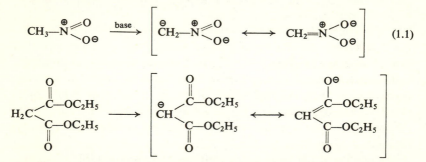

α-carbon atom of a carbonyl compound with base gives the corresponding enolate anion, and it is these anions which are involved in base-catalysed condensation reactions of carbonyl compounds, such as the aldol condensation, and in bimolecular nucleophilic displacements (alkylations) (1.2). The enolate anions must be distinguished from the enols themselves, which are always present in equilibrium with the carbonyl compound in presence of acidic or basic catalysts (1.3). The enols are concerned in certain acid-catalysed condensations of carbonyl compounds. Most monoketones and esters contain only small amounts

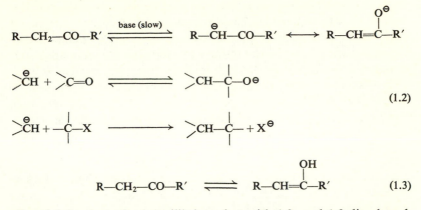

$$R-CH_2-CO-R' \xrightleftharpoons{\text{base (slow)}} R-\overset{\ominus}{C}H-CO-R' \longleftrightarrow R-CH=\overset{\overset{O^{\ominus}}{|}}{C}-R'$$

$$\overset{\ominus}{>}\!\!C\!H + >\!C\!=\!O \rightleftharpoons >\!C\!H\!-\!\overset{|}{\underset{|}{C}}\!-\!O^{\ominus} \tag{1.2}$$

$$\overset{\ominus}{>}\!\!C\!H + -\!\overset{|}{\underset{|}{C}}\!-\!X \longrightarrow >\!C\!H\!-\!\overset{|}{\underset{|}{C}}\!- + X^{\ominus}$$

$$R-CH_2-CO-R' \rightleftharpoons R-CH=\overset{\overset{OH}{|}}{C}-R' \tag{1.3}$$

of enol (<1 per cent) at equilibrium, but with 1,2- and 1,3-dicarbonyl compounds much higher amounts of enol (>50 per cent) may be present.

The formation of the enolate anion results from an equilibrium reaction between the carbonyl compound and the base. A competing equilibrium involves the enolate anion and the solvent. Thus, with diethyl malonate in solvent SolH in presence of base $B^{\ominus}$, we have

$$CH_2(CO_2C_2H_5)_2 + B^{\ominus} \rightleftharpoons \overset{\ominus}{C}H(CO_2C_2H_5)_2 + BH$$

$$\overset{\ominus}{C}H(CO_2C_2H_5)_2 + SolH \rightleftharpoons CH_2(CO_2C_2H_5)_2 + Sol^{\ominus}, \tag{1.4}$$

and to ensure an adequate concentration of the enolate anion at equilibrium clearly both the solvent and the conjugate acid of the base must be much weaker acids than the active methylene compound. The correct choice of base and solvent is thus of great importance if the subsequent alkylation, or other, reaction is to be successful. Reactions must normally be effected under anhydrous conditions since water is a much stronger acid than the usual activated methylene compounds and, if present, would instantly protonate any carbanion produced. Another point of importance is that the solvent must not be a much stronger acid than the conjugate acid of the base, otherwise the equilibrium

$$B^{\ominus} + SolH \rightleftharpoons BH + Sol^{\ominus} \tag{1.5}$$

will lie too far to the right and lower the concentration of $B^{\ominus}$. For example, sodamide can be used as base in liquid ammonia or in benzene, but, obviously, not in ethanol. Base–solvent combinations commonly used to convert active methylene compounds into the corresponding anions include sodium methoxide, sodium ethoxide and sodium or potassium t-butoxide in solution in the corresponding alcohol, or as suspensions in ether, benzene or dimethoxyethane. Sodium t-amyloxide is particularly useful since it is soluble in ether and benzene (Conia, 1963). Metallic sodium or potassium, or sodium hydride, in suspension in benzene, ether or dimethoxyethane, sodamide in suspension in an inert solvent or in solution in liquid ammonia, and solutions of sodium or potassium triphenylmethyl in ether or benzene are also widely used with the less 'active' compounds.

1.2. Alkylation, alkenylation and arylation of relatively acidic methylene groups. In order to effect a reasonably rapid reaction it is, of course, necessary to have a relatively high concentration of the carbanion. Because of their relatively high acidity (see Table 1.1) compounds in which a C–H bond is activated by a nitro group or by two or more carbonyl, ester or cyano groups can be converted largely into their anions with an anhydrous alcoholic solution of a metal alkoxide, such as sodium ethoxide or potassium t-butoxide. An alternative procedure is to prepare the enolate in benzene or ether using finely divided sodium or potassium metal or sodium hydride, which react irreversibly with compounds containing active methylene groups with formation of the metal salt and evolution of hydrogen. β-Diketones can often be converted into their enolates with alkali metal hydroxides or carbonates in aqueous alcohol or acetone.

Recent studies have shown that much faster alkylation of enolate anions can often be achieved in dimethylformamide, dimethyl sulphoxide, 1,2-dimethoxyethane or hexamethylphosphoramide than in the usual protic solvents. This appears to be due to the fact that the former solvents do not solvate the enolate anion and thus do not diminish its reactivity as a nucleophile. At the same time they are able to solvate the cation, separating it from the cation–enolate ion pair and leaving a relatively free enolate ion which would be expected to be a more reactive nucleophile than the ion pair (Parker, 1962).

Alkylation of enolate anions is readily effected with alkyl halides or other alkylating agents. Both primary and secondary alkyl, allyl or benzyl halides may be used successfully, but with tertiary halides poor yields of alkylated product often result because of competing dehydrohalogenation of the halide. It is often advantageous to proceed by way of the toluene-*p*-sulphonate or methanesulphonate rather than a halide.

The sulphonates are excellent alkylating agents, and can usually be obtained from the alcohol in a pure condition more readily than the corresponding halides. Epoxides have also been used as alkylating agents. Attack of the enolate anion on the alkylating agent takes place by a bimolecular nucleophilic displacement (S_N2) process and thus results in inversion of configuration at the carbon atom of the alkylating agent.

$$CH_2(CO_2C_2H_5)_2 \xrightarrow[\text{(2) } C_6H_5.CH_2Cl]{\text{(1) } C_2H_5ONa, C_2H_5OH}$$

$$C_6H_5.CH_2.CH.(CO_2C_2H_5)_2 \quad (85\%)$$

$$C_6H_5.CH_2.CH(CO_2C_2H_5)_2 \xrightarrow[\text{(2) } CH_3OCH_2Cl]{\text{(1) Na, ether}} \begin{array}{c} CH_3OCH_2 \\ \diagdown \\ C_6H_5CH_2 \end{array} C(CO_2C_2H_5)_2 \quad (78\%)$$

$$CH_3.CO.CH_2.CO.CH_3 \xrightarrow[\substack{CH_3COCH_3 \\ \text{reflux}}]{CH_3I, K_2CO_3} CH_3.CO.\underset{\underset{CH_3}{|}}{CH}.CO.CH_3 \quad (75\%)$$

$$(1.6)$$

$$p\text{-}CH_3.C_6H_4SO_2.O \quad \xrightarrow[C_2H_5OH]{\substack{CH_2(CO_2C_2H_5)_2 \\ C_2H_5ONa}} \quad$$

With secondary and tertiary allylic halides or sulphonates reaction of an enolate anion may give mixtures of products formed by competing attack at the allylic position (1.7).

$$C_2H_5\text{—}\underset{\underset{Cl}{|}}{CH}\text{—}CH{=}CH_2 \xrightarrow[C_2H_5ONa, C_2H_5OH]{CH_2(CO_2C_2H_5)_2}$$

$$C_2H_5\text{—}\underset{\underset{CH(CO_2C_2H_5)_2}{|}}{CH}\text{—}CH{=}CH_2 + C_2H_5.CH{=}CH.CH_2.CH(CO_2C_2H_5)_2 \quad (1.7)$$

$$(10\% \text{ of product})$$

Alkylation of active methylene compounds with $\alpha\omega$-polymethylene dihalides, and intramolecular alkylation of ω-haloalkylmalonic esters provides a useful method for synthesising three- to seven-membered rings. Non-cyclic products are frequently formed at the same time by competing intermolecular reactions and conditions have to be carefully chosen to suppress their formation (1.8).

$$Br(CH_2)_5Br + CH_2(CO_2C_2H_5)_2 \xrightarrow[C_2H_5OH]{C_2H_5ONa}$$

$$+ (C_2H_5O_2C)_2CH(CH_2)_5CH(CO_2C_2H_5)_2$$

A difficulty sometimes encountered in the alkylation of active methylene compounds is the formation of unwanted dialkylated products. During the alkylation of diethyl sodiomalonate the monoalkyl derivative formed initially is in equilibrium with its anion as indicated in the first equation of (1.9). In ethanol solution dialkylation does not take place to any appreciable extent because ethanol is sufficiently acidic to reduce the concentration of the anion of the alkyl derivative, but not that of the more acidic diethyl malonate itself, to a very low value.

$$RCH(CO_2C_2H_5)_2 + \overset{\ominus}{C}H(CO_2C_2H_5)_2 \; \rightleftharpoons$$

$$\overset{\ominus}{R}C(CO_2C_2H_5)_2 + CH_2(CO_2C_2H_5)_2 \quad (1.9)$$

$$\overset{\ominus}{R}C(CO_2C_2H_5)_2 + C_2H_5OH \; \rightleftharpoons \; RCH(CO_2C_2H_5)_2 + C_2H_5O\overset{\ominus}{}$$

However, replacement of ethanol by an inert solvent favours dialkylation, and dialkylation also becomes a more serious problem with the more acidic alkylcyanoacetic esters, and in alkylations with very reactive compounds such as allyl or benzyl halides or sulphonates. An improved method of alkylation of β-dicarbonyl compounds, by reaction of the thallium(I) salts with an alkyl iodide or bromide, which is said to lead to high yields of monoalkylated product without any dialkylation, has recently been described by Taylor, Hawks and McKillop (1968) (1.10).

$$\underset{Tl^\oplus}{CH_3.CO.\overset{\ominus}{C}H.CO_2C_2H_5} \xrightarrow[\substack{benzene,\\ boil}]{CH_3I} \underset{\underset{CH_3}{|}}{CH_3.CO.CH.CO_2C_2H_5} \quad (1.10)$$

quantitative

Dialkylation may, of course, be effected deliberately if required by carrying out two successive operations, using either the same or a different alkylating agent in the two steps. It is often found that active methylene compounds with a secondary or tertiary alkyl substituent in the α-position to the activating group undergo further alkylation only with difficulty. This is partly due to increased steric hindrance to approach of the base for proton abstraction and partly, in the case of carbonyl-activated groups at any rate, to steric interference with the attainment of a transition state for proton removal that allows continuous overlap of the *p* orbitals concerned (1.11). This difficulty may be overcome by

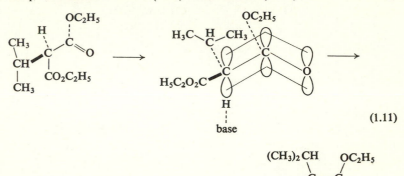

(1.11)

use of a stronger base in a less acidic solvent, such as potassium t-butoxide in t-butanol, or, if the choice is available, by introducing the branched-chain substituent last.

Under ordinary conditions aryl or vinyl halides are not sufficiently reactive to react with enolate anions, although aryl halides with strongly electronegative substituents in the *ortho* and *para* positions can be used. 2,4-Dinitrochlorobenzene, for example, with ethyl cyanoacetate gives ethyl (2,4-dinitrophenyl)cyanoacetate in 90 per cent yield. Aryl derivatives may also be obtained from aryl halides under more vigorous conditions. Reaction of bromobenzene with diethyl malonate, for example, takes place readily in presence of an excess of sodium amide in liquid ammonia, to give diethyl phenylmalonate in 50 per cent yield. The reaction is not a direct nucleophilic displacement, however, but takes place by an elimination–addition sequence in which benzyne is an intermediate (1.12). The reaction can also be effected intramolecularly

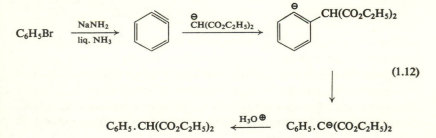

(1.12)

and provides a good route to some cyclic systems (Gilchrist and Rees, 1969) (1.13). Vinyl derivatives of active methylene compounds can be

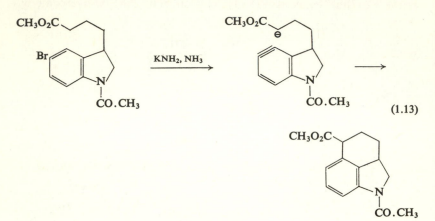

(1.13)

obtained indirectly from alkylidene derivatives as illustrated in (1.14). Kinetically controlled alkylation of the delocalised anion takes place at the α-position, even though γ-alkylation would give a more stable

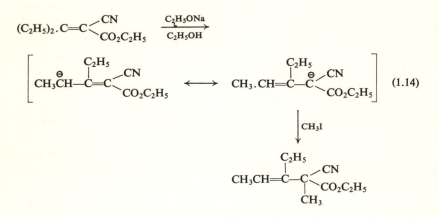

(1.14)

product. A similar course is followed in the kinetically controlled protonation of such anions.

A wasteful side-reaction which frequently occurs in the alkylation of 1,3-dicarbonyl compounds is the formation of the *O*-alkylated product. Thus, reaction of the sodium salt of cyclohexan-1,3-dione with n-butyl bromide gives 37 per cent of 1-n-butoxycyclohexen-3-one and only 15 per cent of 2-n-butylcyclohexan-1,3-dione. This difficulty can be largely overcome by heating the thallium(I) salt of the β-dicarbonyl compound, easily obtained by addition of thallium(I) ethoxide to a solution of the β-dicarbonyl compound in an inert solvent, with an excess of alkyl iodide or bromide (Taylor, Hawks and McKillop, 1968). Under these conditions high yields of monoalkylated products are obtained without any concomitant O-alkylation or dialkylation. Thus, reaction of the thallium salt of acetoacetic ester with methyl iodide gives the C-methyl derivative in quantitative yield, and 2-ethoxycarbonylcyclopentanone with isopropyl iodide similarly forms the 2-isopropyl derivative in 96 per cent yield.

In general, O-alkylation competes significantly with C-alkylation only with reactive methylene compounds in which the equilibrium concentration of enol is relatively high, as in 1,3-dicarbonyl compounds and phenols. Phenols, of course, generally undergo predominant O-alkylation, but under some conditions significant amounts of C-alkylated products may be formed (Kornblum, Seltzer and Haberfield, 1963). A useful method for C-alkylation of phenols by way of the corresponding Mannich base is illustrated in the following synthesis of 13-methylpodocarpic acid (Wenkert, Stenberg and Beak, 1961) (1.15).

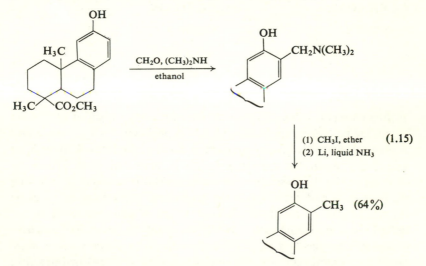

Alkylation of malonic ester, cyanacetic ester and β-ketoesters is useful in synthesis because the alkylated products on hydrolysis and decarboxylation afford carboxylic acids and ketones. From alkylated malonic or cyanacetic esters substituted acetic acids are obtained, and alkylated acetoacetic esters give substituted acetones (1.16). Alkaline hydrolysis

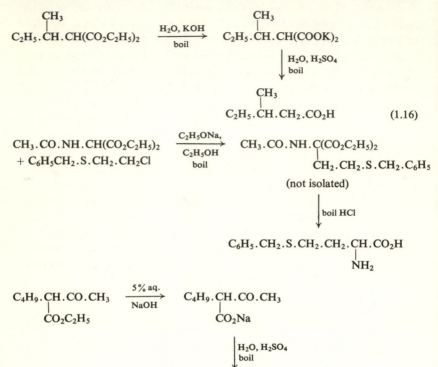

$$C_2H_5.\overset{\overset{\displaystyle CH_3}{|}}{C}H.CH(CO_2C_2H_5)_2 \quad \xrightarrow[\text{boil}]{\text{H}_2\text{O, KOH}} \quad C_2H_5.\overset{\overset{\displaystyle CH_3}{|}}{C}H.CH(COOK)_2$$

$$\downarrow \begin{smallmatrix} \text{H}_2\text{O, H}_2\text{SO}_4 \\ \text{boil} \end{smallmatrix}$$

$$C_2H_5.\overset{\overset{\displaystyle CH_3}{|}}{C}H.CH_2.CO_2H \qquad (1.16)$$

$$CH_3.CO.NH.CH(CO_2C_2H_5)_2 \\ + C_6H_5CH_2.S.CH_2.CH_2Cl \quad \xrightarrow[\text{boil}]{\overset{\text{C}_2\text{H}_5\text{ONa,}}{\text{C}_2\text{H}_5\text{OH}}} \quad CH_3.CO.NH.\overset{\overset{\displaystyle C(CO_2C_2H_5)_2}{|}}{} \\ CH_2.CH_2.S.CH_2.C_6H_5$$

(not isolated)

$$\downarrow \text{boil HCl}$$

$$C_6H_5.CH_2.S.CH_2.CH_2.\overset{\overset{\displaystyle CH.CO_2H}{|}}{}\\ NH_2$$

$$C_4H_9.\overset{\overset{\displaystyle CH.CO.CH_3}{|}}{}\\ CO_2C_2H_5 \quad \xrightarrow[\text{NaOH}]{5\% \text{ aq.}} \quad C_4H_9.\overset{\overset{\displaystyle CH.CO.CH_3}{|}}{}\\ CO_2Na$$

$$\downarrow \begin{smallmatrix} \text{H}_2\text{O, H}_2\text{SO}_4 \\ \text{boil} \end{smallmatrix}$$

$$C_4H_9.CH_2.CO.CH_3$$

of β-ketoesters is often complicated by competing attack of hydroxide ion on the ketone group, leading to fission products. This is particularly liable to occur when the α- position is disubstituted, for if there is an α-hydrogen atom the carbonyl group is protected from attack by enolisation (1.17). These cleavage reactions can be avoided by effecting the hydrolysis and decarboxylation under acid conditions, or by using benzyl, t-butyl or tetrahydropyranyl esters in place of ethyl esters. Benzyl esters are readily cleaved by hydrogenolysis (see p. 300) and acid catalysed cleavage of t-butyl or tetrahydropyranyl esters takes place easily. In a useful extension of the malonic ester synthesis ketones have been obtained by oxidation of the malonic acid with lead tetraacetate followed by hydrolysis of the resulting *gem*-diacetate (Tufariello and Kissel, 1966) (1.18).

11

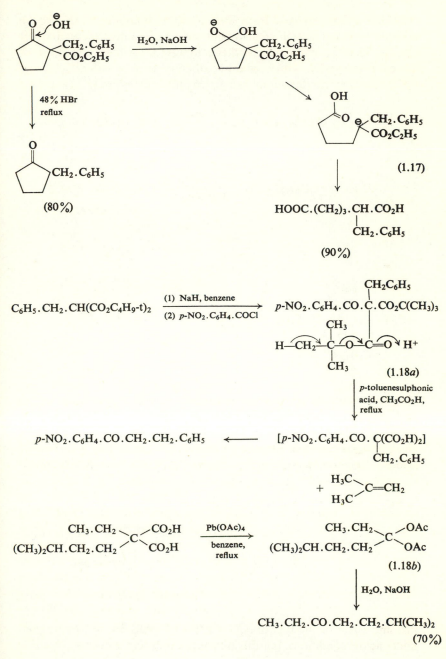

Enolate anions derived from β-dicarbonyl compounds can also be acylated by reaction with an acid chloride or acid anhydride in an inert solvent. These reactions are synthetically useful for the preparation of 'mixed' β-keto-esters which would be obtainable only with difficulty by a Claisen ester condensation using two different esters (1.19). Under

$$CH_3.CO.CH_2.CO_2C_2H_5 \xrightarrow[\text{benzene}]{Na} CH_3.CO.\overset{\ominus}{C}H.CO_2C_2H_5$$

(1.19)

$$\downarrow \begin{array}{l} C_6H_5.CO.Cl, \\ \text{benzene} \end{array}$$

$$C_6H_5.CO.CH_2.CO_2C_2H_5 \xleftarrow[\text{H}_2\text{O, NH}_4\text{Cl}]{\text{NH}_4\text{OH,}} \begin{array}{l} CH_3.CO.CH.CO_2C_2H_5 \\ \quad\quad\quad | \\ \quad\quad\quad CO.C_6H_5 \end{array}$$

$$CH_2(CO_2C_2H_5) \xrightarrow[\text{benzene}]{\text{Mg, ethanol}} C_2H_5O\overset{\oplus}{M}g.\overset{\ominus}{C}H(CO_2C_2H_5)_2$$

$$\downarrow \begin{array}{l} \text{n-C}_7\text{H}_{15}\text{CO.Cl,} \\ \text{benzene} \end{array}$$

$$\text{n-C}_7\text{H}_{15}.CO.CH_2.CO_2C_2H_5 \xleftarrow[\text{CH}_3\text{CO}_2\text{H}]{0.2\% \text{ H}_2\text{SO}_4} \text{n-C}_7\text{H}_{15}.CO.CH(CO_2C_2H_5)_2$$

(43%)

$$+ \text{n-C}_7\text{H}_{15}.CO.CH_3$$

(52%)

alkaline conditions acylmalonic esters are cleaved with removal of the acyl group, but acid-catalysed cleavage yields the β-keto-ester. With more vigorous acid hydrolysis the methyl ketone is formed in excellent yield, and the sequence provides a method for converting a carboxylic acid into the homologous methyl ketone (1.20). Completely substituted

$$p\text{-NO}_2.C_6H_4.CO.Cl + NaCH(CO_2C_2H_5)_2 \xrightarrow{\text{benzene}}$$

$$p\text{-NO}_2.C_6H_4.CO.CH(CO_2C_2H_5)_2$$

$$\downarrow \begin{array}{l} 1\% \text{ H}_2\text{SO}_4 \\ \text{propionic acid,} \\ \text{boil} \end{array} \quad (1.20)$$

$$p\text{-NO}_2.C_6H_4.CO.CH_3 \quad (99\%)$$

malonic esters of the type $R.CO.CR'(CO_2C_2H_5)_2$ are not hydrolysed under acidic conditions, but this difficulty can be circumvented by using the corresponding benzyl esters, which are easily cleaved by hydrogenolysis, thus providing a general route from carboxylic acids $R.CO_2H$ to ketones $R.CO.CH_2.R'$. The sodiobenzyl esters are readily prepared *in situ* from the ethyl esters by ester exchange with benzyl alcohol (1.21).

$$n\text{-}C_4H_9.\underset{\underset{C_2H_5}{|}}{CH}.CO.Cl + C_6H_5CH_2O_2C(CH_2)_{10}\overset{\ominus}{CH}(CO_2CH_2C_6H_5)_2$$

$$\downarrow \text{benzene}$$

$$\begin{array}{l} n\text{-}C_4H_9.CH(C_2H_5).CO \\ \phantom{n\text{-}C_4H_9.}C_6H_5CH_2O_2C(CH_2)_{10} \end{array}\!\!\!\!\!\!\!\!C(CO_2CH_2C_6H_5)_2 \qquad (1.21)$$

$$\downarrow \begin{array}{l} (1)\ Pd,\ H_2 \\ (2)\ -CO_2 \end{array}$$

$$n\text{-}C_4H_9.\underset{\underset{C_2H_5}{|}}{CH}.CO.(CH_2)_{11}.CO_2H \quad (78\%)$$

1.3. γ-Alkylation of 1,3-dicarbonyl compounds. Alkylation of a 1,3-diketone at one of the 'flanking' methyl or methylene groups instead of at the doubly activated CH_2 does not usually take place to any significant extent under ordinary conditions. It can be accomplished selectively and in good yield by way of the corresponding *dianion*, itself prepared from the diketone and two equivalents of sodium or potassium amide in liquid ammonia, by reaction with one equivalent of alkylating agent (Harris and Harris, 1969). Thus, acetylacetone is converted into nona-2,4-dione in 82 per cent yield, and benzoylacetone gives 1,6-diphenyl-penta-1,3-dione in 77 per cent yield. Ketoacids and triketones can also be obtained, by reaction of the dianions with carbon dioxide or with esters (1.22).

The reaction can be applied equally well to β-keto-aldehydes and β-keto-esters, and provides a useful alternative route to 'mixed' Claisen ester condensation products. Reactions with β-keto-aldehydes are generally effected by treating the monosodium salt of the aldehyde with alkali amide, to prevent self condensation of the aldehyde (1.23).

With unsymmetrical diketones, which could apparently give rise to two different dianions, it is found in practice that in most cases only one is formed, and a single alkylation product results. Thus, with hexa-2,4-dione alkylation at the methyl group greatly predominates over that at

$$CH_3.CO.CH_2.CO.CH_3 \xrightarrow[\text{liquid NH}_3]{\text{2 equivs. KNH}_2}$$

$$CH_3CO.\overset{\ominus}{CH}.CO.\overset{\ominus}{CH}_2 \longleftrightarrow CH_3.\overset{\overset{\displaystyle O\ominus}{|}}{C}{=}CH{-}\overset{\overset{\displaystyle O\ominus}{|}}{C}{=}CH_2$$

$$\downarrow \begin{array}{l}\text{(1) n-C}_4\text{H}_9\text{Br, ether}\\ \text{(2) H}_3\text{O}^+\end{array} \qquad\qquad (1.22)$$

$$CH_3.CO.CH_2.CO.CH_2.C_4H_9\text{-n} \quad (82\%)$$

$$C_6H_5.CO.CH_2.CO.CH_3 \xrightarrow[\text{liq. NH}_3]{\text{2KNH}_2} C_6H_5.CO.\overset{\ominus}{CH}.CO.\overset{\ominus}{CH}_2$$

$$\downarrow \begin{array}{l}\text{(1) CO}_2\\ \text{(2) H}_3\text{O}^+\end{array}$$

$$C_6H_5.CO.CH_2.CO.CH_2.CO_2H$$

$$CH_3.CO.CH{=}CH\overset{\oplus}{O}\ \overset{\ominus}{Na} \xrightarrow[\text{NH}_3]{\text{KNH}_2} CH_2{=}\overset{\overset{\displaystyle O\ominus}{|}}{C}{-}CH{=}CH{-}O\ominus$$

$$\downarrow \text{n-C}_4\text{H}_9\text{Br} \qquad\qquad (1.23)$$

$$\text{n-C}_4\text{H}_9.CH_2.CO.CH_2.CHO \xleftarrow{\text{H}_3\text{O}\oplus} \text{n-C}_4\text{H}_9.CH_2.\overset{\overset{\displaystyle O}{\|}}{C}{-}CH{=}CH\overset{\ominus}{O}$$

(72%: isolated as the
copper complex)

the methylene group, and 2-acetylcyclohexanone and 2-acetylcyclo-
pentanone are both alkylated exclusively at the methyl group (1.24).

$$CH_3.CH_2.CO.CH_2.CO.CH_3 \xrightarrow[\substack{\text{(2) CH}_3\text{I}\\ \text{(3) H}_3\text{O}\oplus}]{\text{(1) 2NaNH}_2,\ \text{NH}_3}$$

$$CH_3.CH_2.CO.CH_2.CO.CH_2CH_3 \quad + \quad (CH_3)_2CH.CO.CH_2.CO.CH_3$$

(89% of product)   (11% of product)

$$(1.24)$$

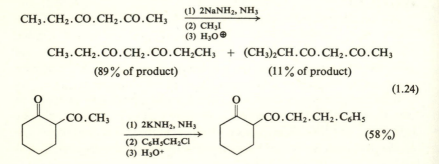

(58%)

In general, the ease of alkylation follows the order $C_6H_5CH_2$— >
CH_3— > —CH_2—.

1.4. Alkylation of β-keto-sulphoxides and -sulphones. Aromatic esters
and aliphatic esters which do not undergo easy proton transfer, react
with two equivalents of methylsulphinyl carbanion, easily available
from dimethyl sulphoxide and sodium hydride, to form anions of
β-keto-sulphoxides (1.25). These β-keto-sulphoxides are valuable syn-

$$CH_3.SO.CH_3 + NaH \xrightarrow[\substack{\text{dimethyl} \\ \text{sulphoxide}}]{\text{warm in}} CH_3.SO.\overset{\ominus}{C}H_2\overset{\oplus}{Na} + H_2O \tag{1.25}$$

$$R.CO_2R' + 2CH_3SO.CH_2^{\ominus} \longrightarrow [R.CO.CHSO.CH_3]^{\ominus} + OR'^{\ominus} + (CH_3)_2SO$$

$$\downarrow H_3O^{\oplus}$$

$$R.CO.CH_2.SO.CH_3$$

thetic intermediates. They are readily alkylated by reaction of the sodium
salts with primary alkyl halides, and on reduction with aluminium
amalgam in aqueous tetrahydrofuran, or with zinc and acetic acid, they
are readily converted into the corresponding ketones. The sequence of
reactions provides another convenient and flexible method for the
conversion of carboxylic acid esters into a wide variety of ketones
(Corey and Chaykovsky, 1965; Gassmann and Richmond, 1966) (1.26).

$$C_2H_5O_2C.(CH_2)_6.CO_2C_2H_5 \xrightarrow[\substack{CH_3.SO.CH_3 \\ \text{then } H_3O^{\oplus}}]{CH_3.SO.CH_2^{\ominus}}$$

$$CH_3.SO.CH_2.CO.(CH_2)_6.CO.CH_2.SO.CH_3$$

$$\downarrow \text{Al–Hg, H}_2\text{O, THF}$$

$$CH_3.CO.(CH_2)_6.CO.CH_3 \quad (82\%)$$

$$C_6H_5.CO_2C_2H_5 \xrightarrow[\substack{CH_3.SO.CH_3 \\ \text{then } H_3O^+}]{CH_3.SO.CH_2^{\ominus}} C_6H_5.CO.CH_2.SO.CH_3 \tag{1.26}$$

$$\downarrow \substack{(1)\ 2\ \text{mol NaH, dimethylformamide} \\ (2)\ CH_3I}$$

$$\overset{\displaystyle CH_3}{\underset{\displaystyle CH_3}{C_6H_5.CO.\overset{|}{\underset{|}{C}}.SO.CH_3}}$$

$$C_6H_5.CO.CH(CH_3)_2 \xleftarrow[CH_3.CO_2H]{Zn, C_2H_5OH}$$

Other promising reagents which undergo a similar series of reactions are the dimethylaminosulphonyl carbanion, from *N,N*-dimethyl-methanesulphonamide, and the methylsulphonyl carbanion, available from reaction of dimethyl sulphone and sodium hydride in dimethyl sulphoxide. In many reactions β-keto-sulphones, prepared by reaction of the methylsulphonyl carbanion with esters, or by oxidation of the sulphoxides, are superior to the β-keto-sulphoxides, particularly where subsequent oxidations or reductions are to be effected elsewhere in the molecule (House and Larson, 1968) (1.27). Reaction of β-keto-

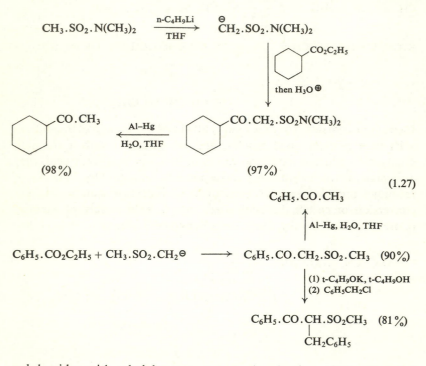

sulphoxides with ethyl bromoacetate and reduction of the alkylated products provides a convenient method for converting carboxylic acid esters into keto-esters containing three more carbon atoms. By pyrolysis of the sulphoxides αβ-unsaturated esters are obtained (Russell and Ochrymowycz, 1969) (1.28). Similarly, Michael addition of the β-keto-sulphoxide to ethyl acrylate affords an intermediate which is readily converted into a δ-keto-ester, or an unsaturated ester, with four more carbon atoms than the starting ester (1.29).

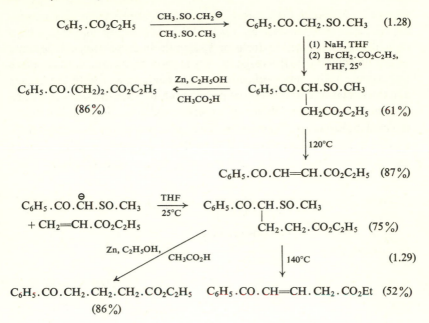

1.5. Alkylation of ketones. Alkylation of monofunctional carbonyl compounds, aldehydes, ketones and esters, is more difficult than that of the 1,3-dicarbonyl compounds discussed above. As can be seen from the Table 1.1, a methyl or methylene group which is activated by only one carbonyl, ester or cyano group requires a stronger base than sodium ethoxide or methoxide to convert it into the enolate anion in high enough concentration to be useful for subsequent alkylation. Alkali metal salts of tertiary alcohols, such as t-butanol or t-amyl alcohol, in solution or suspension in the corresponding alcohol or in an inert solvent, have been used with success, but suffer from the disadvantage that they are not sufficiently basic to convert the ketone completely into the enolate anion, thus allowing the possibility of an aldol condensation between the anion and unchanged carbonyl compound. An alternative procedure, which largely obviates this difficulty, is to use a much stronger base which will convert the compound completely into the anion. Typical bases of this type are sodium and potassium amide, sodium hydride, lithium diethylamide and sodium and potassium triphenyl-methyl, in such solvents as ether, benzene, dimethoxyethane or dimethyl-formamide. The alkali metal amides are often used in solution in liquid ammonia. Although these bases can convert ketones essentially quantitatively into their enolate anions, aldol condensation may again be a difficulty with sodium hydride or sodamide in inert solvents, because of the insolubility of the reagents. Formation of the anion takes place only slowly in the heterogeneous reaction medium and both the ketone and the enolate ion are present at some point. But this difficulty can often be lessened by proper choice of experimental conditions (1.30). Intramolecular alkylation of ketones can be used to prepare cyclic compounds, as in the following synthesis of 9-methyl-1-decalone, and in the key step in the synthesis of the sesquiterpene hydrocarbon seychellene (1.31).

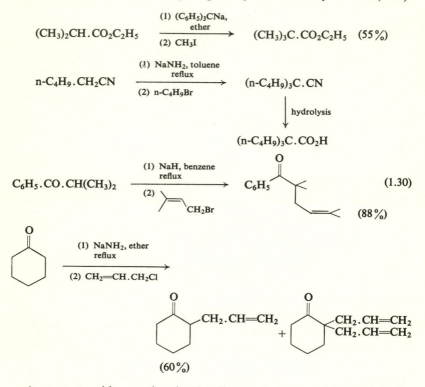

A common side reaction in the direct alkylation of ketones is the formation of di- and polyalkylated products through interaction of the original enolate with the monoalkylated product (1.32). This difficulty can sometimes be avoided to some extent by adding a solution of the enolate in dimethoxyethane to a large excess of the alkylating agent. The enolate may thereby be rapidly consumed before equilibration with

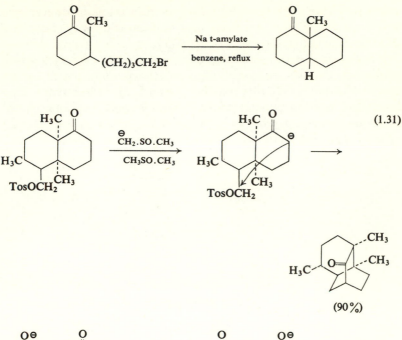

(1.31)

(90%)

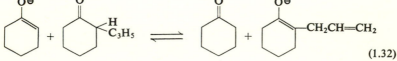

(1.32)

the alkylated ketone can take place. Nevertheless, formation of poly-substituted products is a serious problem in the direct alkylation of ketones and often results in decreased yields of the desired monoalkyl compound.

Alkylation of symmetrical ketones and of ketones which can enolise in one direction only can, of course, give only one mono-*C*-alkylated product. *O*-Alkylation usually does not take place to any appreciable

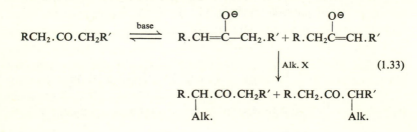

(1.33)

extent with simple ketones. With unsymmetrical ketones, however, two different monoalkylated products may be formed by way of the two possible structurally isomeric enolates (1.33).

If one of the isomeric enolate anions is stabilised by conjugation with another group such as cyano, nitro or ethoxycarbonyl then for all practical purposes only this stabilised anion is formed and alkylation takes place at the position activated by both groups. Even an α-phenyl or an α-vinyl group provides sufficient stabilisation of the resulting anion to direct substitution into the adjacent position (1.34).

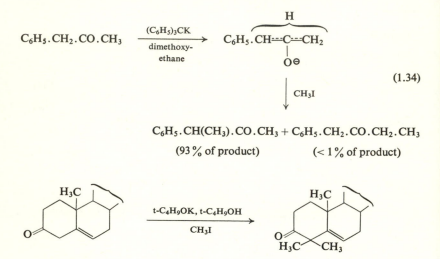

$$C_6H_5.CH_2.CO.CH_3 \xrightarrow[\text{dimethoxy-}\atop\text{ethane}]{(C_6H_5)_3CK} C_6H_5.CH{=}{=}C{=}{=}CH_2$$

(1.34)

$$C_6H_5.CH(CH_3).CO.CH_3 + C_6H_5.CH_2.CO.CH_2.CH_3$$

(93 % of product)   (< 1 % of product)

Alkylation of unsymmetrical ketones bearing only α-alkyl substituents, however, generally leads to mixtures containing both α-alkylated products. The relative amounts of the two products, formed through the enolate anions, depends on the structure of the ketone and may also be influenced by experimental factors such as the nature of the cation and the solvent (see Table 1.2). In the presence of the free ketone or an equivalent proton donor such as a protic solvent, equilibration of the enolate ions can take place (1.35). But if the enolates are prepared from the ketone with one equivalent of a strong base, such as potassium triphenylmethyl or lithium diethylamide, which converts the ketone essentially quantitatively into the anion, equilibration cannot occur, and the composition of mixtures of enolate ions formed under such conditions of kinetic control may differ substantially from that of mixtures formed under equilibrium conditions. In general, enolate mixtures

TABLE 1.2 *Composition of mixtures of enolate anions generated from the ketone and a triphenylmethyl metal derivative in dimethoxyethane* (House, 1967)

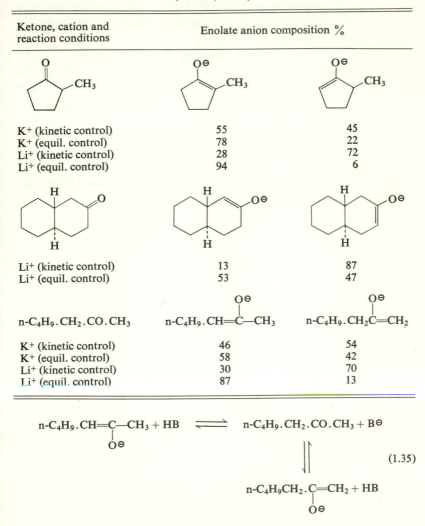

Ketone, cation and reaction conditions	Enolate anion composition %	
K⁺ (kinetic control)	55	45
K⁺ (equil. control)	78	22
Li⁺ (kinetic control)	28	72
Li⁺ (equil. control)	94	6
Li⁺ (kinetic control)	13	87
Li⁺ (equil. control)	53	47
K⁺ (kinetic control)	46	54
K⁺ (equil. control)	58	42
Li⁺ (kinetic control)	30	70
Li⁺ (equil. control)	87	13

$$n\text{-}C_4H_9.CH{=}\underset{\underset{O^{\ominus}}{|}}{C}{-}CH_3 + HB \rightleftharpoons n\text{-}C_4H_9.CH_2.CO.CH_3 + B^{\ominus}$$

(1.35)

$$n\text{-}C_4H_9CH_2.\underset{\underset{O^{\ominus}}{|}}{C}{=}CH_2 + HB$$

formed under kinetic conditions contain more of the less highly sub-stituted enolate than the equilibrium mixture, indicating that the less hindered α-protons are removed more rapidly by the strong base (House, 1967). This is especially true of five- and six-membered cyclic ketones, where the more highly substituted enolate is favoured at equilibrium. However, whichever method is used, mixtures of both structurally isomeric enolates are generally obtained, and mixtures of products result on alkylation. Di- and tri-alkylation products may also be formed by equilibration of the monoalkylated product with the original anion and it is not always easy to isolate the pure monoalkylated derivative in good yield from the resulting complex mixtures.

Two main methods have been used to improve selectivity in the alkylation of unsymmetrical ketones. One widely used procedure is to introduce temporarily an activating group at one of the α-positions to stabilise the corresponding enolate anion; this group is removed later after the alkylation has been effected. Common activating groups used for this purpose are the ethoxycarbonyl, ethoxyoxalyl and formyl groups. Thus, the unsaturated ketone (I) can be converted into the 2-methyl derivative by the steps shown, whereas direct alkylation leads to reaction at C-4 (p. 26), (1.36).

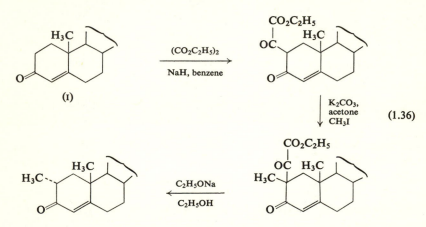

(1.36)

Another useful technique is to block one of the α-positions by intro-duction of a (removable) substituent which *prevents* formation of the corresponding enolate. A widely used method is acylation with ethyl formate and transformation of the resulting formyl or hydroxymethylene substituent into a group that is stable to base, such as an enamine, an enol ether or an enol thioether (1.37). An example of this procedure is shown below in the preparation of 9-methyl-1-decalone from *trans*-1-decalone. Direct alkylation of this compound gives mainly the 2-alkyl derivative (1.38). A useful alternative procedure is alkylation of the

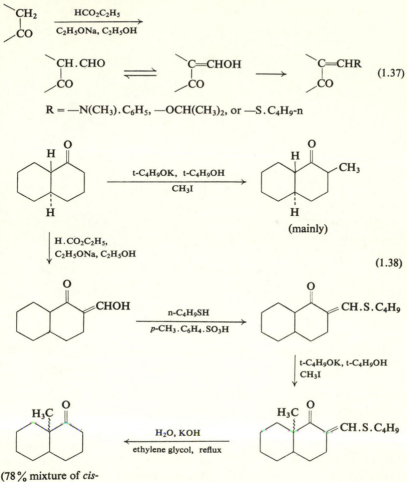

$$R = -N(CH_3).C_6H_5, \quad -OCH(CH_3)_2, \quad \text{or} \quad -S.C_4H_9\text{-}n$$

(1.37)

(mainly)

(1.38)

(78% mixture of *cis-*
and *trans*)

dianion prepared from the formyl derivative with potassium amide (p. 13), (1.39). Specific enolate anions may also be obtained from unsymmetrical ketones by reaction of the structurally specific enol acetates with two equivalents of methyl lithium in dimethoxyethane (House and Trost, 1965). Under these conditions the enol acetate is converted into the lithium enolate without any isomerisation to the alternative structure. Reaction with an alkyl halide then affords a specific monoalkylated derivative, accompanied, usually, by some dialkylated product (1.40).

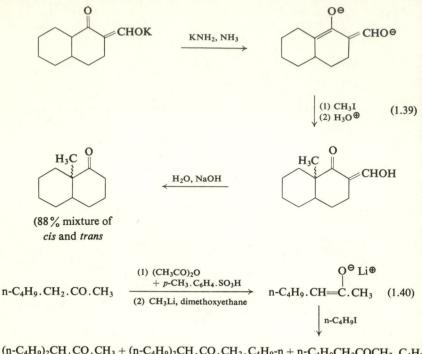

$$(1.39)$$

(88% mixture of
cis and *trans*

$$\text{n-C}_4\text{H}_9 . \text{CH}_2 . \text{CO} . \text{CH}_3 \quad \xrightarrow[\text{(2) CH}_3\text{Li, dimethoxyethane}]{\substack{\text{(1) (CH}_3\text{CO})_2\text{O} \\ + \ p\text{-CH}_3 . \text{C}_6\text{H}_4 . \text{SO}_3\text{H}}} \quad \text{n-C}_4\text{H}_9 . \text{CH}{=}\overset{\displaystyle O^{\ominus} \ \text{Li}^{\oplus}}{\underset{|}{\text{C}}} . \text{CH}_3 \quad (1.40)$$

$$\downarrow \text{n-C}_4\text{H}_9\text{I}$$

$$(\text{n-C}_4\text{H}_9)_2\text{CH} . \text{CO} . \text{CH}_3 + (\text{n-C}_4\text{H}_9)_2\text{CH} . \text{CO} . \text{CH}_2 . \text{C}_4\text{H}_9\text{-n} + \text{n-C}_4\text{H}_9\text{CH}_2\text{COCH}_2 . \text{C}_4\text{H}_9\text{-}$$
$$\quad\quad (55\%) \quad\quad\quad\quad\quad\quad (10\%) \quad\quad\quad\quad\quad\quad\quad (6\%)$$

The success of this procedure is dependent on the availability of the pure enol acetates. The more highly substituted acetates are generally readily available by acid catalysed equilibration of the mixture obtained from the ketone with isopropenyl acetate or by reaction of the ketone with acetic anhydride and a catalytic amount of perchloric or toluene-*p*-sulphonic acid. The less highly substituted acetates are more troublesome to obtain and generally have to be separated from the mixture formed in the isopropenyl acetate reaction. A useful method for separating some isomeric enols by gas-chromatography or distillation of the trimethylsilyl ethers has been described by Stork and Hudrlik (1968). The trimethylsilyl ethers are readily converted into the corresponding enolate anions, without rearrangement, by Grignard reagents or lithium alkyls.

Specific enolate ions can also be obtained by reduction of α*β*-un-saturated ketones, or of α-bromo- or α-acetoxyketones, with lithium in liquid ammonia (cf. p. 338). Alkylation then affords an α-alkyl derivative of the *saturated* ketone, which may not be the same as that obtained by direct base-catalysed alkylation of the saturated ketone. For example, alkylation of 2-decalone in presence of base generally leads to 3-alkyl derivatives, but by proceeding from the corresponding $\Delta^{1(9)}$-enone the 1-alkyl compound can be obtained readily (Stork, Rosen, Golman, Coombs and Tsuji, 1965) (1.41). The success of this method depends

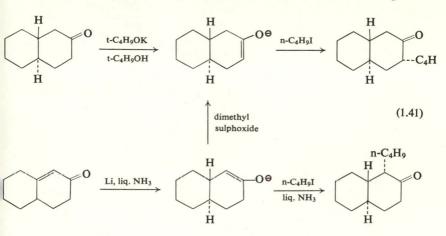

(1.41)

on the fact that in liquid ammonia the alkylation step is faster than equilibration of the initially formed enolate anion to the more stable isomer.

Alkylation of $\alpha\beta$-unsaturated ketones with alkali metal salts of tertiary alcohols as base follows a different course to give either the α-alkyl-$\alpha\beta$-unsaturated ketone or the $\alpha\alpha$-dialkyl-$\beta\gamma$-unsaturated ketone depending on the reaction conditions, although in many cases mixtures of the two products are obtained (Conia, 1963; Ringold and Malhotra, 1962). The reaction has been widely used in the synthesis of α-alkylated enones in the steroid series, and for the introduction of *gem*-dimethyl groups in the synthesis of natural products (1.42).

Reaction proceeds by initial formation of the resonance stabilised anion which undergoes kinetically controlled alkylation at the α-position (compare p. 8) to give the monoalkyl-$\beta\gamma$-unsaturated ketone. The α-proton in this compound is more readily removed by reaction with base than that in either the starting material or the final product, since it is activated by the carbonyl group and an ethylenic double bond. The resulting resonance stabilised anion, in presence of excess of alkylating agent is again alkylated at the α-position to give the $\alpha\alpha$-dialkyl-$\beta\gamma$-unsaturated ketone. If further alkylation does not occur, however, the

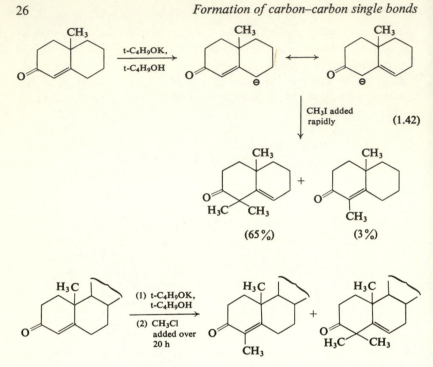

(1.42)

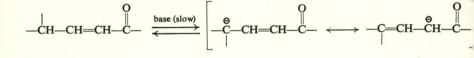

(1.43)

thermodynamically more stable $\alpha\beta$-unsaturated ketone gradually accumulates (1.43). In accordance with this scheme it is found that dialkylation is diminished by slow addition of the alkylating agent or by use of a less reactive alkylating agent (for example, methyl chloride instead of methyl iodide in the example), thus allowing isomerisation of the $\beta\gamma$-unsaturated ketone to the less acidic $\alpha\beta$-isomer to take place before alkylation.

$\alpha\beta$-Unsaturated ketones can also be converted into β-alkyl derivatives of the corresponding *saturated* ketone by conjugate addition of Grignard reagents, as discussed on p. 41.

The stereochemistry of the product obtained in the alkylation of cyclic ketones is important in synthesis but is not always easy to predict. If the product still contains a hydrogen atom attached to the carbon which has been alkylated, equilibration under the alkaline conditions of the reaction generally leads to formation of the more stable isomer, irrespective of the initial direction of attack on the enolate anion. Thus, in the methylation of cholestanone the product is the 2-α-methyl compound in which the methyl group has the more stable equatorial conformation.

In the alkylation of relatively rigid ketones the product obtained in those cases in which equilibration of the system is not a possibility is usually that formed by attack of the alkylating agent on the less-hindered side of the enolate anion. Thus, methylation of norcamphor (II) with sodium triphenylmethyl and methyl iodide leads almost entirely to the *exo* methyl derivative, even though equilibration studies show that the *endo* isomer is (slightly) more stable. Further alkylation of this methyl derivative with methylpentenyl chloride then gives the *endo*-methyl-*exo*-methylpentenyl derivative, again by attack of the alkylating agent from the less hindered *exo* side of the enolate ion (1.44). Reversing the

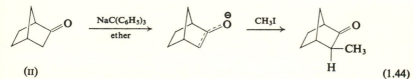

(II)   (1.44)

order of the alkylations leads to the epimeric product, showing that the stereochemistry is controlled by the direction of attack on the enolate ion and not by the size of the alkyl group.

If the enolate anion and the alkylated product are reasonably flexible, both conformational factors and steric hindrance in the enolate anion may influence the steric course of the alkylation process. But it is often difficult to say beforehand what the stereochemical result in any particular case will be, and the various factors involved are not yet completely understood. Thus, while methylation of 2-n-butylthiomethylene-1-decalone leads to a mixture of *cis-* and *trans-*9-methyl derivatives in which the *cis* isomer predominates (see p. 23), introduction of a C-6–C-7 double bond alters the course of the reaction and leads mainly to the *trans* isomer, with an *axial* methyl group. This might be attributed to reduction of 1,3-diaxial interaction of a C-7 hydrogen atom and the methyl group in the transition state were it not for the fact that introduction of a double bond at other positions in the ring system, which also diminishes the number of 1,3-diaxial interactions, does not favour the formation of the *trans* fused product.

Stereoelectronic effects (Corey and Sneen, 1956) have been invoked to rationalise the stereochemistry of alkylation of some cyclohexanone derivatives (House, 1967). Alkylation of an enolate anion should be energetically most favourable when the developing carbonyl group remains in a plane perpendicular to the developing C–C bond between the enolate and the alkylating agent. With this geometry the stabilising interaction of the π orbital of the carbonyl group with the p orbital at the α-carbon continues at a maximum while bonding with the alkylating agent is in progress (1.45). The formation of *cis-* and *trans-*4-t-butyl-2-

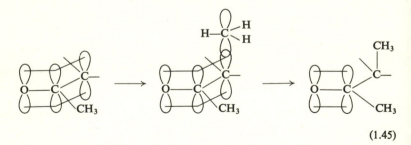

(1.45)

ethylcyclohexanone by alkylation of the enolate anion derived from 4-t-butylcyclohexanone is considered to take place by way of the partial chair and partial twisted boat transition states shown (1.46), in which the ethyl group is introduced perpendicular to the plane of the $>C=O$ bond.

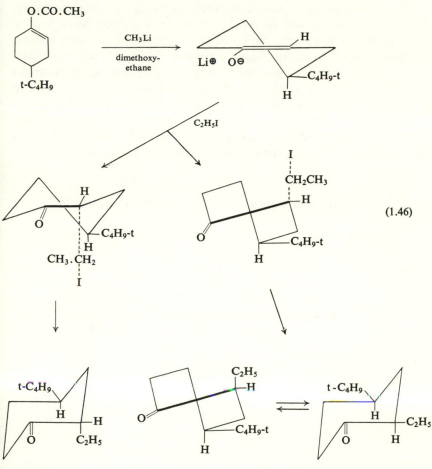

$$(1.46)$$

(51–54% of monoalkylated product)

(46–49% of monoalkylated product)

1.6. The enamine reaction. The enamine reaction, originally interpreted by Robinson (1916) and recently re-introduced by Stork and his co-workers (Stork, Brizzolana, Landesman, Szmuszkovicz and Terrell, 1963; Szmuszkovicz, 1963; Cook, 1968) provides a valuable alternative method for the selective alkylation and acylation of aldehydes and ketones.

Enamines are $\alpha\beta$-unsaturated amines and are simply obtained by reaction of an aldehyde or ketone with a secondary amine in presence of a dehydrating agent such as anhydrous potassium carbonate or, better, by heating in benzene solution in presence of a catalytic amount of toluene-*p*-sulphonic acid, with azeotropic removal of the water formed. Open chain ketones do not form enamines readily under these conditions, but the latter may be obtained by way of the corresponding ketimines (Pfau and Ribière, 1970). The amines found most generally useful in forming enamines are pyrrolidine, morpholine and piperidine in decreasing order of reactivity. Ketones generally give the enamine directly, but with aldehydes the N-analogue of an acetal is first formed and is converted into the enamine on distillation (1.47). All the steps of the

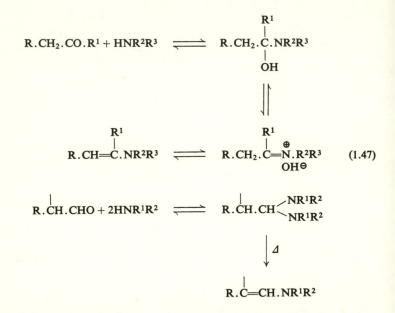

$$R.CH_2.CO.R^1 + HNR^2R^3 \rightleftharpoons R.CH_2.\underset{\underset{OH}{|}}{\overset{\overset{R^1}{|}}{C}}.NR^2R^3$$

$$R.CH=\overset{\overset{R^1}{|}}{C}.NR^2R^3 \rightleftharpoons R.CH_2.\overset{\overset{R^1}{|}}{\underset{OH^\ominus}{C}}=\overset{\oplus}{N}.R^2R^3 \qquad (1.47)$$

$$R.\overset{|}{CH}.CHO + 2HNR^1R^2 \rightleftharpoons R.\overset{|}{CH}.CH\overset{NR^1R^2}{\underset{NR^1R^2}{<}}$$

$$\downarrow \varDelta$$

$$R.\overset{|}{C}=CH.NR^1R^2$$

reaction are reversible, and enamines are readily hydrolysed by water to reform the carbonyl compound. All reactions of enamines must therefore be conducted under strictly anhydrous conditions, but once reaction has been effected the modified carbonyl compound is easily liberated from the product by addition of water to the reaction mixture.

The usefulness of enamines in synthesis is due to the fact that there is some negative charge on the β-carbon atom which can therefore act as a nucleophile in reactions with alkyl and acyl halides and with electro-

$$\underset{}{\overset{}{>}}\mathrm{C}\!\!=\!\!\mathrm{C}\!\!-\!\!\ddot{\mathrm{N}}\!\!<\quad\longleftrightarrow\quad>\!\!\overset{\ominus}{\mathrm{C}}\!\!-\!\!\mathrm{C}\!\!=\!\!\overset{\oplus}{\mathrm{N}}\!\!< \tag{1.48}$$

philic olefins (1.48). Reaction with alkyl halides, for example, leads irreversibly to *C*-alkylated and *N*-alkylated products. Subsequent hydrolysis of the *C*-alkylated imminium salt gives the alkylated ketone; the *N*-alkylated product is usually water soluble and unaffected by the hydrolysis (1.49).

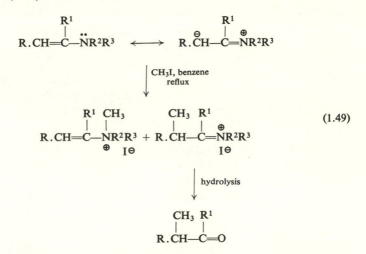

This procedure has a number of advantages over direct base-catalysed alkylation of aldehydes and ketones. Since no base or other catalyst is required there is less tendency for wasteful self-condensation reactions of the carbonyl compound, and even aldehydes can be alkylated and acylated in good yield. A particularly valuable feature of the reaction is that monoalkylated products are readily obtained without concurrent formation of di- and tri-alkylated compounds. The enamine of the monoalkylated ketone is formed by equilibration with the parent enamine, but reacts much more slowly with the alkyl halide than the parent enamine. Another advantage of the enamine reaction is that in the alkylation of unsymmetrical ketones the product formed by reaction at the *less* substituted α-carbon atom is formed in preponderant amount, in contrast to base-catalysed alkylation which usually gives a mixture of products. For example, reaction of the pyrrolidine enamine of 2-methylcyclohexanone with methyl iodide gives 2,6-dimethylcyclohexanone almost exclusively. This selectivity derives from the fact that the enamine from an unsymmetrical ketone consists mainly of the isomer in which the double bond is directed toward the least substituted carbon. The 'more substituted' enamine is destabilised by steric inhibition of the resonance involving the nitrogen lone pair and the double bond, caused by interference of the substituent (CH_3— in the example) (1.50) and the α-methylene group of the amine.

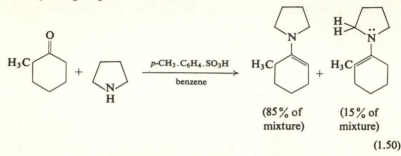

(85% of
mixture)   (15% of
mixture)

(1.50)

Alkylation of enamines with alkyl halides generally proceeds in only poor yield, because the main reaction is *N*-alkylation rather than *C*-alkylation. Good yields of alkylated products are obtained using reactive halides such as benzyl or allyl halides, and it is believed that in these cases there is migration of the substituent group from nitrogen to carbon. This may take place in some cases by an intramolecular pathway, resulting in rearrangement of allyl substituents or, in other cases, by dissociation of the *N*-alkyl derivative followed by irreversible *C*-alkylation (1.51).

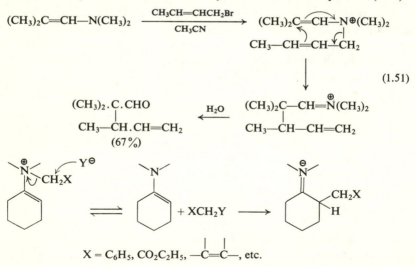

$X = C_6H_5$, $CO_2C_2H_5$, $-C≡C-$, etc.

Fortunately, a modification of the original procedure provides a method which leads to high yields of alkylated aldehydes and ketones even with simple aliphatic alkyl halides (Stork and Dowd, 1963). It is found that many *imines* formed from aliphatic primary amines and enolisable aldehydes and ketones undergo essentially complete enolisation on treatment with one equivalent of ethylmagnesium bromide in boiling tetrahydrofuran. The magnesium salts so formed react readily with primary and secondary alkyl halides to give, after hydrolysis, high yields of monoalkylated carbonyl compound. Even aldehydes are readily alkylated by this method. Allylic halides give products without rearrangement of the allyl group, and the new alkyl substituent is again introduced at the less substituted α-carbon of an unsymmetrical ketone (1.52).

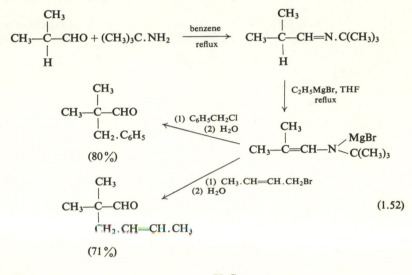

$$H_3C \!\!>\!\! CH.CO.CH_3 \quad \xrightarrow[+ \ n\text{-}C_4H_9Br]{\text{imine salt}} \quad H_3C \!\!>\!\! CH.CO.(CH_2)_4.CH_3 \quad (70\%)$$

Alkylation of enamines of aldehydes and ketones can also be effected with electrophilic olefins, such as αβ-unsaturated ketones, esters or nitriles to give high yields of monoalkylated carbonyl compounds, and the sequence provides a useful alternative to base-catalysed Michael addition. In these reactions *N*-alkylation is reversible and good yields of monoalkylated products are usually obtained. Reaction again takes place at the least substituted α-carbon atom (1.53).

According to Fleming and Harley-Mason (1964) the initial reaction

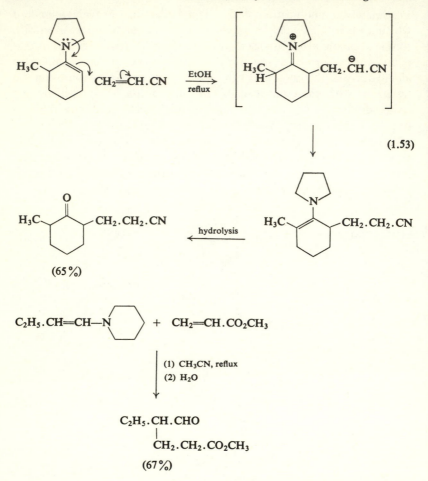

(1.53)

(65%)

(67%)

of an enamine with acrylonitrile is to form a cyclobutane derivative
which may be isolated from reaction at room temperature but which is
converted into the substituted enamine on warming. With acrylic
aldehydes and $\alpha\beta$-unsaturated ketones a dipolar addition product is
first formed which affords the substituted enamine on heating, but at
room temperature leads, reversibly, to a dihydropyran (Colonna, Fattuta,
Risaliti and Russo, 1970; Fleming and Karger, 1967).

Enamines also react readily with acid chlorides or anhydrides to give
products which, on hydrolysis, afford β-diketones or β-keto-esters.
Reaction at the nitrogen atom is again reversible, and good yields of

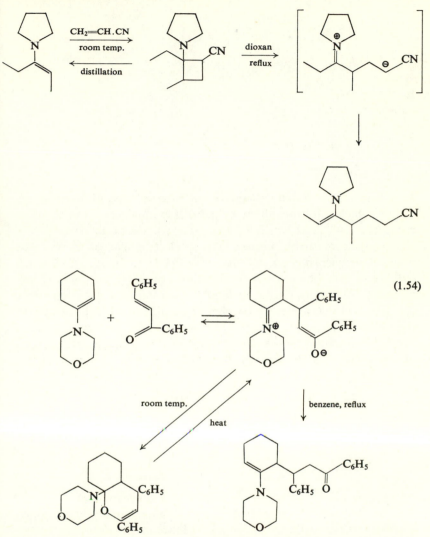

(1.54)

C-acylated products are obtained. The morpholine enamine of cyclo-hexanone, for example, and n-heptanoyl chloride give 2-n-heptanoyl-cyclohexanone in 75 per cent yield (1.55). In these reactions triethylamine is often added to neutralise the hydrogen chloride formed which would otherwise combine with the enamine; alternatively two equivalents of enamine may be used.

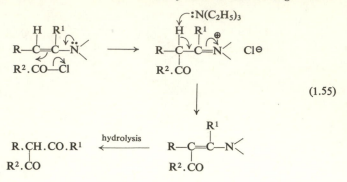

$$(1.55)$$

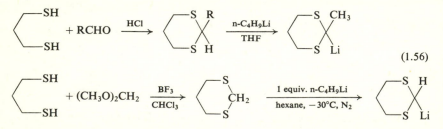

1.7. Reactions of bisthio carbanions. A general class of synthetic reactions which will probably be of increasing importance in the future
consists of processes which by some means reverse temporarily the
characteristic reactivity, nucleophilic or electrophilic, of an atom or
group. One such process, recently developed by Corey and Seebach
(1965) results in the transformation of the normally electrophilic carbon
of a carbonyl group into a nucleophilic carbon. The key step is the
conversion of the carbonyl group into a 1,3-dithiane, which, because of
the stabilising effect of the two electronegative sulphur atoms, is easily
converted into the corresponding carbanion at C-2, that is at the carbon
of the original carbonyl group, by reaction with butyl lithium (1.56).

$$(1.56)$$

The resulting lithium compounds are stable in solution at low temperatures, and undergo the whole range of reactions shown by other organolithium compounds. After reaction the 1,3-dithiane system can be
reconverted into the carbonyl group by hydrolysis with acid in presence
of mercuric ion (Seebach, 1969).

Thus, primary and secondary alkyl halides (iodides are best) react
readily to form 2-alkyl-1,3-dithianes, which on hydrolysis afford
aldehydes or ketones, and the sequence provides a method for converting
an aldehyde into a ketone, or an alkyl halide into the homologous alde-

hyde. Two alkyl groups can be introduced by two successive reactions without isolation of intermediates, and this sequence has been applied in a convenient new synthesis of 3- to 7-membered cyclic ketones (Seebach, Jones and Corey, 1968) (1.57).

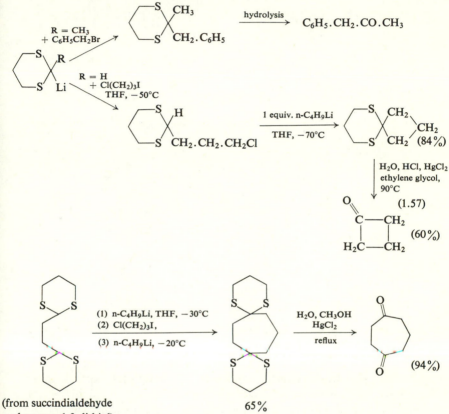

A variety of other reactions has been effected. Epoxides, for example, react readily to form mercaptals of β-hydroxyaldehydes or ketones, aldehydes and ketones give derivatives of α-hydroxyaldehydes or ketones, and acid chlorides and esters give mercaptals of 1,2-dicarbonyl compounds (1.58). In the last example the αβ-unsaturated ketone gives the 1,2- addition product.

An alternative method for preparing 2-lithio derivatives of 1,3-dithianes by addition of organolithium compounds to trimethylene-

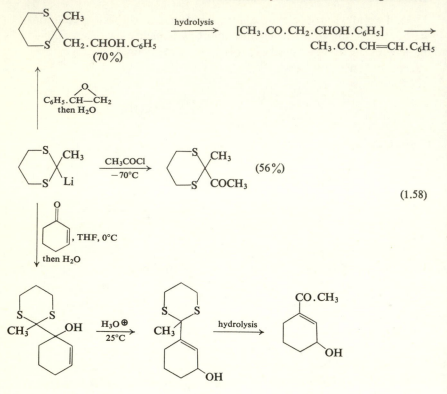

(1.58)

dithiocarbene or to 2-methylene-1,3-dithiane has been described by Carlson and Helquist (1969).

1.8. The dihydro-1,3-oxazine synthesis of aldehydes and ketones.
Another very promising method for the synthesis of aldehydes and ketones from alkyl halides has recently been described by Meyers and his co-workers (1969), and is illustrated in (1.59) in general terms. The starting material is a dihydro-1,3-oxazine derivative (available commercially or by a simple preparation from 2,4-dimethylpentan-2,4-diol and a nitrile), the 'activated' alkyl group of which is readily metallated by treatment with n-butyllithium at −80°C. The resulting anions react rapidly with a wide variety of alkyl halides (bromides and iodïdes) to give high yields of substituted dihydro-oxazines which are smoothly reduced in quantitative yield with sodium borohydride to the tetrahydro derivatives. These compounds are readily cleaved in dilute acid to give

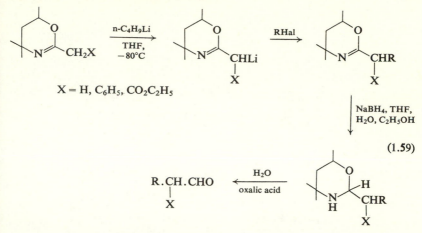

X = H, C_6H_5, $CO_2C_2H_5$

(1.59)

the aldehyde. By this method halides are converted into aldehydes with two more carbon atoms, and the sequence thus complements the synthesis from lithio-1,3-dithiane described above (p. 36) which gives an aldehyde with one more carbon. A useful feature of the reaction is that by reduction with borodeuteride instead of borohydride C-1 labelled aldehydes can be obtained. By appropriate choice of starting materials a variety of different kinds of aldehydes and ketones can be synthesised. Thus, successive alkylation with two molecules of alkyl halide leads to αα-disubstituted aldehydes, and with one molecule of an αω-dihalide alicyclic aldehydes can be obtained. Aldehydes and ketones afford precursors of αβ-unsaturated aldehydes, and with epoxides γ-hydroxyaldehydes are formed. Hydrolysis at the dihydro-oxazine stage leads to carboxylic acids (1.60).

The scope of the reaction has been extended even further by the observation that 2-vinyl- and 2-isopropenyl-dihydro-1,3-oxazines react with Grignard or organolithium reagents to form compounds which can be converted into substituted propionaldehydes and α-methyl unsymmetrical ketones (1.61).

An even more general synthesis of ketones is illustrated by the following synthesis of 1-phenylpentan-3-one (Meyers and Smith, 1970) (1.62). The oxazine synthesis of aldehydes and ketones complements the method using organoboranes (p. 229) which is difficult to apply in some cases because of the complexity of the olefins required.

Some other recent methods for synthesising aldehydes have been reviewed by Carnduff (1966).

$$X = C_6H_5 \xrightarrow[\text{(3) } H_3O^\oplus]{\substack{\text{(1) } CH_3I \\ \text{(2) } NaBH_4}} \underset{|}{\overset{CH_3}{C_6H_5.CH.CHO}} \quad (70\%)$$

$$X = CO_2C_2H_5 \xrightarrow[\text{(3) } CH_3I, \text{ etc.}]{\substack{\text{(1) } CH_3I \\ \text{(2) } n\text{-}C_4H_9Li}} \overset{CH_3}{\underset{CO_2C_2H_5}{CH_3-C-CHO}} \quad (66\%)$$

$$X = H \xrightarrow[\text{(2) } n\text{-}C_4H_9Li, \text{ etc.}]{\text{(1) } BrCH_2CH_2Cl} OHC\!-\!\triangleleft \quad (69\%) \qquad (1.60)$$

$$X = H \xrightarrow[\text{(2) } NaBH_4 \quad \text{(3) } H_3O^\oplus]{\substack{\text{(1) } C_6H_5.CH=CH.CHO \\ \text{then } H_3O^\oplus}} C_6H_5(CH=CH)_2.CHO \quad (61\%)$$

$$X = H \xrightarrow{\text{ethylene oxide, etc.}} \underset{CHO \ CH_2OH}{CH_2-CH_2} \rightleftharpoons \underset{HO.CH \underset{O}{} CH_2}{CH_2-CH_2} \quad (63\%)$$

$$X = H \xrightarrow{Br(CH_2)_5Br} \text{[oxazine]} \xrightarrow[\text{(3) } H_3O^\oplus]{\substack{\text{(1) } Mg, THF \\ \text{(2) } D_2O}} HO_2C(CH_2)_5CH_2D \quad (80\%)$$

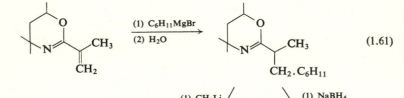

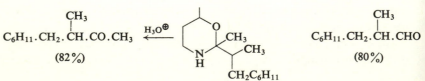

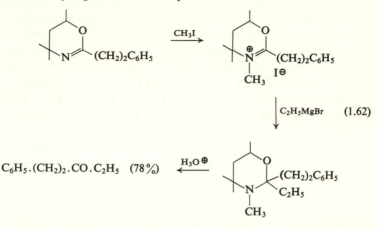

$$C_6H_5.(CH_2)_2.CO.C_2H_5 \quad (78\%) \quad \xleftarrow{H_3O^{\oplus}}$$

1.9. 1,4-Addition of organometallic compounds to αβ-unsaturated ketones. Lithium dialkyl- and diarylcuprates.

1,4-Addition of Grignard reagents to αβ-unsaturated ketones and esters to give β-alkyl and -aryl derivatives of the corresponding saturated carbonyl compound has been known for many years. Early experiments gave erratic results, but it has been found that conjugate addition is strongly favoured by the presence of some metal ions, particularly cuprous ions. High yields of 1,4-addition products can be obtained from reaction of αβ-unsaturated ketones and Grignard reagents in presence of small amounts (1 mol. per cent) of cuprous ion, or of the ether-soluble cuprous iodide-tri-n-butylphosphine complex (Kharasch and Tawney, 1941). Conjugate addition of Grignard reagents to αβ-unsaturated esters is also catalysed by cuprous ions (Munch-Peterson, 1958), (1.63). Another excellent method for effecting conjugate addition to αβ-unsaturated ketones by means of organoboranes is discussed on p. 234.

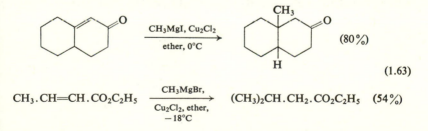

$$CH_3.CH{=}CH.CO_2C_2H_5 \quad \xrightarrow[\substack{Cu_2Cl_2,\ ether,\\ -18°C}]{CH_3MgBr,} \quad (CH_3)_2CH.CH_2.CO_2C_2H_5 \quad (54\%)$$

The factors affecting the stereochemistry of conjugate addition to cyclic ketones are not completely understood, but it seems that in most cases the reactions are sterically controlled. The organomagnesium compound attacks the unsaturated ketone from the least hindered side and in a direction perpendicular to the plane of the conjugated system to form, where possible, a new axial bond (Marshall and Anderson, 1966; House and Thompson, 1963). The steric bulk of the Grignard reagent influences the proportion of the stereoisomers formed. Thus, in the conjugate addition of methylmagnesium iodide to the $\alpha\beta$-unsaturated ketone (III) the product consists mainly of the isomer in which the methyl group is axial, but with the more bulky isopropylmagnesium bromide the axial and equatorial epimers are formed in nearly equal amount (1.64).

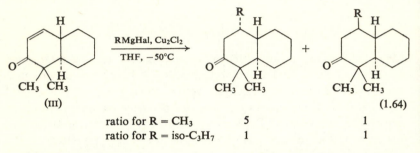

| ratio for R = CH₃ | 5 | 1 |
| ratio for R = iso-C₃H₇ | 1 | 1 |

The rôle of the cuprous ions in these reactions is to form an ether-soluble organocopper species which reacts very rapidly with the olefinic bond of the conjugated ketone. The effective reagent with methylmagnesium halides, methylcopper, is formed and regenerated rapidly by interaction of cuprous salt and Grignard reagent, but since methylcopper itself rapidly forms an insoluble polymer, it probably exists in solution as a solvated complex (House, Respess and Whitesides, 1966) (1.65).

$$Cu^{\oplus} + CH_3MgI \longrightarrow CuCH_3 + MgI^{\oplus}$$

$$CH_3.CH{=}CH.CO.CH_3 + CuCH_3 \longrightarrow (CH_3)_2CH.CH{=}C.CH_3$$
$$\underset{\displaystyle O^{\ominus}Cu^{\oplus}}{|}$$

$$\Big\downarrow MgI^{\oplus} \qquad (1.65)$$

$$(CH_3)_2CH.CH{=}C.CH_3 + Cu^{\oplus}$$
$$\underset{\displaystyle OMgI}{|}$$

A number of promising reagents, based on methylcopper, have been developed for specific addition to the double bond of $\alpha\beta$-unsaturated ketones. One is lithium dimethylcuprate, $Li^{\oplus}Cu^{\ominus}(CH_3)_2$, which is readily obtained by reaction of two equivalents of methyllithium with cuprous iodide in ether, probably as a solvated complex (1.66). Two other useful

$$CH_3Li + Cu_2I_2 \longrightarrow CH_3Cu \longrightarrow (CH_3Cu)_n \qquad (1.66)$$

$$\big\updownarrow CH_3Li, \text{ ether}$$

$$Li^{\oplus}Cu^{\ominus}(CH_3)_2[(C_2H_5)_2O]_2$$

ether-soluble complexes of methylcopper are obtained by reaction of methyllithium with the tri-n-butylphosphine complex of cuprous iodide or with trimethyl phosphite in ether in the presence of cuprous iodide (1.67). All of these reagents react selectively at the β-carbon atom of

$$CH_3Li + (n\text{-}C_4H_9)_3P.Cu_2I_2 \xrightarrow{\text{ether}} CH_3Cu(n\text{-}C_4H_9)_3P[(C_2H_5)_2O]_2$$

$$CH_3Li + Cu_2I_2 + 3(CH_3O)_3P \longrightarrow CH_3.Cu[(CH_3O)_3P]_3 \qquad (1.67)$$

$\alpha\beta$-unsaturated ketones, although the latter two only do so in presence of various inorganic salts (House and Fischer, 1968). $\alpha\beta$-Unsaturated acids and esters are unaffected, and isolated carbonyl groups react comparatively slowly. Thus, while the cyclohexylideneacetone derivative (IV) reacted readily with lithium dimethylcuprate by conjugate addition, the related ethyl ester and carboxylic acid did not react at all (1.68).

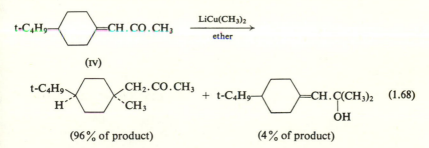

(96% of product)   (4% of product)

Copper derivatives of allylic or acetylenic carbanions show less tendency to undergo conjugate addition than the alkyl, vinyl or aryl compounds (House and Fischer, 1969).

The mechanism of the transfer of the methyl group from methylcopper to the β-position of conjugated ketones is still uncertain. It has been suggested that reaction proceeds through a cyclic six-membered transition state (v) to give an enolate anion stereospecifically, but House, Respess and Whitesides (1966) have found that this cannot be the case because a mixture of stereoisomeric enol acetates was obtained on direct acetylation of the enolate obtained on conjugate addition of methylmagnesium iodide to *trans*-pent-3-en-2-one (1.69). It is now suggested that reaction

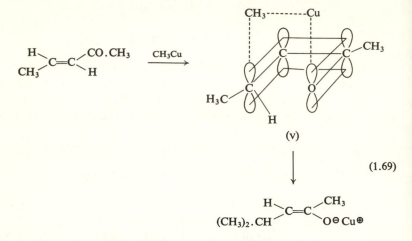

(v)

(1.69)

takes place by one electron transfer from the organocopper(I) species, followed by transfer of an alkyl radical to the β-carbon atom (1.70).

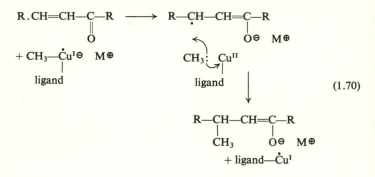

(1.70)

For successful reaction, the alkyl-Cu^I must be complexed with at least one ligand (e.g. $I^\ominus$, $CN^\ominus$, $CH_3{}^\ominus$) which places a net negative charge on the copper complex, thus accounting for the effect of added salts in the reactions of $CH_3Cu(n-C_4H_9)_3P$ and $CH_3Cu[(CH_3O)_3P]_3$.

In favourable cases allylic alcohols R—CH=CH—CHR′OH (R = H or Ar) can be alkylated on the double bond by reaction with primary alkyllithium reagents in presence of tetramethylenediamine. Thus, allyl alcohol and n-propyllithium give 2-methylpentanol in 73 per cent yield (Felkin, Swierczewski and Tambuté, 1969), (1.71).

$$CH_2{=}CH.CH_2OH \xrightarrow[\text{tetramethylenediamine}]{\begin{array}{c}\text{(1) pentane}\\ \text{(2) } H_2O\end{array}} \underset{\underset{\text{n-}C_3H_7}{|}}{CH_3.CH.CH_2OH} \quad (73\%) \quad (1.71)$$

$$+ \text{ n-}C_3H_7Li$$

1.10. Coupling of organonickel and organocopper complexes.

Formation of carbon–carbon bonds by reaction of organic halides with organometallic compounds, as in the Wurtz reaction or the reaction of Grignard reagents with halides, is of only limited use in synthesis. Yields are often poor, owing to intervening side reactions such as α- and β-eliminations, and in 'mixed' reactions, as with RHal and R′M, substantial amounts of RR and R′R′ are formed besides the desired RR′ because of fast halogen–metal interchange. Improved yields of coupled products can be obtained from aryl and s-alkyl Grignard reagents in presence of thallium(I) bromide. Under these conditions *p*-tolylmagnesium bromide was converted into 4,4′-dimethyldiphenyl in 90 per cent yield, and pent-2-ylmagnesium bromide gave 3,4-dimethyloctane in 50 per cent yield (McKillop, Elsom and Taylor, 1968).

A new and promising method for the selective combination of unlike groups using organonickel reagents has recently been described by Corey and Semmelhack (1967). Reaction of allylic bromides with excess of nickel carbonyl in dry benzene gives, in good yield, the crystalline π-allylnickel(I) bromides, which are represented as shown in (1.72).

$$R = H, CH_3, CO_2C_2H_5, \text{ etc.} \quad (1.72)$$

These complexes are relatively inert towards alkyl halides in hydrocarbon or ethereal solvents, but in polar coordinating media such as dimethylformamide or *N*-methylpyrrolidone, a smooth reaction takes place at room temperature between the complex and a wide variety of organic halides (iodides are best) to give the coupled product in high yield (1.73).

$$2R'Hal + \left(R{-}\overset{\ }{\underset{\ }{\text{Ni}}}\overset{Br}{\diagdown}\right)_2 \longrightarrow 2R'{-}CH_2{-}\underset{\underset{\text{}}{}}{\overset{R}{\underset{|}{C}}}{=}CH_2 \quad (1.73)$$

R′ = 4-hydroxycyclohexyl, R = CH_3 $\longrightarrow$ 88%

R′ = C_6H_5 R = CH_3 $\longrightarrow$ 98%

Vinyl and aryl halides react just as well as alkyl halides, and, helpfully, hydroxyl and carbonyl groups do not interfere with the reaction. Carbonyl groups do react with the complexes, but more slowly and at higher temperatures. With dihalides disubstitution products can be obtained with the appropriate quantity of the complex, but with chloro-iodides selective reaction can be effected at the iodo group affording a route to halides useful for further chain extension or other reaction. Aldehydes give $\beta\gamma$-unsaturated alcohols (1.74). The complex from

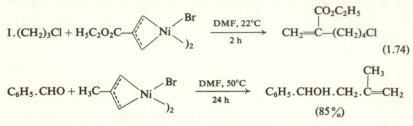

(1.74)

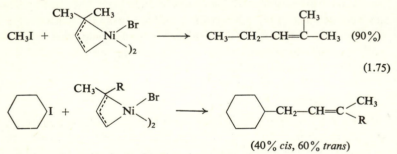

(85%)

DMF $\equiv$ Dimethylformamide

$\alpha\alpha$-dimethylallyl bromide generally undergoes preferential coupling at the primary rather than the tertiary position. In cases where substituents at C-1 and C-3 of the allyl group allow the possibility of geometrically isomeric products both isomers are usually produced (1.75).

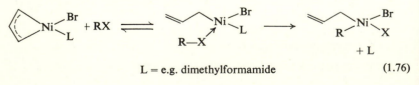

(1.75)

(40% *cis*, 60% *trans*)

R = $(CH_3)_2C$=$CH.CH_2.CH_2$—

Exactly how the new carbon–carbon bond is formed in these reactions is still uncertain, but it is thought that a complex formed by the route shown in (1.76) may be involved.

L = e.g. dimethylformamide

(1.76)

Allyl halides undergo a more complex reaction with π-allylnickel(I) complexes and cannot be used in coupling reactions of the type described above, except where the allyl group in the halide and in the complex is the same. In other cases mixtures of products are obtained (Corey, Semmelhack and Hegedus, 1968).

The reaction has been usefully extended to the synthesis of cyclic 1,5-dienes which are very easily obtained in remarkably high yield by reaction of $\alpha\omega$-bisallyl bromides with nickel carbonyl (1.77).

$$(CH_2)_n \Big\langle {{CH\!=\!CH.CH_2Br} \atop {CH\!=\!CH.CH_2Br}} \xrightarrow[\text{DMF}]{\text{Ni(CO)}_4} (CH_2)_n \Big\langle {{CH\!=\!CH.CH_2} \atop {CH\!=\!CH.CH_2}} \Big| \qquad (1.77)$$

$$n = 6 \quad (59\%)$$
$$n = 8 \quad (70\%)$$
$$n = 12 \quad (84\%)$$

The products consist very largely of the *trans, trans* cyclic dienes, irrespective of whether the starting material contained *cis* or *trans* double bonds. This can be simply explained by allylic isomerisation because of allyl halogen–nickel exchange or allylic rearrangement within the organonickel complex.

With the bisallyl bromide where $n = 2$, however, the main reaction was the formation of a *six-* and not an eight-membered ring; 4-vinylcyclohexene was the predominant product. Similarly, for $n = 4$, the product consisted entirely of *cis-* and *trans*-1,2-divinylcyclohexane; no cyclodeca-1,5-diene was detected. Evidently, in these cases, six-membered ring formation is so much favoured over eight- and ten-membered ring formation that the usual strong preference for the joining of primary carbon atoms is overcome.

This cyclisation procedure, because it leads to cyclic 1,5-dienes, makes available a wide variety of cyclic structures which are not readily obtainable by any other method, and has already been used in the synthesis of a number of natural products. The important sesquiterpene hydrocarbon humulene, for example, was readily synthesised using this method (Corey and Hamanaka, 1967).

A more general procedure for coupling unlike organohalides which is not restricted to the coupling of allylic with non-allylic halides as in the reactions discussed above, makes use of the lithium organocuprates, already discussed on p. 43. Recent studies have shown that they are excellent reagents for the specific replacement of iodine or bromine in a variety of organic substrates by alkyl, vinyl or aryl groups, as illustrated in the examples (1.78), (Corey and Posner, 1967, 1968; Whitesides *et al.*, 1969). In general, lithium di-n-alkyl copper reagents react more rapidly with

$$\text{n-C}_{10}\text{H}_{21}\text{I} + \text{Li(CH}_3)_2\text{Cu} \xrightarrow[\text{0°C, 6 h}]{\text{ether}} \text{n-C}_{11}\text{H}_{24} \quad (90\%)$$

$$\text{n-I(CH}_2)_{10}\text{CO}_2\text{H} \xrightarrow[\text{ether, } -40°C]{\text{Li(Et)}_2\text{Cu}} \text{n-C}_{12}\text{H}_{25}\text{CO}_2\text{H} \quad (70\%)$$

$$\text{n-C}_8\text{H}_{17}\text{I} + \text{Li(CH}_2{=}\text{CH})_2\text{Cu} \cdot \text{P(n-Bu)}_3$$

$$\downarrow \text{THF, ether}$$

$$\text{n-C}_8\text{H}_{17}{-}\text{CH}{=}\text{CH}_2 \quad (95\%)$$

(1.78)

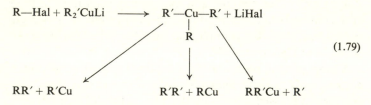

$$(65\%)$$

halides than does lithium dimethylcuprate, but they are less thermally stable. They are also more apt to give rise to halogen–copper exchange as a side reaction, particularly in reactions with aryl iodides, and proper choice of experimental conditions is more important with these reagents. Preliminary experiments indicate that replacement of halogen in alkyl halides takes place with predominant inversion of configuration, as expected for an S_N2-type displacement. With vinyl halides it appears that in most cases the geometry about the double bond is preserved, but at least one example has been recorded in which reaction was not stereospecific.

Little is yet known about the mode of formation of the new carbon–carbon bond. A possible first step is the formation of an intermediate 'trialkylcopper' which can lead to coupling or halogen replacement by metal, depending on the conditions (1.79).

$$\text{R}{-}\text{Hal} + \text{R}_2'\text{CuLi} \longrightarrow \text{R}'{-}\underset{\underset{\text{R}}{|}}{\text{Cu}}{-}\text{R}' + \text{LiHal}$$

(1.79)

$$\text{RR}' + \text{R}'\text{Cu} \qquad \text{R}'\text{R}' + \text{RCu} \qquad \text{RR}'\text{Cu} + \text{R}'$$

Reaction of lithium dialkylcuprates with iodides or bromides is much faster than with non-conjugated carbonyl groups, and this difference has been exploited in a promising new method for cyclisation of ε- and δ-halo-carbonyl compounds. Cyclisation of compounds of this type by an intramolecular Grignard-type reaction is notoriously difficult to effect owing to the lack of a satisfactory method for converting a halo-carbonyl compound into a reactive organometallic derivative. In the examples shown below the cyclic alcohols were obtained in high yield by reaction of the halo-ketone with lithium di-n-butylcuprate or with the dianion derived from nickel tetraphenylporphin under carefully controlled conditions, presumably by way of halogen metal interchange (Corey and Kuwajima, 1970), (1.80).

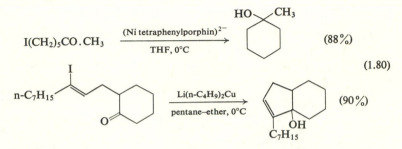

(1.80)

A number of other organic reactions leading to the formation of new carbon–carbon bonds are promoted by copper and copper ions (see pp. 41 and 58) (Bacon and Hill, 1965). Some of these may involve the transient formation of organocopper species. A very useful reaction is the oxidative coupling of terminal acetylenes which is easily effected by agitation of the ethynyl compound with an aqueous mixture of cuprous chloride in an atmosphere of air or oxygen (the Glaser reaction) or by reaction of the ethynyl compound with cupric acetate in pyridine solution. Good yields of coupled products can be obtained from substrates containing a variety of functional groups, and the reaction has been applied to the synthesis of cyclic compounds from αω-diacetylenes (Eglinton and McCrae, 1963), and as a key step in the synthesis of a variety of natural products, for example the fungal polyene, corticrocin (VI) (1.81). The mechanism of the reaction is not entirely clear, but it seems to be generally held that it proceeds by an initial ionisation step, facilitated probably by cuprous ion complex formation, followed by a one-electron transfer involving cupric copper and subsequent dimerisation of the radical formed. There is no doubt, however, that the reaction provides one of the easiest methods of forming carbon–carbon bonds and ranks with the Ullmann, Kolbe and acyloin reactions for molecular duplication.

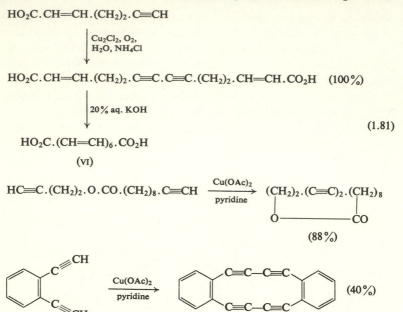

$$HO_2C.CH{=}CH.(CH_2)_2.C{\equiv}CH$$

$$\Big\downarrow \quad \begin{array}{l} Cu_2Cl_2,\ O_2, \\ H_2O,\ NH_4Cl \end{array}$$

$$HO_2C.CH{=}CH.(CH_2)_2.C{\equiv}C.C{\equiv}C.(CH_2)_2.CH{=}CH.CO_2H \quad (100\%)$$

$$\Big\downarrow \quad 20\%\ \text{aq. KOH}$$

(1.81)

$$HO_2C.(CH{=}CH)_6.CO_2H$$

(VI)

$$HC{\equiv}C.(CH_2)_2.O.CO.(CH_2)_8.C{\equiv}CH \xrightarrow[\text{pyridine}]{Cu(OAc)_2}$$

(88 %)

(40 %)

The direct coupling of terminal acetylenes discussed above is obviously not so well suited for the synthesis of unsymmetrical products by coupling of two different acetylenes. Happily, this difficulty has been overcome by the recent discovery that, in presence of a base and a catalytic amount of cuprous ion, terminal acetylenes react rapidly with 1-bromoacetylenes with elimination of hydrogen bromide, to form the unsymmetrical diyne in high yield. The reaction is believed to take the course shown in (1.82). The second step regenerates the cuprous ion

$$R.C{\equiv}CH + Cu^+ \longrightarrow R.C{\equiv}C.Cu + H^\oplus \quad (\text{fast})$$

$$R.C{\equiv}C.Cu + Br.C{\equiv}C.R' \longrightarrow R.(C{\equiv}C)_2.R' + CuBr$$

(1.82)

which is best kept at low concentration to avoid self-coupling of the bromoalkyne. The addition of a base facilitates the reaction by removing the liberated acid and assisting in the solution of the cuprous derivative. This valuable reaction has already been used to synthesise a variety of unsymmetrical diynes, including a number of naturally occurring poly-acetylenic compounds (1.83).

Recent studies indicate that primary and secondary alkyl, vinyl and aryl groups can be coupled by oxidation of the corresponding lithium

$$HO_2C.CH_2.C{\equiv}CH \xrightarrow[20°C,\ C_2H_5NH_2]{Cu^+,\ H_2O} HO_2C.CH_2.C{\equiv}C.C{\equiv}C.C_6H_5$$
$$+ Br.C{\equiv}C.C_6H_5 \qquad\qquad (82\%)$$

$$CH_3.(CH_2)_2.C{\equiv}C.Br + HC{\equiv}C.CH{=}CH.CH_2OH$$

$$\Big\downarrow {\substack{Cu^+,\ CH_3OH \\ H_2O,\ 20°C}} \qquad\qquad (1.83)$$

$$CH_3.(CH_2)_2.(C{\equiv}C)_2.CH{=}CH.CH_2OH \quad (84\%)$$

diorganocuprates (p. 43) with molecular oxygen. Lithium di-s-butyl-cuprate, for example, gave 3,4-dimethylhexane in 82 per cent yield (Whitesides, San Filippo, Casey and Panek, 1967).

1.11. Synthesis of 1,5-dienes from allyl compounds. A useful procedure for the reductive coupling of allylic or benzylic alcohols has been developed by Sharpless, Hanzlik and van Tamelen (1968) and used by them to synthesise a number of 1,5-dienes and polyenes of biosynthetic interest. The method involves treatment of the starting alcohol with a controlled amount of titanium(III) or (IV) chloride, reduction of the titanium alkoxide species, without isolation, with metallic potassium, alkyllithium or sodium naphthalide at −78°C, and final warming of the titanium organic intermediates. 'Mixed' products can be obtained by using an excess of one of the alcohols. The process is believed to involve production of titanium(II) alkoxides or benzoxides, which decompose providing products by way of benzyl or allyl radicals (1.84).

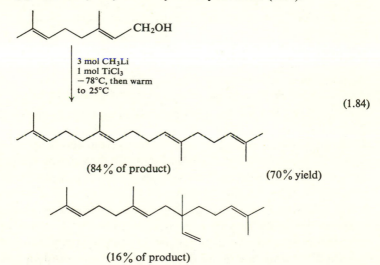

(1.84)

(84 % of product)   (70 % yield)

(16 % of product)

A number of other methods for the synthesis of 1,5-dienes by coupling of allylic units have recently been developed. A promising method which leads to high yields under mild conditions, with virtually complete preservation of the position and geometry of the olefinic bonds, involves reaction between an allylic bromide and an allylic phosphonium salt followed by reduction of the coupled product with lithium and ethylamine. All-*trans*-squalene was obtained from farnesyl bromide in 65 per cent yield by this method, and 'mixed' couplings have also been achieved in good yield (Axelrod, Milne and van Tamelen, 1970). Stork, Grieco and Gregson (1969) have reported that 1,5-dienes can also be obtained in good yield by coupling of allylic Grignard reagents with allylic chlorides. If reaction is conducted in presence of hexamethylphosphoramide very little of the 1,3′-coupled product is said to be formed, although, in fact, the example quoted by the authors does not allow a distinction between coupling at the two possible reactive sites in the allylic Grignard reagent (1.85).

$$\underset{\overset{|}{CH_3}}{(CH_3O)_2CH.CH_2.CH_2.C}{=}CH.CH_2Cl + \underset{\overset{|}{CH_3}}{CH_2}{=}C.CH_2MgCl$$

$$\Big\downarrow \begin{array}{l} THF, \\ hexamethylphosphoramide \end{array} \qquad (1.85)$$

$$\underset{\overset{|}{CH_3}}{(CH_3O)_2CH.CH_2.CH_2C}{=}CH.CH_2.CH_2.\underset{\overset{|}{CH_3}}{C}{=}CH_2 \quad (95\%)$$

Finally, an experimentally simple and effective way of linking allyl residues under mild conditions is provided by the ready electrocyclic rearrangement of vinylsulphonium ylids, themselves generated from sulphonium salts and a base. A sulphide containing a new carbon–carbon bond is first formed and on desulphurisation affords the coupled product (Blackburn, Ollis, Plackett, Smith and Sutherland, 1968; Baldwin, Hackler and Kelly, 1968). By proper choice of the initial sulphide either head-to-tail or tail-to-tail coupling of allyl units can be achieved. Rearrangement also proceeds readily when only one of the sulphide groups is an allyl residue (1.86).

The rearrangements can be regarded as nucleophilic displacement reactions involving an ylid or as sigmatropic reactions involving the covalent form of the ylid intermediate.

With diallyl sulphides more than one product can be formed, and the

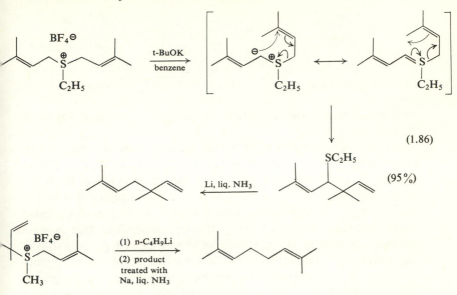

(1.86)

(95%)

course of the rearrangement depends on the alkylation pattern and on the mode of operation.

1.12. Synthetic applications of carbenes and carbenoids. A carbene is a neutral, bivalent carbon intermediate in which a carbon atom is covalently bonded to two other groups and has two valency electrons distributed between two non-bonding orbitals. If the two electrons are spin-paired the carbene is a singlet; if the spins of the electrons are

$$\overset{A}{\underset{B}{\diagdown}} C:$$

parallel it is a triplet (Kirmse, 1964; Hine, 1964).

A singlet carbene is believed to have a bent sp^2 hybrid structure in which the paired electrons occupy the vacant sp^2 orbital. A triplet carbene may be either a bent sp^2 hybrid with an electron in each unoccupied orbital, or a linear sp hybrid with one electron in each of the unoccupied p orbitals. Structures intermediate between the last two are also possible (1.87). The results of experimental observations and molecular orbital calculations indicate that many carbenes have a nonlinear triplet ground state. Exceptions are the dihalogenocarbenes and carbenes with oxygen, nitrogen and sulphur atoms attached to the bivalent carbon, which are probably singlets. The singlet and triplet

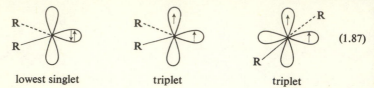

states of a carbene do not necessarily show the same chemical behaviour (Bethell, 1969).

A variety of methods is available for the generation of carbenes, but for synthetic purposes they are usually obtained by thermal or photolytic decomposition of diazoalkanes, or by α-elimination of hydrogen halide from a haloform or of halogen from a *gem*-dihalide by action of base or a metal. In many of these latter reactions it is doubtful whether a 'free' carbene is actually formed. It seems more likely that in these reactions the carbene is complexed with a metal or held in a solvent cage with a salt, or that the reactive intermediate is, in fact, an organometallic compound and not a carbene. Such organometallic or complexed intermediates which, while not 'free' carbenes, give rise to products expected of carbenes are usually called carbenoids (1.88). Carbenes

$$R.CH.N_2 \xrightarrow{h\nu} [RCH\colon] + N_2$$

$$N_2.CH.CO_2C_2H_5 \xrightarrow{\Delta} [\colon CHCO_2C_2H_5] + N_2$$

$$CHCl_3 \xrightarrow{B\ominus} BH + \colon CCl_3\ominus \longrightarrow \colon CCl_2 + Cl\ominus + BH$$

$$R.CHBr_2 + R'Li \longrightarrow R.CHBrLi + R'Br$$

$$\downarrow ?$$

$$[RCH\colon] + LiBr$$

(1.88)

produced by photolysis of diazoalkanes are highly energetic species and indiscriminate in their action, and photolysis is not, therefore, a good method for generating alkylcarbenes for synthesis. Thermal decomposition of diazoalkanes often produces a less energetic, and more selective, carbene, particularly in presence of copper powder or copper salts. Copper–carbene complexes are probably involved in these reactions (see p. 58). Another convenient and widely-used route to alkylcarbenes is by thermal or photolytic decomposition of the lithium or sodium salts of toluene-*p*-sulphonylhydrazones. The diazoalkane is first formed and decomposes under the reaction conditions (1.89). Ketocarbenes and

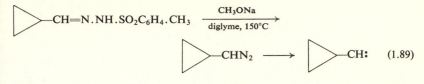

$$\text{alkoxycarbonylcarbenes}$$ are usually produced by heating or photolysing diazoketones and diazoesters.

Carbenes in general are very reactive electrophilic species. Their activity depends to some extent on the method of preparation, on the nature of the substituent groups R and R' in R—$\ddot{\text{C}}$—R', and also on the presence or absence of certain metals or metallic salts (see p. 58). Carbenes undergo a variety of reactions, including insertion into C–H bonds, addition to olefinic and acetylenic bonds, and skeletal rearrangements, some of which are useful in synthesis. Insertion into C–H bonds is not of general synthetic value because mixtures are often produced. Methylene itself attacks primary, secondary and tertiary C–H bonds indiscriminately, although other carbenes may be more selective. With alkylcarbenes intramolecular insertion into C–H bonds is the preferred course of reaction. The major product is usually an olefin, formed by insertion at the β-C–H bond; insertion at the γ-C–H bond gives a cyclopropane as a second product. In general, no intermolecular reactions are observed when intramolecular insertion is possible (1.90). Only in

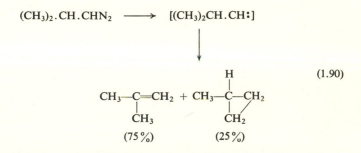

favourable cases where the possibilities of different reactions are limited by geometric factors are insertion reactions of carbenes synthetically useful. For example, camphor toluene-*p*-sulphonylhydrazone, when heated with sodium methoxide in diglyme, gives tricyclene by intramolecular insertion of the derived carbene. A similar reaction was used by Corey, Chow and Scherrer (1957) in a key step in their synthesis of α-santalene (1.91).

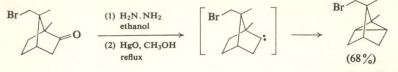

$$(1.91)$$

Probably the most useful application of carbenes in synthesis, however, is in the formation of three-membered rings by addition to olefinic and acetylenic bonds. This is a common reaction of all carbenes which do not undergo intramolecular insertion. Particularly useful synthetically are the halocarbenes, which are readily obtained from a variety of precursors (p. 54). Generated in presence of an olefin they give rise to halocyclopropanes which are valuable intermediates for the preparation of cyclopropanes, allenes, and ring-expanded products (Parham and Schweizer, 1963). Addition of halocarbenes to olefins is a stereospecific *cis* reaction, but this is not necessarily the case with other carbenes. Intramolecular additions are also possible. The toluene-*p*-sulphonyl-hydrazones of $\alpha\beta$-unsaturated aldehydes and ketones, for example, on reaction with sodium methoxide in aprotic media at 160–220° yield alkyl substituted cyclopropenes, presumably by way of the corresponding alkenylcarbene. *gem*-Dihalocyclopropenes are obtained by reaction of dihalocarbenes with disubstituted acetylenes. They have been used to prepare cyclopropenones (1.92).

Addition of carbenes to aromatic systems to form ring-expanded products is a valuable reaction in synthesis. Methylene itself, formed by photolysis of diazomethane, adds to benzene to form cycloheptatriene in 32 per cent yield; a small amount of toluene is also formed by an insertion reaction. Better yields of cycloheptatriene are obtained in presence of copper salts (p. 59). One of the oldest known carbene reactions is the addition of ethoxycarbonylcarbene, from diazoacetic ester, to benzene to form a mixture of cycloheptatrienylcarboxylic esters. Under the conditions of the reaction the intermediate norcaradiene undergoes a Cope rearrangement to give the ring expanded product. With poly-nuclear aromatic hydrocarbons the norcaradiene can sometimes be isolated (1.93).

Dichlorocarbene is insufficiently reactive to attack benzene, but it does add to anthracene, to 1- and 2-methoxynaphthalene and to 9-methoxyphenanthrene. With 9-methoxyphenanthrene the dichloro-carbene addition product has been isolated and pyrolysed to give a dibenzochlorotropone. In most other cases the ring expansion proceeds

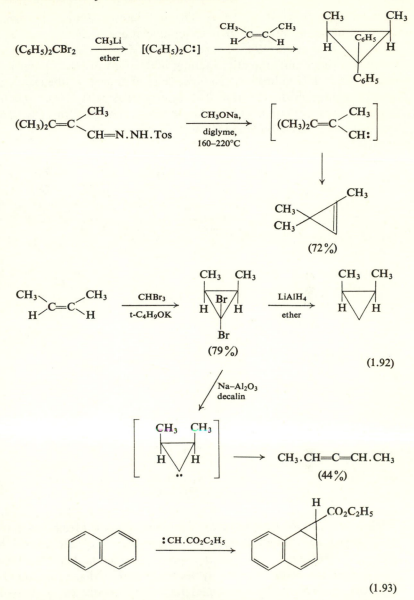

(1.92)

(1.93)

spontaneously. Pyrroles and indoles, for example, in neutral solution, are readily converted into pyridines and quinolines. Under basic conditions substitution products are also formed in a Reimer–Tiemann type reaction (Jones and Rees, 1969). The well-known Reimer–Tiemann reaction, of course, proceeds by electrophilic attack of dichlorocarbene on a phenoxide anion. With indene, since the double bond in the five-membered ring is not aromatic, addition of dichlorocarbene takes place readily. The addition product can be isolated, if desired, or the reaction mixture converted directly into 2-chloronaphthalene by steam distillation. A small amount of chloroazulene is also formed in this reaction by addition to the benzene ring (1.94).

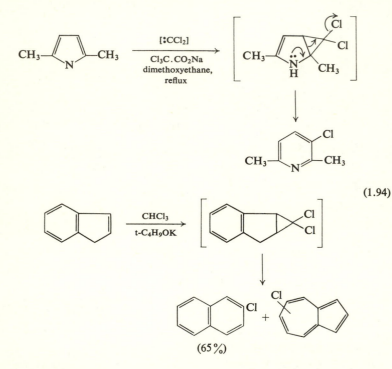

(1.94)

(65%)

The composition of the mixture of products obtained from reactions of carbenes is profoundly altered by the presence of certain transition metals, notably copper and its salts. Under these conditions the intermediates obtained, for example, by decomposition of diazo compounds are more selective than 'free' carbenes. Insertion reactions are suppressed and higher yields of addition products are obtained in reactions with olefins and aromatic compounds (Müller, Kessler and Zeeh, 1966). Thus, benzene reacts readily with diazomethane in the presence of cuprous chloride to form cycloheptatriene in 85 per cent yield. The reaction is general for aromatic systems, substituted benzenes giving a mixture of the corresponding substituted cycloheptatrienes. Related reactions of cyclic and acyclic olefins produce cyclopropanes in good yield and with complete retention of configuration. Intramolecular addition to olefinic double bonds also takes place readily in presence of cuprous salts, and this was exploited by Corey and Achiwa (1969) in a neat synthesis of the hydrocarbon sesquicarene (VII) from *cis,trans*-farnesol (1.95). Unsaturated diazoketones and diazoacetic esters likewise

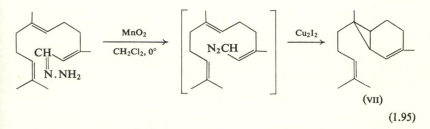

(VII)

(1.95)

form intramolecular addition products in presence of cuprous salts and reactions of this type have been used to prepare bridged bicyclic and 'cage' ketones (1.96). In the absence of copper the main reaction of diazoketones is the Wolff rearrangement (p. 61).

It is very unlikely that 'free' carbenes are involved in any of these catalysed reactions, and cyclopropane formation is believed to involve a copper–carbene–olefin complex similar to that invoked in the Simmons–Smith reaction (1.98). The view that the reactive intermediate is some kind of carbene–copper complex gains strong support from recent kinetic studies and from the observation that catalysed decomposition of ethyl diazoacetate with optically active copper complexes in presence of styrene gave a mixture of cyclopropane derivatives which was optically active (Moser, 1969; Cowell and Ledwith, 1970).

Related to these copper-catalysed reactions of diazoalkanes and diazoketones is the valuable Simmons–Smith reaction (Simmons and Smith, 1959) which is widely used for the synthesis of cyclopropane derivatives from olefins by reaction with methylene iodide and zinc–copper couple. This is a versatile reaction and has been applied with success to a wide variety of olefins. Many functional groups are un-affected and the reaction has been used in the synthesis of a number of naturally-occurring cyclopropane derivatives. Dihydrosterculic acid (VIII), for example, was obtained from methyl oleate in 51 per cent yield. The

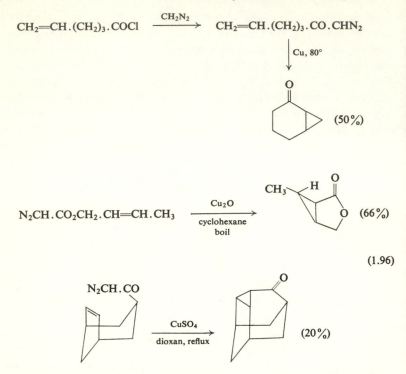

(1.96)

reaction is stereospecific and takes place by *cis* addition of methylene to the least hindered side of the double bond (1.97). The reactive intermediate is believed to be an iodomethylenezinc iodide complex which

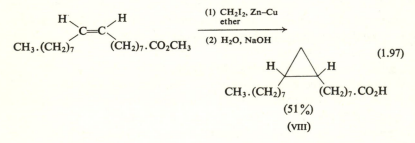

reacts with the olefin in a bimolecular process to give a cyclopropane and zinc iodide (Blanchard and Simmons, 1964) (1.98). A valuable feature of the reaction in synthesis is the stereochemical control exerted on the developing cyclopropane ring by a suitably situated hydroxyl group in the olefin. With allylic and homoallylic alcohols or ethers the rate

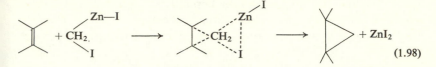

$$(1.98)$$

of the reaction is greatly increased and in five- and six-membered cyclic allylic alcohols the product in which the cyclopropane ring is *cis* to the hydroxyl group is formed stereospecifically. A methoxycarbonyl group in the α- position to the double bond has a similar directing effect (1.99). These effects are ascribed to co-ordination of oxygen to the zinc,

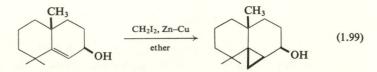

$$(1.99)$$

followed by transfer of methylene to the nearest face of the adjacent double bond. It has recently been found that in eight- and nine-membered cyclic allylic alcohols reaction with methylene iodide and zinc–copper couple takes place stereospecifically to give products in which the cyclopropane ring is *trans* to the hydroxyl group. Presumably, in these reactions the conformation of the ring in the intermediate is such that it is the *trans* face of the double bond which is nearest to the co-ordinated reagent (Poulter, Friedrich and Winstein, 1969).

High yields of cyclopropanes can also be obtained from olefins and ethylidene iodide using diethylzinc as catalyst; zinc–copper couple was ineffective. Reaction is again stereospecific and with cyclic allylic and homoallylic alcohols gives the *cis,anti*-cyclopropane only (Nishimuru, Kawabata and Furukawa, 1969) (1.100).

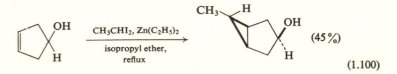

$$(1.100)$$

Carbenes undergo a number of skeletal rearrangements, some of which are useful in synthesis. The most important of these is the Wolff rearrangement of diazoketones to ketenes, which is brought about by heat, light or by action of some metallic catalysts. This reaction is the key step in the well known Arndt–Eistert method for converting a carboxylic acid into its next higher homologue (Bachman and Struve, 1942) (1.101). With cyclic diazoketones the rearrangement leads to ring contraction, and this reaction has been widely used to prepare

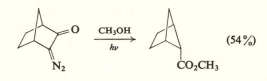

$$R.CO.Cl \xrightarrow{CH_2N_2} R.CO.CHN_2 \longrightarrow \left[\begin{array}{c} O \\ \| \\ C - \ddot{C}H \\ | \\ R \end{array} \right]$$

(1.101)

$$R.CH_2.CO_2R' \cdot \xleftarrow{R'OH} R.CH=C=O$$

derivatives of strained small-ring compounds such as bicyclo[2,1,1]-hexane and benzocyclobutene (Meinwald and Meinwald, 1966) (1.102).

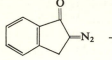

(1.102)

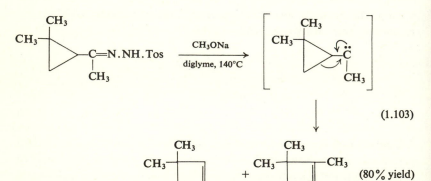

(1.103)

Migration of alkyl groups does not occur easily in alkyl- and dialkyl-carbenes, but cyclopropylcarbenes are exceptional and rearrange to cyclobutenes in high yield (Gutsche and Redmore, 1968; Bird, Frey and Stevens, 1967). The less substituted bond of an unsymmetrical cyclopropane migrates preferentially (1.103).

There is no clear evidence that a carbene is involved in any of these rearrangements. A concerted migration and expulsion of nitrogen is usually a valid alternative (1.104).

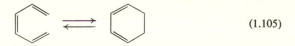

$$\qquad\qquad\qquad\qquad\qquad\qquad\qquad\qquad\qquad\qquad (1.104)$$

1.13. Some photocyclisation reactions.

Few areas of organic chemistry have been more productive of new synthetic reactions in recent years than organic photochemistry (see, for example, Schönberg, 1968). In general, absorption of light by an organic molecule can produce three types of activated molecule not accessible by normal thermal means, namely the electronically excited singlet and triplet states, and, often, a vibrationally 'hot' ground state. Each of these excited states may undergo different chemical reactions in proceeding back to the ground state (Neckers, 1967). The triplet excited state, which generally has a relatively long lifetime, is frequently encountered in photochemical reactions. Because of the high energy of the excited states photochemical reactions often lead to strained structures which would be difficult to obtain by conventional thermal reactions.

Many photochemical transformations are of value in synthesis, and some are referred to elsewhere (see pp. 181, 291). Particularly useful are photoelectrocyclic reactions of the type (1.105), which lead to the

$$\qquad\qquad\qquad\qquad\qquad\qquad\qquad\qquad\qquad\qquad (1.105)$$

interconversion of acyclic conjugated trienes and cyclohexa-1,3-dienes. The formation of an acyclic hexatriene from a cyclohexa-1,3-diene was first noted in the case of the light-induced formation of calciferol from ergosterol, *via* precalciferol. More recent studies show that this ring cleavage reaction takes place with a number of other simple and complex cyclohexa-1,3-dienes and may be fairly general (see Barton, 1959; Corey and Hortmann, 1963). The reaction is reversible and under appropriate conditions certain acyclic 1,3,5-trienes are converted into cyclohexadienes. Thus, irradiation of *trans,cis,trans*-octa-2,4,6-triene in ether affords a stationary state containing 10 per cent of *trans*-1,2-dimethylcyclohexa-3,5-diene. By far the most useful application of this reaction in synthesis, however, is in the conversion of stilbene derivatives, in which two of the double bonds of the 'triene' are contained in benzene rings, into 4a,5a-dihydrophenanthrenes and thence into phenanthrenes (1.106). Stilbene undergoes a rapid *cis–trans* isomerisation under the

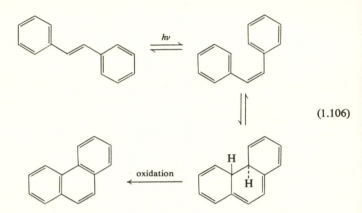

(1.106)

influence of ultraviolet light, and *cis*-stilbene, upon further irradiation, cyclises to give the dihydrophenanthrene. The dihydrophenanthrene has not been isolated but there is convincing evidence for its formation (Moore, Morgan, and Stermitz, 1963). In presence of mild oxidising agents such as oxygen or iodine it is readily converted into phenanthrene, and the sequence of reactions has been widely used to prepare a variety of phenanthrene derivatives containing alkyl, chloro, bromo, methoxy, phenyl, and carboxyl substituents from the appropriately substituted stilbene derivatives. For example, 1,2,7,8-tetramethylphenanthrene was obtained from 2,2′,3,3′-tetramethylstilbene in 50 per cent yield by irradiation of a dilute solution in hexane in presence of iodine; a minor product in this reaction was 1,2,5-trimethylphenanthrene formed by cyclisation in the alternative direction with expulsion of a methyl group. Polycyclic aromatic hydrocarbons can also be obtained by photocyclisation of the appropriate stilbene analogues (Blackburn and Timmons, 1969). Thus 1-α-styrylnaphthalene is converted into chrysene and 2-α-styrylnaphthalene gives benzo[c]phenanthrene. In the last case, cyclisation takes place at the 1-position of the naphthalene nucleus only.

No benz[*a*]anthracene, formed by cyclisation at the 3-position was obtained. The convenience of this synthetic method compensates for the moderate yields obtained in some cases.

Heterocyclic aromatic compounds can also be obtained by this route. Thus, 2-α-styrylpyridine gave 1-azaphenanthrene and 2-α-styrylthiophen cyclised to naphtho[2,1-*b*]thiophen. Azobenzene derivatives and Schiff's bases are cyclised to the corresponding benzocinnoline and phenanthridine derivatives on irradiation in strongly acid solution. The anil (ix), for example, gave calycanine, a degradation product of the alkaloid calycanthine (1.107).

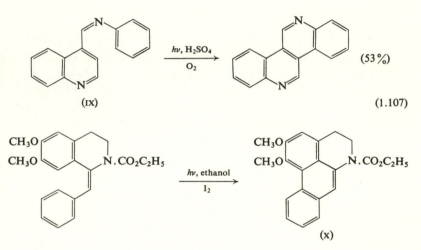

These photocyclisation reactions have proved useful in the synthesis of some natural products. The aporphine alkaloid derivative (x), for example, was prepared by photocyclisation of the appropriate stilbene analogue.

An alternative procedure for the cyclisation of stilbenes makes use of the ready photolysis of the carbon–iodine bond in iodoaromatic compounds. It had been found earlier that photolysis of iodoaromatic compounds in aromatic solvents, particularly benzene, was useful for the preparation of biphenyl and polyphenyl derivatives. In an intramolecular version of the reaction 2-iodostilbene gave phenanthrene in 90 per cent yield by photolysis in hexane solution. This procedure is particularly useful for the preparation of nitrophenanthrenes which cannot be obtained by irradiation of nitrostilbenes (cf. Kupchan and Wormsey, 1965). It seems likely that these reactions proceed by a radical

mechanism and not by way of a dihydrophenanthrene type of intermediate.

Another reaction of practical value in synthesis is the cycloaddition
of olefins to double bonds to form four-membered rings. Most simple
olefins absorb in the far ultraviolet and in the absence of sensitisers
undergo mainly fragmentations and *trans–cis* isomerisation, but conjugated olefins which absorb at longer wavelengths form cycloaddition
compounds readily (Dilling, 1966; Eaton, 1968). Thus butadiene on
irradiation in dilute solution with light from a high pressure mercury
arc forms cyclobutene and bicyclo[1,1,0]butane. Many substituted
acyclic 1,3-butadienes similarly form the corresponding cyclobutenes
on direct irradiation in dilute solution. In presence of a sensitiser reaction may follow a different course. Thus butadiene itself on irradiation
with a sensitiser dimerises to form, mainly, *trans*-1,2-divinylcyclobutane,
and many other acyclic dienes behave in a similar way. Myrcene, on the
other hand, which on direct irradiation in ether solution affords a mixture
of the cyclobutene (XI) and β-pinene, in presence of ketonic sensitisers
gives only the bridged ring compound (XII) (1.08). The sensitised re-

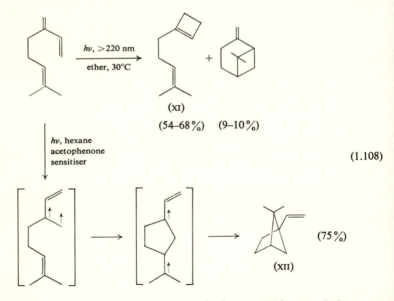

(XI)

(54–68%)   (9–10%)

(1.108)

(75%)

(XII)

action proceeds by way of a triplet excited state. This state behaves as
a diradical and the two bond-forming steps occur consecutively. It is
found in such cases that, where a choice is available, the initial bond

formation takes place to give a five-membered ring rather than a smaller or larger ring.

Many cyclic conjugated dienes undergo 'bridging' reactions on direct irradiation. For example, cyclohepta-1,3-diene forms bicyclo[3,2,0]hept-6-ene, and cyclohepta-1,3,5-triene similarly forms bicycloheptadiene. Bicyclo compounds with suitably disposed double bonds often undergo intramolecular cycloaddition to form cage structures which would be difficult to obtain by any other route. Thus, bicyclo[2,2,1]hepta-2,5-diene on irradiation through quartz in presence of acetone as sensitiser forms quadricyclene (XIII) in 95 per cent yield (1.109).

$\alpha\beta$-Unsaturated carbonyl compounds also readily undergo intra-molecular cycloaddition reactions. One of the first recorded examples was the conversion of carvone into carvone camphor (XIV). Many other cases have since been observed. Thus hexa-1,5-dien-3-one gives the highly strained bicyclo[2,1,1]hexan-2-one (1.110).

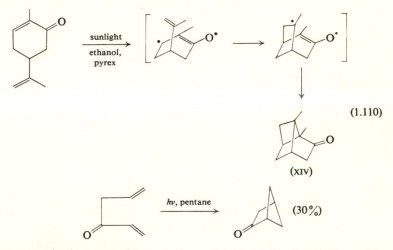

In general, these reactions are brought about by irradiation with light of wavelength greater than 300 nm. Reaction is thought to take place through a triplet excited state of the enone formed by intersystem crossing from the initial $n \rightarrow \pi^*$ excited singlet.

Intermolecular addition of olefins to $\alpha\beta$-unsaturated ketones is also of value in synthesis (Sammes, 1970). Thus, the first step in Corey's synthesis of caryophyllene involved addition of isobutene to cyclohexenone to give the *trans*-cyclobutane derivative (xv). Acetylenes also add to $\alpha\beta$-unsaturated ketones on irradiation, to form cyclobutene derivatives, as in the example (1.111).

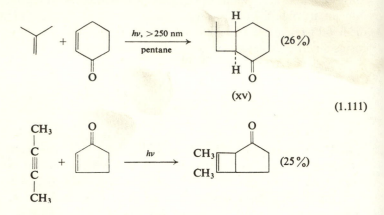

(xv)

(1.111)

Study of the addition of a series of substituted ethylenes to cyclohexenone by Corey, Bass, LeMahieu and Mitra (1964) revealed that the ease of addition was increased by electron-donating substituents on the olefin, following the order shown in (1.112). The reactions proceeded

$$> CH_2{=}C{=}CH_2 \gg CH_2{=}CH.CN$$

(1.112)

with a high degree of orientational specificity and gave mainly the product in which the α-carbon of the enone was attached to the most nucleophilic carbon of the olefin. Addition of *cis*- and *trans*-but-2-ene to cyclohexenone gave the same mixture of products in each case, suggesting that the reactions proceed in a stepwise manner through radical intermediates.

The synthetic usefulness of these cycloaddition reactions has been extended beyond the immediate formation of cyclobutane derivatives in elegant work by de Mayo and his co-workers (Challand, Hikino,

Kornis, Lange, and de Mayo, 1969). Photoaddition of an olefin to the enolised form of a 1,3-diketone results in the formation of a β-hydroxy-ketone, retroaldolisation of which, with ring opening, provides a 1,5-diketone. Thus, irradiation of a solution of acetylacetone in cyclo-hexene affords the 1,5-diketone (xvi) by spontaneous retroaldolisation of the intermediate β-hydroxy ketone (1.113). Cyclic 1,3-diketones may

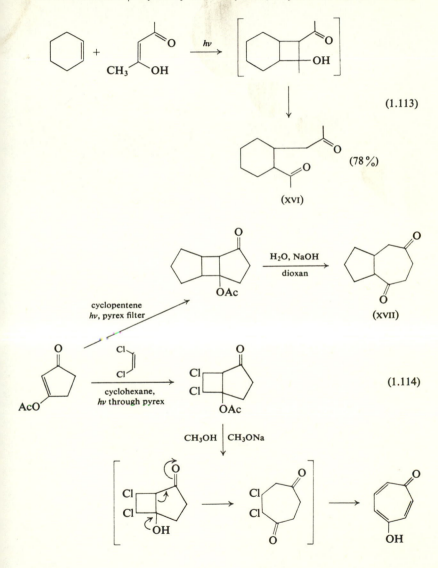

(1.113)

(78%)

(xvi)

(xvii)

(1.114)

also be used, giving rise to products in which the resultant 1,5-diketone is incorporated in a ring. Thus, the diketoperhydroazulene (xvii) was readily obtained from the enol acetate of cyclopentan-1,3-dione and cyclopentene, and an elegant synthesis of γ-tropolone resulted from photo-addition of *cis*-dichloroethylene to the enol acetate followed by reaction of the initial addition product with base.

2 Formation of carbon–carbon double bonds

2.1. β-Elimination reactions. The formation of carbon–carbon double bonds is important in synthesis, not only for the obvious reason that the compound being synthesised may contain a double bond, but also because formation of the double bond followed by reduction may be the most convenient route to a new carbon–carbon single bond.

One of the most commonly used methods for forming carbon–carbon double bonds is by β-elimination reactions of the type

$$-\overset{|}{\underset{\underset{H}{|}}{C}}-\overset{|}{\underset{\underset{X}{|}}{C}}- \longrightarrow -\overset{|}{C}=\overset{|}{C}- + HX \qquad (2.1)$$

where X = e.g. OH, OCOR, halogen, OSO_2Ar, $\overset{+}{N}R_3$, $\overset{+}{S}R_2$. Included among these reactions are acid-catalysed dehydrations of alcohols, solvolytic and base-induced eliminations from alkyl halides and sulphonates, and the well-known Hofmann elimination from quaternary ammonium salts. They proceed by both E1 and E2 mechanisms and synthetically useful reactions are found in each category (Saunders, 1964). Some examples are given in (2.2). These reactions, although much

$$(2.2)$$

$$CH_3.CHBr.CH_2.CH_3 \xrightarrow[C_2H_5OH]{C_2H_5ONa} CH_3.CH\!=\!CH.CH_3 + CH_3.CH_2.CH\!=\!CH_2$$
$$\text{(81 \% of product)} \qquad \text{(19 \% of product)}$$

$$CH_3.\underset{\underset{\oplus S(CH_3)_2}{|}}{C}H.CH_2.CH_3 \xrightarrow[C_2H_5OH]{C_2H_5ONa} CH_3.CH\!=\!CH.CH_3 + CH_3.CH_2.CH\!=\!CH_2$$
$$\text{(26 \% of product)} \qquad \text{(74 \% of product)}$$

$$CH_3.\underset{\underset{\oplus N(CH_3)_3I\ominus}{|}}{C}H.(CH_2)_2.CH_3 \xrightarrow[130°C]{H_2O,\ KOH} CH_2\!=\!CH.(CH_2)_2CH_3 + CH_3.CH\!=\!CHCH_2.CH_3$$
$$\text{(98 \% of product)} \qquad \text{(2 \% of product)}$$

used, leave much to be desired as synthetic procedures. One disadvantage is that in many cases elimination can take place in more than one way, so that mixtures of products are obtained. Mixtures of geometrical isomers may also be formed (see below). The direction of elimination in unsymmetrical compounds is governed largely by the nature of the leaving group, but may be influenced to some extent by the experimental conditions. It is found in general that acid-catalysed dehydration of alcohols, and other E1 eliminations, as well as eliminations from alkyl halides and arylsulphonates with base, gives rise to the most highly substituted olefin as the principal product (the Saytzeff rule), whereas base-induced eliminations from quaternary ammonium salts and from sulphonium salts gives predominantly the least substituted olefin (the Hofmann rule). Exceptions to both rules are observed, however. If there is a conjugating substituent such as CO or C_6H_5 at one β-carbon atom, elimination will take place towards that carbon to give the conjugated olefin, irrespective of the method used. For example, (*trans*-2-phenylcyclohexyl)trimethylammonium hydroxide on Hofmann elimination gives 1-phenylcyclohexene exclusively; none of the 'expected' 3-phenylcyclohexene is detected. Another exception is found in the elimination of hydrogen chloride from 2-chloro-2,4,4-trimethylpentane, which gives mainly the terminal olefin; presumably in this case the intermediate leading to the expected Saytzeff product is destabilised by steric interaction between a methyl group and the t-butyl substituent (2.3). An additional disadvantage of acid-catalysed dehydration of

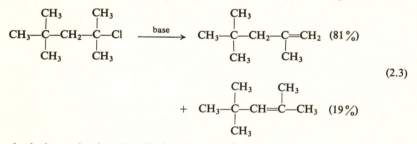

(2.3)

alcohols, and other E1 eliminations, is that since they proceed through an intermediate carbonium ion, elimination is frequently accompanied by rearrangement of the carbon skeleton. Thus, if the α-hydroxy compound in the first equation of (2.2) is replaced by the β-isomer, dehydration is accompanied by ring contraction to give an isopropylidenecyclopentane derivative. Numerous other examples are known in the terpene series, as in the conversion of camphenilol into santene (2.4).

$$\text{(structure)} \xrightarrow{\text{H}_3\text{O}^\oplus} \text{(structure)} \qquad (2.4)$$

The Hofmann reaction with quaternary ammonium salts and base-induced eliminations from alkyl halides and arylsulphonates are generally *anti* elimination processes, that is to say the hydrogen atom and the leaving group depart from opposite sides of the incipient double bond (Saunders, 1964; Cope and Trumbull, 1960) although examples involving *syn* elimination are known (Bailey and Saunders, 1970; de Puy, Naylor and Beckman, 1970). This has been established in a number of ways. For example, reaction of ethanolic potassium hydroxide with *meso*-stilbene dibromide gave *cis*-bromostilbene whereas the (±)-dibromide gave the *trans*-stilbene. Again, the quaternary ammonium salts derived from *threo*- and *erythro*-1,2-diphenylpropylamine were found to undergo stereospecific elimination on treatment with sodium ethoxide in ethanol as shown in (2.5). In agreement with the relatively rigid

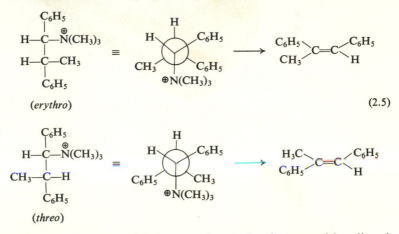

(2.5)

requirements of the transition state, the *erythro* isomer, with eclipsed phenyl groups, reacted more slowly than the *threo*.

For stereoelectronic reasons elimination takes place most readily when the hydrogen atom and the leaving group are in the anti-periplanar arrangement (I), although a number of apparent '*syn*' eliminations have been observed in which hydrogen and the leaving group are eclipsed (II) (2.6). In open-chain compounds the molecule can usually adopt a conformation in which H and R are anti-periplanar, but in cyclic systems this is not always so, and this may have an important bearing on the

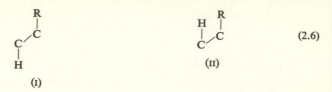

$$\text{(2.6)}$$

(I)   (II)

direction of elimination in cyclic compounds. In cyclohexyl derivatives anti-periplanarity of the leaving groups requires that they be diaxial even if this is the less stable conformation, and on this basis we can understand why menthyl chloride (III) on treatment with ethanolic sodium ethoxide gives only 2-menthene (2.7), while neomenthyl chloride (IV) gives a mixture of 2- and 3-menthene in which the 'Saytzeff product' predominates. The elimination from menthyl chloride is much slower than that from neomenthyl chloride because the molecule has to adopt an unfavourable conformation with axial substituents before elimination can take place.

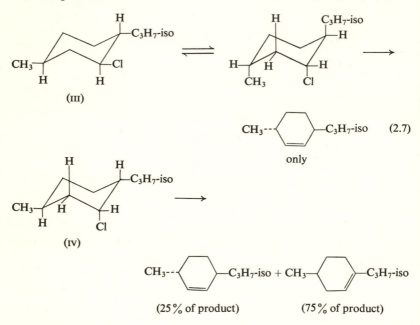

(III)

$$\text{(2.7)}$$

only

(IV)

(25 % of product)   (75 % of product)

In spite of the disadvantages, however, acid-catalysed dehydration of alcohols and base-induced eliminations from halides and aryl-sulphonates are widely used in the preparation of olefins. The bases used in the latter reactions include alkali metal hydroxides and alkoxides, as well as organic bases such as pyridine and triethylamine. Recently, superior results have been obtained using 1,5-diazabicyclo[3,4,0]non-5-ene (VI) (Oediger, Kabbe, Möller and Eiter, 1966). Thus, the chloroester (V) could not be dehydrochlorinated with pyridine or quinoline but with the base (VI) at 90°C the enyne was obtained in 85 per cent yield (2.8).

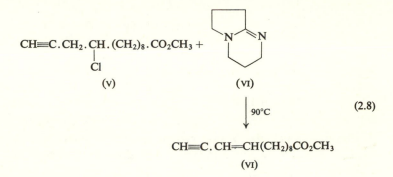

Lithium chloride (or bromide) in dimethylformamide has also been found to be an effective reagent for dehydrohalogenation in some cases (Wendler, Taub and Kuo, 1960).

2.2. Pyrolytic *syn* eliminations. Another important group of olefin-forming reactions, some of which are useful in synthesis, are pyrolytic eliminations (DePuy and King, 1960). Included in this group are the pyrolyses of carboxylic esters and xanthates, which provide valuable alternative methods for dehydration of alcohols without rearrangement, and pyrolysis of amine oxides. For the majority of cases these reactions are believed to take place in a concerted manner by way of a cyclic transition state and, in contrast to the eliminations discussed above, they are necessarily *syn* eliminations, that is the hydrogen atom and the leaving group depart from the same side of the incipient double bond (2.9).

The *syn* character of these pyrolytic eliminations has been demonstrated in a number of ways. Thus, pyrolysis of the *erythro* and *threo* isomers of 1-acetoxy-2-deutero-1,2-diphenylethane gave in each case *trans*-stilbene, but the stilbene from the *erythro* compound retained nearly all its deuterium, whereas the stilbene from the *threo* compound had lost most of its deuterium. Either the hydrogen or the deuterium could be *syn* to the acetoxy group, but the preferred conformations (shown)

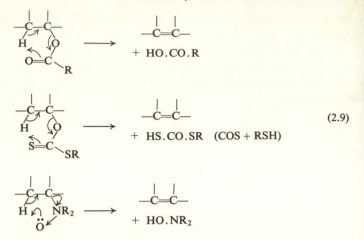

$$(2.9)$$

are those in which the phenyl groups are as far removed from each other as possible (2.10). Similarly, *threo*-dimethyl(3-phenylbut-2-yl)-

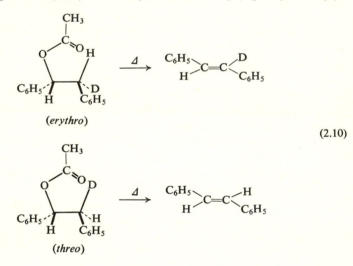

$$(2.10)$$

amine oxide on pyrolysis gave mainly *cis*-2-phenylbut-2-ene, while the *erythro* compound gave the *trans* olefin, and the oxide of *cis*-1-phenyl-2-dimethylaminocyclohexane gave 3-phenylcyclohexene, elimination involving the *syn* and not the *anti* hydrogen even although the latter is activated by the phenyl group.

Pyrolysis of esters to give an olefin and a carboxylic acid is usually effected at a temperature of about 300–500°C and may be carried out by simply heating the ester if its boiling point is high enough, or by passing the vapour through a heated tube. Yields are usually good, and the absence of solvents and other reactants simplifies the isolation of the product. In practice acetates are nearly always employed, but esters of other carboxylic acids can be used. The reaction provides an excellent method for preparing pure terminal olefins from primary acetates and, because it does not involve either acidic or basic reagents, it is especially useful for preparing strained, highly reactive dienes and trienes. An example is the preparation of 4,5-dimethylenecyclohexene, which is obtained from 4,5-diacetoxymethylcyclohexene without extensive re-arrangement to *o*-xylene (2.11). With secondary and tertiary acetates,

$$CH_3.CH_2.CH_2.CH_2OAc \xrightarrow[N_2]{500°C} CH_3.CH_2.CH=CH_2 \quad (100\%)$$

(2.11)

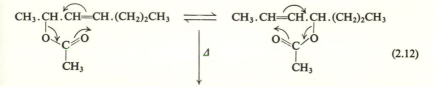

where elimination can take place in more than one direction, mixtures of products are usually obtained (see page 72).

The high temperature required for elimination is a disadvantage in these reactions, and in some cases the olefin formed may not be stable under the reaction conditions. Thus, pyrolysis of 1-cyclopropylethyl acetate affords mainly cyclopentene by rearrangement of the initially formed 1-vinylcyclopropane. In some cases also rearrangement of the ester may take place before elimination, leading to mixtures of products. This is especially liable to occur with allylic esters as in the example (2.12).

$$CH_3.CH.CH=CH.(CH_2)_2CH_3 \rightleftharpoons CH_3.CH=CH.CH.(CH_2)_2CH_3$$

$$\downarrow \Delta$$

(2.12)

$$CH_2=CH.CH=CH.(CH_2)_2CH_3 + CH_3.CH=CH.CH=CH.CH_2CH_3$$

Pyrolysis of amine oxides (the Cope reaction) (Cope and Trumbull, 1960) and of methyl xanthates (the Chugaev reaction) (Nace, 1962) takes place at much lower temperatures (100–200°C) than that of carboxylic esters, with the result that further decomposition of sensitive olefins can often be avoided in these reactions. On the other hand separation of the olefin from the other products of the reaction may be more troublesome, particularly in the Chugaev reaction, where sulphur-containing impurities have often to be removed by distillation from sodium. Pyrolysis of amine oxides generally takes place under particularly mild conditions, and this often allows the generation of a new olefinic bond without migration into conjugation with other unsaturated systems in the molecule, as in the synthesis of penta-1,4-diene (2.13). If an allyl or benzyl group is attached to the nitrogen atom rearrangement to give an *O*-substituted hydroxylamine may compete with elimination.

$$
\text{t-C}_4\text{H}_9\overset{\overset{\displaystyle \text{CH}_3}{|}}{\underset{\underset{\displaystyle \text{H}}{|}}{\text{C}}}\text{—O.CS.S.CH}_3 \xrightarrow{\ 170°\text{C}\ } \text{t-C}_4\text{H}_9.\text{CH}{=}\text{CH}_2 \quad (71\%)
$$

$$
\underset{\underset{\displaystyle \text{O}}{\downarrow}}{\text{CH}_2{=}\text{CH}.(\text{CH}_2)_3.\text{N}(\text{CH}_3)_2} \xrightarrow{\ 140°\text{C}\ } \begin{array}{l}\text{CH}_2{=}\text{CH}.\text{CH}_2.\text{CH}{=}\text{CH}_2 \quad (61\%) \\ + (\text{CH}_3)_2\text{NOH} \qquad\qquad (2.13)\end{array}
$$

$$
\underset{\underset{\displaystyle \text{O}}{\downarrow}}{\text{CH}_2{=}\text{CH}.\text{CH}_2.\text{N}(\text{C}_2\text{H}_5)_2} \xrightarrow{\ 150°\text{C}\ } \begin{array}{l}(\text{C}_2\text{H}_5)_2\text{N}.\text{O}.\text{CH}_2.\text{CH}{=}\text{CH}_2 \quad (59\%) \\ + \text{CH}_2{=}\text{CH}_2\end{array}
$$

Like the base-induced eliminations discussed above, pyrolytic *syn* eliminations have the disadvantage that with derivatives of unsymmetrical secondary and tertiary alcohols and with unsymmetrical amine oxides, mixtures of products are formed. With open chain compounds, the three methods give closely similar results. If there is a conjugating substituent in the β-position elimination takes place to give the conjugated olefin, but otherwise the composition of the product is determined mainly by the number of hydrogen atoms on each β-carbon. For example, pyrolysis of s-butyl acetate affords a mixture containing 57 per cent but-1-ene

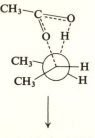

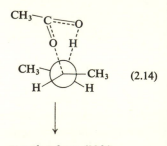

(2.14)

cis-but-2-ene (15%)   *trans*-but-2-ene (28%)

and 43 per cent but-2-ene, in close agreement with the $3:2$ distribution predicted on the basis of the number of β-hydrogen atoms. Of the but-2-enes the *trans* isomer is formed in larger amount, presumably because there is less steric interaction between the two methyl substituents in the transition state which leads to the *trans* olefin (2.14). It is found in general that with aliphatic esters, xanthates or amine oxides which could form either a *cis* or *trans* olefin on pyrolysis, the more stable *trans* isomer is formed in larger amount (2.15).

$$CH_3.CH_2.\underset{\underset{O.CS.SCH_3}{|}}{CH}.CH_2.CH_2CH_3 \xrightarrow{250°C} CH_3.CH_2.CH{=}CH.CH_2.CH_3$$

(17% *cis*)

(33% *trans*) (2.15)

$$+ CH_3.CH{=}CH.(CH_2)_2.CH_3$$

(16% *cis*)

(34% *trans*)

$$CH_3.CH_2.\underset{\underset{\underset{O}{\downarrow}}{\overset{|}{N(CH_3)_2}}}{CH}.CH_3 \xrightarrow[85-150°C]{91\%} CH_3.CH{=}CH.CH_3 + CH_3CH_2CH{=}CH_2 \quad (67\%)$$

(11.7% *cis*)

(21.0% *trans*)

$$C_6H_5.\underset{\underset{CH_3}{|}}{CH}{-}\underset{\underset{\underset{O}{\downarrow}}{\overset{|}{N(CH_3)_2}}}{CH}.CH_3 \xrightarrow{110-115°C} C_6H_5.\underset{\underset{CH_3}{|}}{C}{=}CH.CH_3 + 7\% \text{ unconjugated}$$

(93% of product)

threo $\longrightarrow$ 94% *cis*, 0·1% *trans*

erythro $\longrightarrow$ 4% *cis*, 89% *trans*

With alicyclic compounds some restrictions are imposed by the conformation of the leaving groups and the necessity to form the cyclic intermediate. Thus, the cyclohexyl acetate (VII) in which the leaving group is axial, in contrast to its isomer (VIII), does not form a double bond in the direction of the ethoxycarbonyl group, even though it would be conjugated, because the necessary cyclic transition state is not sterically possible. With the xanthate (IX), however, in which the leaving group is equatorial, the six-membered cyclic transition state can lead equally well to abstraction of a hydrogen atom from either β-carbon, and an equimolecular mixture of two products results (2.16).

Pyrolysis of alicyclic amine oxides does not necessarily lead to product mixtures similar in composition to those obtained from the corresponding

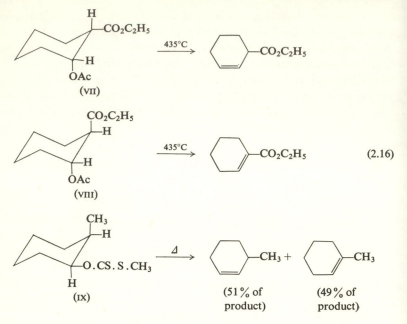

$$(2.16)$$

(51% of   (49% of
product)   product)

acetates and xanthates (De Puy and King, 1960). A notable difference is found in the pyrolysis of 1-methylcyclohexyl derivatives. The acetates and xanthates afford mixtures containing 1-methylcyclohexene and methylenecyclohexane in a ratio of 3:1, whereas pyrolysis of the oxide of 1-dimethylamino-1-methylcyclohexane gives methylenecyclohexane almost exclusively. The reason for this is thought to be that in the oxide the coplanar five-membered cyclic transition state only allows abstraction of hydrogen from the methyl substituent, whereas with the more flexible six-membered transition state of the ester pyrolysis, hydrogen abstraction is also possible from the ring. With larger more flexible rings, where the requisite planar transition state is more easily attained, the cycloalkene is the preponderant product from amine oxides as well.

As a general rule formation of compounds with a double bond exocyclic to a ring is not favoured in pyrolysis of esters, presumably because of their relative instability with respect to other isomers. For instance, 1-cyclohexylethyl acetate on pyrolysis gives mainly vinylcyclohexane.

Sulphoxides also undergo a ready *syn* elimination on pyrolysis to form olefins. These reactions also appear to take place by way of a concerted cyclic pathway. Since sulphoxides are readily obtained by reaction of primary alkyl bromides or toluene-*p*-sulphonates with the anion of dimethyl sulphoxide, the sequence provides, for example, a convenient method for preparing terminal olefins from primary alcohols containing one fewer carbon atom (Entwistle and Johnstone, 1965; Kingsbury and Cram, 1960) (2.17).

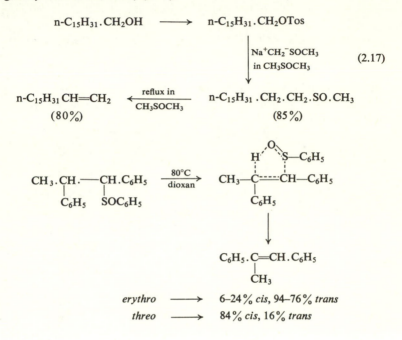

$$\text{erythro} \longrightarrow \text{6–24\% cis, 94–76\% trans}$$
$$\text{threo} \longrightarrow \text{84\% cis, 16\% trans}$$

2.3. The Wittig and related reactions. The reaction occurring between a phosphorane, or phosphonium ylid, and an aldehyde or ketone to form a phosphine oxide and an olefin, has become known as the Wittig reaction after the German chemist Georg Wittig who first showed the value of this procedure in the synthesis of olefins (Trippett, 1963; Maercker, 1965; Johnson, 1966) (2.18). The reaction is easy to carry

$$R_3P{=}CR^1R^2 + {}^{R^3}_{R^4}{>}C{=}O \longrightarrow R_3P{=}O + {}^{R^1}_{R^2}{>}C{=}C{<}^{R^3}_{R^4} \quad (2.18)$$

out and proceeds under mild conditions to give, in general, high yields of the olefinic compound. A particularly valuable feature of the Wittig procedure is that, in contrast to the elimination and pyrolytic reactions discussed above, it gives rise to olefins in which the position of the newly-formed double bond is unambiguous, as illustrated in the examples (2.19).

$$(C_6H_5)_3P{=}CH_2 + n\text{-}C_7H_{15}COCH_3 \xrightarrow{\text{ether}} n\text{-}C_{17}H_{15}.\underset{\underset{CH_3}{|}}{C}{=}CH_2$$

$$(2.19)$$

$$(C_6H_5)_3P{=}CH.CH_3 + CH_3.CO.CO_2C_2H_5 \xrightarrow[\text{reflux}]{\text{ether}} CH_3.CH{=}\underset{\underset{CH_3}{|}}{C}.CO_2C_2H_5$$

Phosphoranes are resonance-stabilised structures in which there is some overlap between the carbon p orbital and one of the d orbitals of phosphorus as shown in (2.20). Reaction with a carbonyl compound

$$\underset{R^2}{\overset{R^1}{>}}C{=}PR_3 \quad \longleftrightarrow \quad \underset{R^2}{\overset{R^1}{>}}\overset{\ominus}{C}{-}\overset{\oplus}{P}R_3 \qquad (2.20)$$

phosphorane form   ylid form

takes place by attack of the carbanionoid carbon of the ylid form on the electrophilic carbon of the carbonyl group with the formation of a betaine which collapses to the products by way of a four-membered cyclic transition state (2.21). Depending on the reactants either the

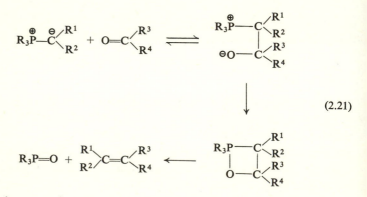

$$(2.21)$$

first or the second step may be rate-determining. It has never been observed that the last step is the slowest, and it is uncertain whether the four-membered ring compound is a true intermediate or a transition state. Evidence for the formation of a betaine in the first stage of the reaction is provided by the isolation of compounds of this type in certain cases.

The reactivity of the phosphorane depends on the nature of the groups R, R^1 and R^2. In practice, R is nearly always phenyl. Alkylidene trialkylphosphoranes, in which the formal positive charge on the phosphorus is lessened by the inductive effect of the alkyl groups, are more reactive than alkylidene triphenylphosphoranes in the initial addition to a carbonyl group to form a betaine, but by the same token decomposition of the betaine becomes more difficult, and alkylidene trialkylphosphoranes are superior to the triphenyl compounds only in certain special cases. In the alkylidene part of the phosphorane, if R^1 or R^2 is an electron-withdrawing group (e.g. CO or CO_2R) the negative charge in the ylid becomes delocalised into R^1 or R^2 and the nucleophilic character, and reactivity towards carbonyl groups, is decreased. Reagents of this type are much more stable and less reactive than those in which R^1 and R^2 are alkyl groups, and with them the rate-determining step in reactions with carbonyl groups is the initial addition to form the betaine. The more electrophilic the carbonyl group the more readily does the reaction proceed. Thus, in the reaction of fluorenylidenetriphenylphosphorane with benzaldehyde derivatives introduction of electron-withdrawing substituents into the benzaldehyde resulted in an increase in reaction rate; the reverse held for electron-releasing substituents (2.22). Similarly the preparatively important carbalkoxy-

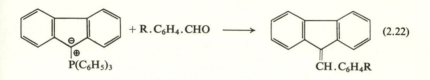

$$\text{(2.22)}$$

benzaldehyde	yield %
R = *o*-NO$_2$	96
p-Cl	93
H	84
p-OCH$_3$	37
p-N(CH$_3$)$_2$	0

methylenetriphenylphosphoranes react fairly readily with aldehydes, but give only poor yields in reaction with the less reactive carbonyl group of ketones (2.23).

In the majority of Wittig reagents R^1 and R^2 are alkyl groups which have little effect on the carbanionoid character of the molecule. These reagents are markedly nucleophilic and react readily with carbonyl and other polar groups. Addition of the ylid to carbonyl groups takes place

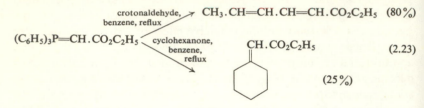

$$\text{(C}_6\text{H}_5)_3\text{P}\!\!=\!\!\text{CH}.\text{CO}_2\text{C}_2\text{H}_5 \underset{\substack{\text{cyclohexanone,}\\\text{benzene,}\\\text{reflux}}}{\overset{\substack{\text{crotonaldehyde,}\\\text{benzene, reflux}}}{\big<}} \begin{array}{l} \text{CH}_3.\text{CH}\!\!=\!\!\text{CH}.\text{CH}\!\!=\!\!\text{CH}.\text{CO}_2\text{C}_2\text{H}_5 \quad (80\%) \\[2em] \text{CH}.\text{CO}_2\text{C}_2\text{H}_5 \end{array}$$

(2.23)

(25%)

rapidly and decomposition of the betaine now becomes the rate-determining step of the reaction. Since the polarity of the carbonyl group is of little consequence when the second step is rate-determining, aldehydes and ketones usually react equally well with these reagents.

Alkylidenephosphoranes can be prepared by a number of methods, but in practice they are usually obtained by action of base on alkyltriphenylphosphonium salts, which are themselves readily available from an alkyl halide and triphenylphosphine (2.24). The phosphonium salt

$$(\text{C}_6\text{H}_5)_3\text{P} + \text{R}^1\text{R}^2\text{CHX} \longrightarrow (\text{C}_6\text{H}_5)_3\overset{\oplus}{\text{P}}\!\!-\!\!\text{CHR}^1\text{R}^2 \ \ \text{X}^\ominus$$

$$\text{base} \Big\updownarrow$$

(2.24)

$$(\text{C}_6\text{H}_5)_3\text{P}\!\!=\!\!\text{CR}^1\text{R}^2 + \text{HB}$$

can usually be isolated and crystallised but the phosphorane is generally prepared in solution and used without isolation. Formation of the phosphorane is reversible, and the strength of base necessary and the reaction conditions depend entirely on the nature of the ylid. A common procedure is to add the stoichiometric amount of an ethereal solution of phenyl- or n-butyllithium to a solution or suspension of the phosphonium salt in ether, benzene or tetrahydrofuran, followed, after an appropriate interval, by the carbonyl compound. Lithium and sodium alkoxides, in solution in the corresponding alcohol or in dimethylformamide, are being used increasingly. This method is simpler than the procedure using organolithium compounds, and has the added advantage that it can be used to prepare phosphoranes from compounds containing functional groups, such as the ethoxycarbonyl group, which would react with organolithium compounds. By this method, also, good yields of condensation products can be obtained from unstable ylids by generating the ylid in presence of the carbonyl compound with which it is to react. Another convenient reagent is the sodium salt of dimethyl sulphoxide which is easily formed by warming dimethyl sulphoxide with sodium hydride. Reaction is effected in excess of dimethyl sulphoxide as solvent by successive addition of the phosphonium salt and the carbonyl compound (Corey and Chaykovsky, 1962).

Reactions involving non-stabilised ylids must be conducted under anhydrous conditions and in an inert atmosphere, because these ylids react both with oxygen and with water. Water effects hydrolysis, with formation of a phosphine oxide and a hydrocarbon; the most electro-negative group is cleaved (2.25). With oxygen, reaction leads in the

$$(C_6H_5)_3P{=}CH.C_6H_5 \quad \xrightarrow{\ H_2O\ } \quad (C_6H_5)_3\overset{\oplus}{P}.CH_2C_6H_5 \ \ OH^{\ominus}$$

$$\downarrow \qquad\qquad (2.25)$$

$$(C_6H_5)_3P{=}O + C_6H_5CH_3$$

first place to triphenylphosphine oxide and a carbonyl compound which undergoes a Wittig reaction with unoxidised ylid to form a symmetrical olefin (2.26). This method has been used to prepare symmetrical olefins.

$$(C_6H_5)_3P{=}CHC_6H_5 \quad \xrightarrow{\ O_2\ } \quad (C_6H_5)_3PO + C_6H_5CHO$$

$$\downarrow {\scriptstyle (C_6H_5)_3P{=}CH.C_6H_5} \qquad (2.26)$$

$$C_6H_5.CH{=}CH.C_6H_5 \ \ (55\%)$$

For example vitamin A was converted into β-carotene *via* axerophthyl-enetriphenylphosphorane.

A very useful alternative method for the preparation of resonance-stabilised phosphoranes for use in the Wittig reaction proceeds from phosphonate esters, themselves readily available from alkyl halides and triethyl phosphite. Reaction with a suitable base gives the corresponding anions (as x) which are strongly nucleophilic and react readily with the carbonyl group of aldehydes and ketones to form an olefin and a water soluble phosphate ester. Thus, the anion (x) from ethyl bromoacetate and triethyl phosphite reacts rapidly with cyclohexanone at room temperature to give ethyl cyclohexylideneacetate in 70 per cent yield (2.27). In general, the phosphonate ester reaction is the method of choice for the interaction of resonance-stabilised phosphoranes with aldehydes and ketones, and this procedure has been widely used for the preparation of $\alpha\beta$-unsaturated esters. It is unsuitable for the preparation of olefins

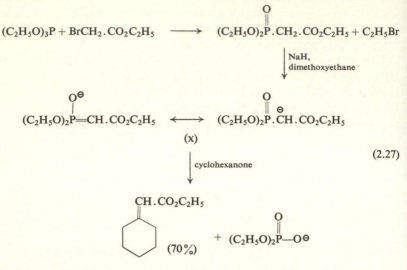

$$(C_2H_5O)_3P + BrCH_2.CO_2C_2H_5 \longrightarrow (C_2H_5O)_2\overset{O}{\overset{\|}{P}}.CH_2.CO_2C_2H_5 + C_2H_5Br$$

$$\Big\downarrow \text{NaH, dimethoxyethane}$$

$$(C_2H_5O)_2\overset{O^{\ominus}}{\overset{|}{P}}=CH.CO_2C_2H_5 \longleftrightarrow (C_2H_5O)_2\overset{O}{\overset{\|}{P}}.\overset{\ominus}{CH}.CO_2C_2H_5$$

$$(x)$$

$$\Big\downarrow \text{cyclohexanone}$$

(2.27)

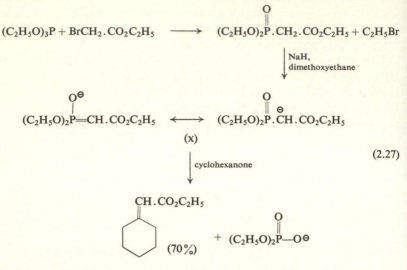

from non-stabilised reagents because decomposition of the betaine through a four-membered cyclic transition state, which is rate-determining in these cases, is hindered by the decreased positive character of the phosphorus.

Because of the mild conditions under which it proceeds, its versatility and the unambiguous position of the double bond formed, the Wittig reaction is an almost ideal olefin synthesis. Its only disadvantage is that as originally described it is not subject to steric control. Where the structure of the olefin allows it a mixture of *cis* and *trans* isomers is usually produced in which, with stabilised ylids, the *trans* isomer generally predominates although non-stabilised ylids may give more of the *cis* olefin. Benzylidenetriphenylphosphorane and benzaldehyde, for example, give a product containing 25 per cent *cis*- and 75 per cent *trans*-stilbene.

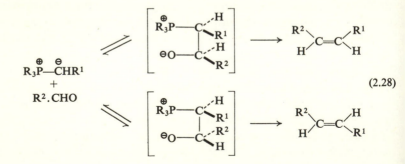

(2.28)

This preference for the *trans* isomer is a reflection of the fact that the *threo* form of the intermediate betaine, which leads to the *trans* olefin, is less sterically hindered than the *erythro* form (2.28). Recent work has demonstrated, however, that the steric course of the reaction can be substantially altered by varying the reaction conditions, and by proper choice of experimental conditions of a high degree of steric control can be achieved. Reactions of stabilised ylids with aldehydes in dimethyl-formamide containing lithium iodide are said to give increased amounts of *cis* olefin (Bergelson, Barsukov and Shemyakin, 1967). Under these conditions Bergelson and Shemyakin (1964) have effected the synthesis of a number of unsaturated fatty acids with *cis* olefinic bonds. For example, reaction of ω-ethoxycarbonyl-n-octylidenetriphenylphosphor-ane with pelargonic aldehyde in dimethylformamide solution gave the expected olefin which was 95 per cent ethyl oleate (2.29). These variations

$(C_6H_5)_3P\!=\!CH.(CH_2)_7.CO_2C_2H_5$

$\quad + \; C_8H_{17}.CHO$

$\xrightarrow[\text{dimethyl formamide}]{\text{NaI}}$

$$C_8H_{17}.CH\!=\!CH.(CH_2)_7CO_2C_2H_5 \quad (2.29)$$

$$(95\% \;\; cis)$$

in the ratio of *cis* and *trans* products can be ascribed to the influence of the solvents and additives on the relative stabilities and relative rates of decomposition of the *threo*- and *erythro*-betaines. The various factors involved are discussed by Bergelson, Barsukov and Shemyakin (1967) and by Schlosser and Christmann (1967). See also Reucroft and Sammes (1971).

Almost pure *trans* olefins can be obtained by equilibration of the intermediate betaines by converting them into betaine ylids by the action of phenyllithium. This is exemplified in the synthesis of *trans*-oct-3-ene, which is represented by Schlosser and Christmann (1966) as shown in (2.30). Interconversion of the stereoisomeric betaine ylids (XI) is very rapid and the equilibrium strongly favours the form (XI*b*) which affords the *threo*-betaine on protonation. Subsequent treatment with butanol gives almost pure *trans* olefin.

This technique has been extended by Schlosser and Christmann (1969) and by Corey and Yamamoto (1970) to provide a method for the stereo-specific synthesis of trisubstituted olefins of the types in (2.31). For example, the *cis*-allylic alcohol (XIV) was obtained from n-heptaldehyde and ethylidenetriphenylphosphorane by the steps shown in (2.32).

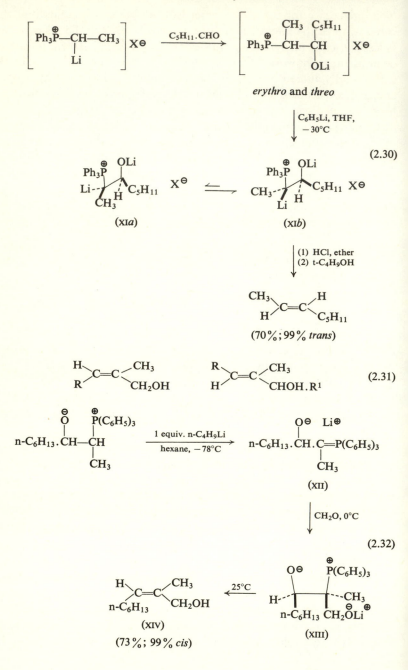

Reaction of the betaine formed by addition of n-heptaldehyde to the phosphorane with one equivalent of n-butyllithium gave a β-oxido-phosphonium ylid which with paraformaldehyde gave stereospecifically the derivative (XIII), and thence the allylic alcohol (XIV) which was almost entirely the *cis* isomer. Reaction of β-oxidophosphonium ylids with other aldehydes gave secondary alcohols which, in the cases studied, had the *trans* configuration. Thus, sequential treatment of ethylidene-triphenylphosphorane with benzaldehyde, n-butyllithium and again benzaldehyde gave the alcohol (XV), (2.33). Tracer studies have shown that the oxygen atom in the product originates in the first aldehyde used. Reaction of the oxidophosphonium ylid (XII) with acetaldehyde thus affords specifically the alcohol (XVI) in 67 per cent yield.

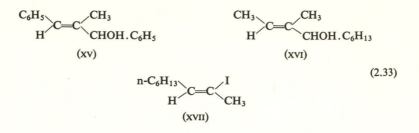

(2.33)

Alkylation of β-oxidophosphonium ylids with alkyl halides, as a possible route to trialkylethylenes, was unsuccessful except with methyl iodide, in which case the reaction was not stereospecific. But successive reaction with mercuric acetate and iodine gave stereospecifically iodides of the type (XVII) which could, presumably, be converted into trialkyl-ethylenes by reaction with organocopper reagents (p. 47) (Corey, Shulman and Yamamoto, 1970).

The value of the Wittig reaction is clearly shown by the fact that it has already been used in the synthesis of very many olefins, including a considerable number of natural products (Maercker, 1965). It is a versatile synthesis and can be used for the preparation of mono-, di-, tri-, and tetra-substituted ethylenes, and also of cyclic compounds. The carbonyl components may contain a wide variety of other functional groups such as hydroxyl, ether, ester, halogen and terminal acetylene, which do not interfere with the reaction. $\alpha\beta$-Unsaturated carbonyl compounds generally react only at the carbonyl group, in contrast to their behaviour with Grignard reagents. The mild conditions of the Wittig reaction make it an ideal method for the synthesis of sensitive olefins such as carotenoids and other poly-unsaturated compounds (see below).

One especially useful application of the Wittig reaction is in the formation of exocyclic double bonds. The Wittig reaction is the only method for converting a cyclic ketone into a exocyclic olefin. The Grignard method gives the endocyclic isomer almost exclusively. Thus, cyclohexanone and methylenetriphenylphosphorane give methylene-cyclohexane, and the same reaction has been used to prepare a variety of methylene steroids. Inhoffen's synthesis of vitamins D_2 and D_3 involved the Wittig reaction in three of the steps of which the first is shown in (2.34). Unsaturated aldehydes and ketones can also be used,

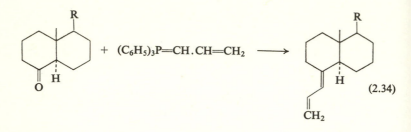

$$+ \quad (C_6H_5)_3P{=}CH.CH{=}CH_2 \longrightarrow \qquad\qquad (2.34)$$

and the interaction of a bifunctional carbonyl compound with two equivalents of a phosphorane has been employed in the preparation of polyenes and polyenynes. Alternatively, a bifunctional phosphorane can be condensed with two molecules of a carbonyl compound. Both procedures have been extensively used in the synthesis of carotenoids.

With bifunctional carbonyl compounds and bisphosphoranes cyclic compounds can be obtained, as in the synthesis of 1,2,5,6-dibenzocyclo-octatetraene (2.35).

In compounds that contain both ester and carbonyl groups the latter react preferentially as long as the Wittig reagent is not present in excess.

A valuable group of Wittig reagents is derived from α-haloethers. They react with aldehydes and ketones to form vinyl ethers which on acid hydrolysis are converted into aldehydes containing one more carbon atom. Thus, for example, cyclohexanone is converted into formylcyclohexane as shown in the equation (2.36).

A promising new route to substituted olefins has recently been described by Corey and Kwiatkowsky (1968). It is based on readily available phosphonbis-N,N-dialkylamides of type (XVIII) and has a number of advantages over the Wittig procedure. In particular, it can be used for

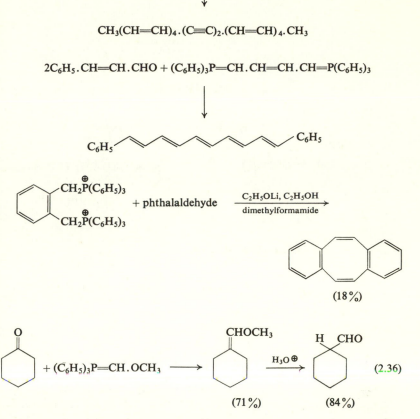

$OHC . CH=CH . (C≡C)_2 . CH=CH . CHO + 2(C_6H_5)_3P=CH(CH=CH)_2CH_3$

(2.35)

$CH_3(CH=CH)_4 . (C≡C)_2 . (CH=CH)_4 . CH_3$

$2C_6H_5 . CH=CH . CHO + (C_6H_5)_3P=CH . CH=CH . CH=P(C_6H_5)_3$

C_6H_5 ⟋⟍⟋⟍⟋⟍⟋⟍ C_6H_5

+ phthalaldehyde $\xrightarrow[\text{dimethylformamide}]{C_2H_5OLi, C_2H_5OH}$

(18%)

+ $(C_6H_5)_3P=CH . OCH_3$ ⟶ $\xrightarrow{H_3O⊕}$ (2.36)

(71%)   (84%)

the stereoselective synthesis of *cis* and *trans* isomers. By reaction with n-butyllithium the phosphonamides form lithio derivatives which react readily with aldehydes and ketones to give β-hydroxyphosphonamides. On heating in benzene solution these hydroxyphosphonamides decompose to form olefins in high yield. This method offers wide scope for the synthesis of olefins, for not only can a range of aldehydes and ketones be used, but also a variety of phosphonamides, prepared directly from the appropriate alkyl halide or elaborated from (XVIII) as in the example shown (2.37). Furthermore, diastereoisomeric β-hydroxy-

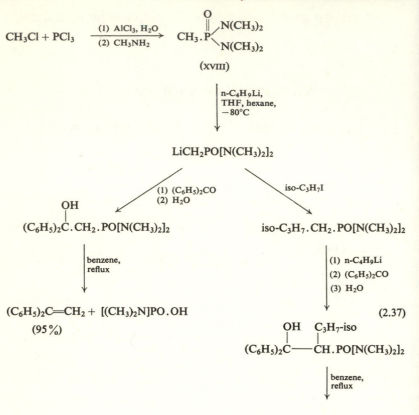

phosphonamides decompose stereospecifically, each giving a single olefin. Separation of pure diastereoisomers by fractional crystallisation and separate decomposition thus affords a route to the *cis* and *trans* forms of a particular olefin. This is illustrated below for the synthesis of *cis*- and *trans*-1-phenylpropene. In this case one diastereoisomer of the hydroxyphosphonamide was separated by fractional crystallisation of the mixture formed in the reaction of the lithio derivative with benzaldehyde, but it was found easier to obtain the other by reduction of the corresponding β-keto-phosphonamide, itself readily available by reaction of the lithiophosphonamide with methyl benzoate (2.38). This exemplifies another convenient route to the synthetically useful β-hydro-oxyphosphonamides. Recent studies have shown that the thermal decomposition of β-hydroxyphosphonamides involves preferential *syn*

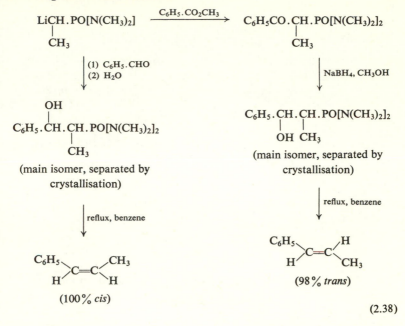

(2.38)

elimination, probably by way of a four-membered cyclic transition state similar to that invoked in the Wittig reaction (Corey and Cane, 1969).

β-Hydroxysulphinamides (specifically sulphinanilides and sulphin-*p*-toluidides) formed by reaction of lithio reagents of type (XIX) with aldehydes and ketones, also undergo ready thermal decomposition to form olefins (Corey and Durst, 1968).

$$[LiCH_2.SO.N.C_6H_5]^{\ominus}Li^{\oplus}$$

(XIX)

Recently the reactions of a number of sulphur ylids, particularly dimethylsulphoxonium methylide (XX) and dimethylsulphonium methylide (XXI) have been investigated. They do not behave in the same way as Wittig reagents, and in their reactions with carbonyl compounds they form oxiranes and not olefins. αβ-Unsaturated carbonyl compounds give either oxiranes or cyclopropyl ketones, depending on the reagent. Yields are generally high and these reactions provide a good synthetic route to oxiranes and cyclopropyl ketones (Corey and Chaykovsky, 1965; Corey and Jautelat, 1967; Johnson, Hruby and Williams, 1964). Some reactions are illustrated in (2.39). The ylids are prepared by proton

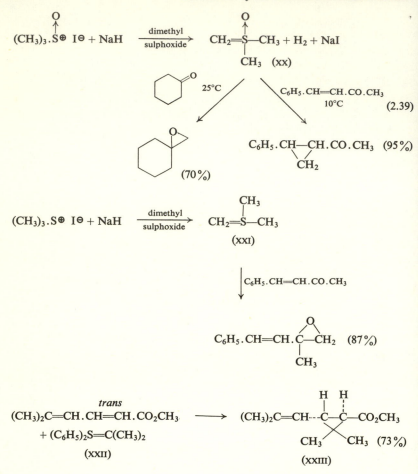

abstraction from sulphoxonium salts and sulphonium salts. The sulphoxonium ylid (xx) is more stable than the sulphonium ylid (xxi) and attacks the olefinic double bond of $\alpha\beta$-unsaturated carbonyl compounds rather than the carbonyl group. In reactions with cyclohexanones, moreover, the sulphoxonium ylid forms a new equatorial carbon–carbon bond, whereas the sulphonium ylid preferentially gives an axial carbon–carbon bond. The diphenylsulphonium isopropylide reagent (xxii) is particularly valuable for synthesising the *gem*-dimethylcyclopropyl unit, which is found in many natural products, as exemplified in the synthesis of ($\pm$)-methyl *trans*-chrysanthemate (xxiii).

2.4. Stereoselective synthesis of tri- and tetra-substituted ethylenes. *cis*-
and *trans*-1,2-Disubstituted ethylenes are readily available by reduction
of the corresponding acetylenes. Various suitable methods of effecting
these reactions are discussed later, in the relevant sections of the book
(see pp. 307, 345). Recently a number of stereoselective syntheses of
tri- and tetra-substituted ethylenes have been developed in the course of
work aimed at the synthesis of biologically and biosynthetically important
polyisoprenoids. The first of these methods due to Cornforth, Cornforth
and Mathew (1959) was subsequently applied by them in their synthesis
of all-*trans*-squalene. The critical step in this method is the reaction
of a Grignard reagent with an α-chloro-aldehyde or -ketone. The most
reactive conformation of an α-chloroaldehyde or α-chloroketone is that
in which the C=O and C—Cl dipoles are antiparallel. It was found that
addition of a Grignard reagent to this conformation is highly stereoselective and takes place mainly from the side of the carbonyl group
which is less sterically hindered by the groups R^1 and R^2 on the α-carbon
atom, leading predominantly to the chlorohydrin in which the incoming
group R^4 and the larger of the groups R^1 and R^2 are *anti* to each other
(2.40). The resulting chlorohydrin is converted, by a series of stereo-

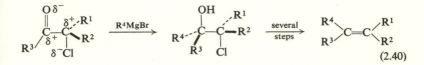

(2.40)

selective conversions, into the olefin in which three of the groups on the
double bond are derived from the chlorocarbonyl compound, and the
other from the Grignard reagent. The method is exemplified in (2.41)
for the synthesis of 3-methyl-*trans*-pent-2-ene. A similar series of
reactions beginning with the addition of methylmagnesium iodide to
2-chloropentan-3-one gave 3-methyl-*cis*-pent-2-ene.

An entirely different procedure which makes possible the stereospecific
conversion of a propargylic alcohol into a 2- or 3-alkylated allylic
alcohol has been described by Corey, Katzenellenbogen and Posner
(1967) (2.42). This procedure is based on a remarkable, specific, conversion of propargylic alcohols into β- or γ-iodoallylic alcohols by
reduction with a modified lithium aluminium hydride reagent and subsequent reaction of the crude reduction product with iodine. If the
reduction is effected in presence of sodium methoxide, the final product

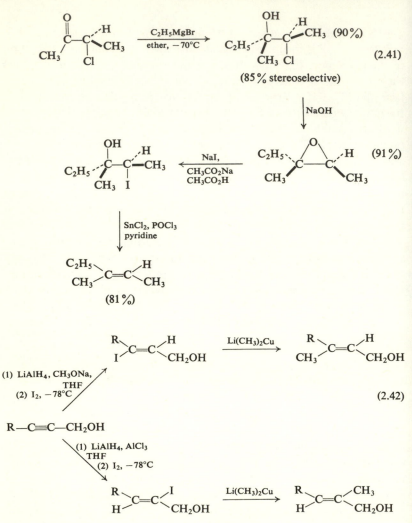

is exclusively the γ-iodoallylic alcohol, whereas reduction with lithium aluminium hydride and aluminium chloride gives finally the β-iodoallylic alcohol. Reaction of the resulting iodo compounds with lithium organo-cuprates (p. 47) then affords the corresponding substituted allylic alcohols. This method is applicable to a variety of synthetic problems in which the stereospecific introduction of trisubstituted olefinic linkages is involved. For example, it has been used by Corey and his co-workers

(1968) in their synthesis of juvenile hormone, the relevant steps of which are shown in (2.43) .

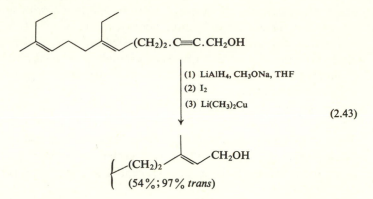

(2.43)

Another synthesis which leads to $\beta\beta$-dialkylacrylic esters from $\alpha\beta$-acetylenic esters has been described by Corey and Katzenellenbogen (1969) and Siddall, Biskup and Fried (1969) (2.44). Lithium alkylcopper

$$R^1—C\equiv C—CO_2CH_3 + LiR_2Cu \longrightarrow$$

$$\underset{(xxiv)}{R^1\diagdown \atop R^2\diagup}C=C\diagup^{CO_2CH_3}_{\diagdown H} \quad + \quad \underset{(xxv)}{R^1\diagdown \atop R^2\diagup}C=C\diagup^{H}_{\diagdown CO_2CH_3} \quad (2.44)$$

reagents (p. 43) prepared *in situ* from alkyllithium and cuprous iodide add rapidly to $\alpha\beta$-acetylenic esters to give the stereoisomeric acrylic esters (xxiv) and (xxv) in high yield. The stereochemistry of the products is highly dependent on the reaction temperature and the nature of the solvent, and high yields of the *cis* addition compound (xxiv) are obtained by conducting the reaction at $-78°C$ in tetrahydrofuran solution. In contrast to the procedure with propargyl alcohols described above, this reaction yields an olefin in which the substituents in the acetylenic precursor are *cis* to each other in the olefinic product.

At $0°C$, a mixture of the two isomeric olefins is obtained, probably resulting from equilibration of the copper intermediates, which becomes rapid at about $-20°C$ (2.45). These intermediates react with alkyl

$$\underset{R^2\diagup}{R^1\diagdown}C=C\diagup^{CO_2CH_3}_{\diagdown Cu} \quad\rightleftharpoons\quad \underset{R^2\diagup}{R^1\diagdown}C=C\diagup^{Cu}_{\diagdown CO_2CH_3} \quad (2.45)$$

halides to give tetra-substituted ethylenes, but unfortunately only at temperatures which allow isomerisation, so that mixtures of the two geometrical isomers are obtained. But by oxidation with oxygen in presence of an excess of lithium dialkylcuprate at −78°C the tetra-substituted ethylene is formed stereospecifically in moderate yield. Similarly with iodine at −78°C the corresponding iodide is obtained and could presumably be converted into an alkyl derivative by reaction with lithium dialkylcuprate.

A promising stereoselective synthesis of olefins by reaction of allylic acetates with lithium dialkylcuprates has recently been described by Anderson, Henrick and Siddall (1970). Reaction can take two paths leading either to direct or allylic replacement of acetate by the alkyl group of the lithium dialkylcopper reagent (2.46). In general, secondary

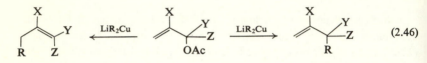

$$ (2.46) $$

allylic acetates (Z = H) react by allylic replacement to give, with a high degree of selectivity, substituted ethylenes in which the groups X and Y are *cis* to each other. Tertiary allylic acetates undergo mainly direct replacement of the acetate group (2.47).

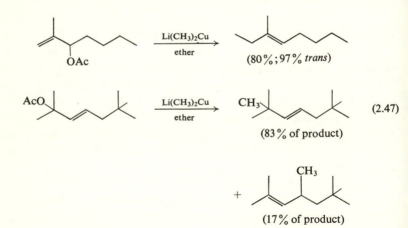

$$ (2.47) $$

With propargylic acetates allenes are formed (Rona and Crabbé, 1968) (2.48).

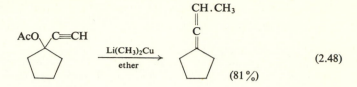

$$(2.48)$$

A method for obtaining trisubstituted olefins of the type (XXIX, R = alkyl) is based on work by Julia, Julia and Tchen (1961) who earlier developed a novel synthesis of homoallylic bromides by rearrangement of cyclopropylcarbinol systems. Thus, the secondary cyclopropyl-carbinol (XXVI, $R^1 = CH_3$, $R^2 = H$) on treatment with 48 per cent hydrobromic acid gave the *trans* compound (XXVII, $R^1 = CH_3$, $R^2 = H$) in 80 per cent yield with 90–95 per cent stereoselectivity (2.49). With

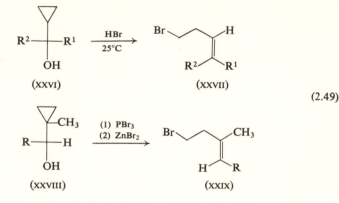

$$(2.49)$$

tertiary carbinols, unfortunately, the reaction was not so stereoselective; the carbinol (XXVI, $R^1 = $ n-C_4H_9, $R^2 = CH_3$) gave a mixture of the *trans*-bromide (XXVII, $R^1 = $ n-C_4H_9, $R^2 = CH_3$) and its *cis* isomer in a ratio of 3:1. With methylcyclopropylcarbinols of the type (XXVIII), however, treatment with phosphorus tribromide followed by anhydrous zinc bromide in ether affords trisubstituted ethylenes (XXIX) in good yield and with a high degree of stereoselectivity (Brady, Ilton and Johnson, 1968). Thus the alcohol (XXX) was smoothly converted into the diene (XXXI) in 85–90 per cent yield (2.50). A useful feature of this procedure is that under the mild conditions of the reaction acid-catalysed migration of double bonds does not take place. In the reaction with (XXX) there was no isomerisation to give an isopropylidene derivative. The reason for the high stereoselectivity can be seen by considering the transition states for the two modes of reaction as seen in the Newman projection

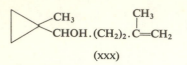

(XXX)

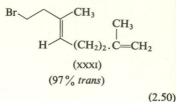

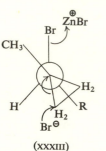

(XXXI)

(97% *trans*)

(2.50)

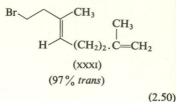

(XXXII)   (XXXIII)

formulae (XXXII) and (XXXIII). Because of non-bonded interactions between the protons of the cyclopropyl ring and the substituent R in (XXXIII) the conformation (XXXII) is preferred and the usual antiparallel electronic reorganisation leads to the olefin in which R and CH_3 are *cis* to each other.

Another useful method for the stereospecific synthesis of olefins proceeds from acetylenes by way of vinylalanes and vinylboranes. Vinylalanes are readily available by hydroalumination of acetylenes with, for example, diisobutylaluminium hydride. With disubstituted acetylenes the reaction proceeds by *cis* addition to give *cis*-vinylalanes; with terminal acetylenes *trans*-vinylalanes are formed. Reaction of these vinylalanes with halogens proceeds with retention of configuration to give the corresponding vinyl halides, the halogen atoms of which are replaceable by alkyl groups by reaction with an organocopper reagent. Thus, iodination of the vinylalane from the reaction of hex-1-yne with diisobutylaluminium hydride produces the isomerically pure *trans*-hex-1-enyl iodide, while hex-3-yne gives *cis*-3-iodohex-3-ene (2.51). As a complement to the above reaction with terminal acetylenes it is found that *cis*-vinyl halides can be obtained by addition of dicyclohexylborane (p. 211) to 1-bromo- and 1-iodoacetylenes followed by protonolysis of

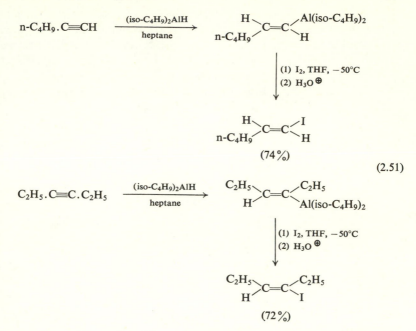

(2.51)

the product with acetic acid. Under these conditions hex-1-yne gave *cis*-1-iodohex-1-ene in 95 per cent yield (Zweifel and Arzoumanian, 1967), (2.52). Brown, Bowman, Misumi and Unni (1967) have also found that

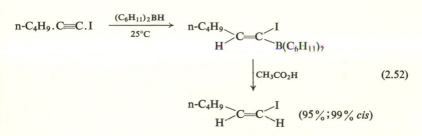

(2.52)

addition of bromine to *trans*-alk-1-enyldisiamylboranes, derived from the reaction of terminal acetylenes with di(3-methylbut-2-yl)borane (p. 213), gives either *cis*- or *trans*-1-bromoalk-1-enes, depending on the procedure used to eliminate the elements of disiamylboron bromide from the dibromide intermediate. Hydrolysis yields the *cis* derivative, while thermal decomposition in refluxing carbon tetrachloride gives the *trans* compound. Since the reaction with bromine presumably proceeds through the usual bromonium ion mechanism to give *trans* addition to the double bond of the vinylborane, solvolysis must involve an *anti* elimination and thermal decomposition a *syn* elimination, as illustrated in (2.53). Using these procedures oct-1-yne, for example, was converted

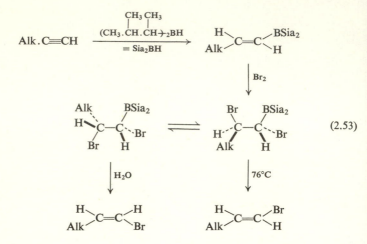

(2.53)

into *cis*-1-bromooct-1-ene in 68 per cent yield, and into *trans*-1-bromo-oct-1-ene in 50 per cent yield.

Reaction of vinylalanes with methyllithium affords the corresponding aluminates which are converted into αβ-unsaturated carboxylic acids in excellent yields on reaction with carbon dioxide. Similarly, para-formaldehyde and other aldehydes afford allylic alcohols (Zweifel and Steele, 1967*a*) (2.54).

In contrast to the reaction with diisobutylaluminium hydride discussed above, hydroalumination of disubstituted acetylenes with lithium hydridodiisobutylmethylaluminate, obtained from diisobutylaluminium hydride and methyllithium, results in *trans* addition to the triple bond, thus opening the way to a convenient synthesis of isomeric series of olefins from a disubstituted acetylene. Carbonation gives αβ-unsaturated acids, reaction with paraformaldehyde or acetaldehyde affords allylic alcohols and iodine gives vinyl iodides, all isomeric with the products obtained in the reaction sequences using diisobutylaluminium hydride discussed above (Zweifel and Steele, 1967*b*). Thus the isomeric α-methylcrotonic acids are conveniently obtained from but-2-yne as illustrated in (2.54). In all these reactions the isobutyl groups of the hydroaluminating agent are converted into isobutane in the hydrolysis step, and do not interfere with the isolation of the products.

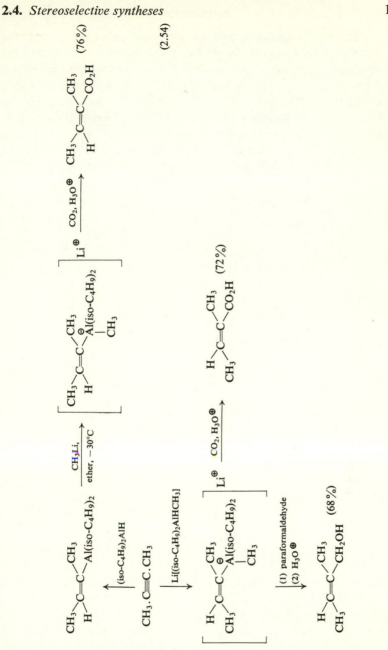

(2.54)

Vinylboranes, obtained by hydroboration of acetylenes (p. 221) do not behave in the same way as vinylalanes. Iodination does not afford the corresponding iodide, but results in stereospecific transfer of one alkyl group from boron to the adjacent carbon, thus providing another stereospecific synthesis of substituted ethylenes. Thus, addition of iodine and sodium hydroxide to the vinylborane obtained by hydroboration of hex-1-yne with dicyclohexylborane, affords *cis*-1-cyclohexylhex-1-ene in 75 per cent yield. The reaction is thought to take the course indicated in (2.55), (Zweifel, Arzoumanian and Whitney, 1967).

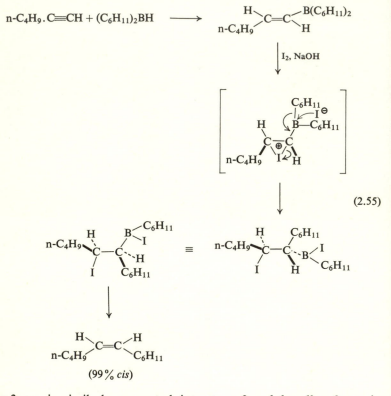

(2.55)

Hex-3-yne is similarly converted into *trans*-3-cyclohexylhex-3-ene in 85 per cent yield. The necessity for *anti* elimination of the iodine and the boron group ensures the stereoselectivity of the olefin-forming step.

This procedure has been extended to the synthesis of conjugated *cis,trans*-dienes from acetylenes (Zweifel, Polston and Whitney, 1968). For example, *cis,trans*-4,5-diethylocta-3,5-diene is obtained in 68 per cent yield by way of the trivinylborane prepared by addition of diborane to hex-3-yne (2.56). A disadvantage of this direct procedure is that only

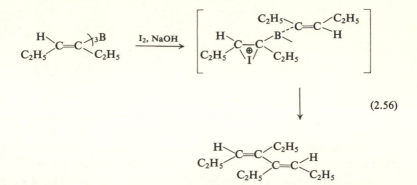

(2.56)

two of the three vinyl groups of the borane are incorporated into the final product. An attempt to overcome this difficulty by use of the divinyl-thexylborane, prepared from the acetylene and 2,3-dimethylbut-2-ylborane (p. 213) was frustrated by the fact that the thexyl group migrated preferentially. Oxidation of the divinylthexylborane with trimethylamine oxide, however, results in selective oxidation of the thexyl group and the resulting divinylborinates, treated with iodine and sodium hydroxide, rearrange to give the conjugated *cis,trans* diene in good yield (2.57).

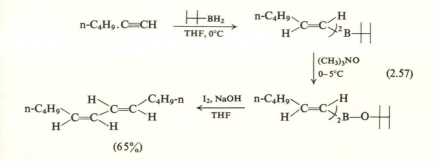

(2.57)

trans,trans-Conjugated dienes can be obtained by reaction of disubstituted acetylenes with diisobutylaluminium hydride in a 2:1 ratio, followed by hydrolysis of the intermediate dienylalane. Thus, *trans,trans*-3,4-dimethylhexa-2,4-diene is obtained from but-2-yne in 79 per cent yield (2.58).

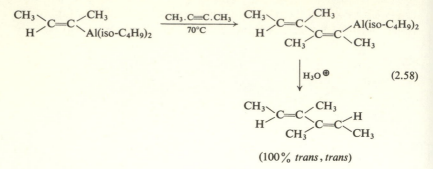

(2.58)

(100% *trans*, *trans*)

2.5. Fragmentation reactions. A number of other methods for forming carbon–carbon double bonds have less general application than those discussed above, but are of value in particular circumstances. One of these makes use of the fragmentation of the monotoluene-*p*-sulphonates or methanesulphonates of suitable cyclic 1,3-diols on treatment with base (Grob and Schiess, 1967; Clayton, Henbest and Smith, 1957) (2.59).

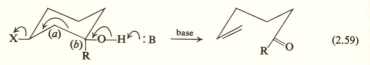

(2.59)

X = leaving group, e.g.: $-OSO_2C_6H_4.CH_3$-*p*, $-OSO_2CH_3$

A feature of these reactions is that when the C–X bond and the $C_{(a)}$–$C_{(b)}$ bond have the *trans* antiparallel arrangement the reaction proceeds very readily by a concerted pathway to give an olefin the stereochemistry of which is governed solely by the relative orientation of groups in the cyclic precursor. For example, the decalin derivative (xxxiv) in which the tosyloxy group and the adjacent angular hydrogen atom are *cis* affords *trans*-cyclodec-5-enone in high yield, whereas the isomer (xxxv) in which the tosyloxy substituent and the hydrogen atom are *trans* affords the *cis*-cyclodecenone (that is, the relative orientation of the hydrogen atoms in the precursor is retained in the olefin). In these derivatives a *trans* antiparallel arrangement of the breaking bonds is easily attained, but this is not so in the isomer (xxxvi) and this compound on treatment with base gives a mixture of products containing only a very small amount of the *trans*-cyclodecenone (2.60). This reaction has been used to prepare a variety of *cis*- and *trans*-cyclodecenone and cyclo-nonenone derivatives, notably in the course of the synthesis of caryo-

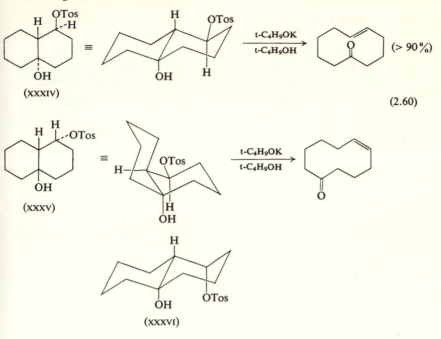

$$(2.60)$$

phyllene. (Corey, Mitra and Uda, 1964). Recently Marshall (1969), has extended the reaction to the preparation of cyclodecadienes by fragmentation of appropriately substituted decalylboranes, easily available by hydroboration of the appropriate olefin (2.61), (see p. 211). Another promising development is the application of fragmentation reactions

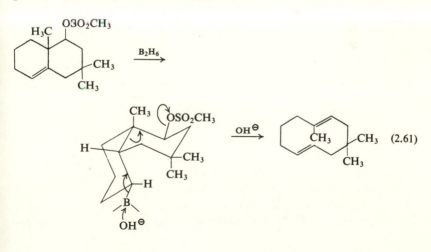

$$(2.61)$$

to the stereospecific synthesis of trisubstituted acyclic olefins. Control of olefin geometry is thereby transposed to control of relative stereochemistry in cyclic systems. The ketone (xxxix), for example, an intermediate in a synthesis of juvenile hormone was obtained stereospecifically from the bicyclic compound (xxxvii) using two successive fragmentation steps (Zurflüh, Wall, Siddall and Edwards, 1968), (2.62). The geometry

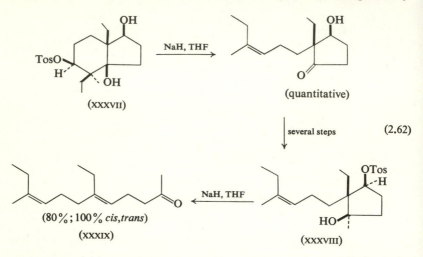

(quantitative)

(xxxvii)

several steps   (2.62)

(80%; 100% *cis,trans*)

(xxxix)

(xxxviii)

of the intermediates (xxxvii) and (xxxviii) is such as to allow easy fragmentation at each stage.

2.6. Oxidative decarboxylation of carboxylic acids.   Another useful method for generating olefinic double bonds is by oxidative decarboxylation of vicinal dicarboxylic acids (2.63). This transformation can be

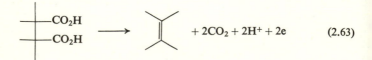

$$\text{—CO}_2\text{H} \longrightarrow \qquad + 2\text{CO}_2 + 2\text{H}^+ + 2e \qquad (2.63)$$

brought about in a number of ways. The familiar procedure with lead-tetraacetate in boiling benzene gives poor and variable yields and is not applicable to bicyclic diacids with nearby double bonds (Criegee, 1965). It has been found by Cimarusti and Wolinsky (1968), however, that much improved yields are obtained when reaction is effected in presence of oxygen (2.64). An even better procedure is by electrolysis of the

$$\text{(2.64)}$$

acid in pyridine solution in presence of trimethylamine. This method gives good yields and since reaction proceeds under mild conditions this is an attractive procedure for preparing highly strained unsaturated small and bridged ring compounds. Dewar benzene, for example, is best prepared by this route from bicyclo[2,2,0]hex-2-en-5,6-dicarboxylic acid (Radlick, Klem, Spurlock, Sims, van Tamelen and Whitesides, 1968), (2.65). Another useful method which proceeds smoothly under mild conditions is by thermal or photolytic decomposition of di-t-butyl peresters, which are readily obtained from the diacid chlorides and t-butyl hydroperoxide. This photolytic process can be used for the synthesis of thermally labile alkenes (Cain, Vukov and Masamune, 1969). The

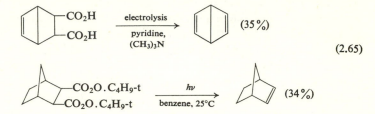

$$\text{(2.65)}$$

vicinal dicarboxylic acids required as starting materials in these reactions are readily available by Diels–Alder or photosensitised addition reactions of maleic anhydride with dienes or olefins.

Related to these reactions is the oxidative decarboxylation of mono-carboxylic acids with lead tetraacetate. Under ordinary conditions the course of the reaction depends on the structure of the acid and poor yields and mixtures of products are often obtained. In presence of catalytic amounts of cupric acetate, however, preparatively useful yields of alkenes are formed from primary and secondary carboxylic

$$CH_3.(CH_2)_7.CO_2H + Pb(OAc)_4 \xrightarrow[\substack{\text{benzene,} \\ h\nu, 30°C}]{Cu(OAc)_2} CH_3.(CH_2)_5.CH{=}CH_2$$

quantitative

$$\text{(2.66)}$$

acids, either photolytically or thermally, in boiling benzene (Bacha and Kochi, 1968). Thus, nonanoic acid is converted into oct-1-ene in quantitative yield and cyclobutanecarboxylic acid gives cyclobutene in 77 per cent yield (2.66).

The following radical chain mechanism is proposed.

$$RCO_2Pb^{III} \longrightarrow R^{\cdot} + CO_2 + Pb^{II}$$
$$R^{\cdot} + Cu^{II} \longrightarrow olefin + H^+ + Cu^I \qquad (2.67)$$
$$Cu^I + RCO_2Pb^{IV} \longrightarrow Cu^{II} + RCO_2Pb^{III}, etc.$$

2.7. Decomposition of toluene-*p*-sulphonylhydrazones. Alkenes are also readily obtained from aliphatic and alicyclic ketones with at least one α-hydrogen atom, by reaction of the corresponding toluene-*p*-sulphonyl-hydrazones with two equivalents of an alkyl lithium, preferably methyl lithium (Shapiro and Heath, 1967; Kaufman, Cook, Schechter, Bayless and Friedman, 1967). Reaction proceeds under mild conditions without rearrangement of the carbon skeleton and in general leads to the least substituted double bond, where there is a choice. αβ-Unsaturated tosyl-hydrazones are likewise converted into dienes, and the reaction has been particularly useful in the preparation of difficultly accessible olefins such as bicyclo[2,1,1]hex-2-ene (2.68). It is essential in these reactions

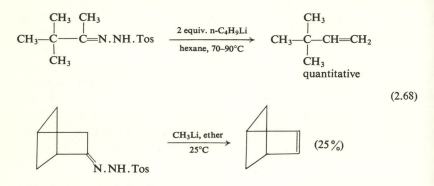

$$(2.68)$$

to use at least two equivalents of the alkyllithium. With smaller amounts competitive carbenic and carbonium ion processes intervene and mixtures are obtained.

Reaction appears to proceed by way of a carbanion intermediate. The reaction path shown in (2.69) is suggested. Aromatic tosylhydrazones

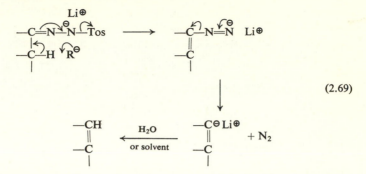

$$(2.69)$$

with no α-hydrogen atom form nucleophilic substitution products on reaction with alkyllithium, instead of olefins. Fluorenone tosylhydrazone, for example, with methyllithium, gives 9-methylfluorene. Variable amounts of substitution products have also been obtained in the reaction of some alicyclic derivatives (see Herz and Gonzalez, 1969).

2.8. Stereospecific synthesis from 1,2-diols. A valuable new stereospecific and position-specific synthesis of olefins from 1,2-diols follows the general scheme shown in (2.70). This process proceeds with complete

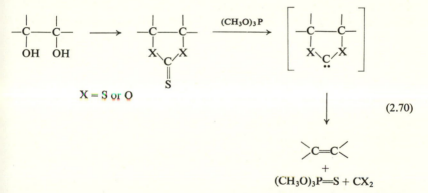

$$(2.70)$$

stereospecificity by a *syn* elimination pathway. It allows the stereospecific synthesis of strained cycloolefins and, together with the *trans*-hydroxylation reaction (p. 259) provides a general and unambiguous method for the interconversion of *cis* and *trans* olefins (Corey, Carey and Winter, 1965). Thus, *meso*-1,4-diphenylbutan-2,3-diol was converted into *cis*-1,4-diphenylbut-2-ene in 96 per cent yield, while the *dl*-compound gave *trans*-olefin. *cis*-Cyclooctene was converted into *trans*-cyclooctene as shown in (2.71). An alternative procedure which avoids the disadvantages of Corey's method (prolonged reaction at high temperature and relative inaccessibility of reagents) was found by Hines, Peagram, Whitham and

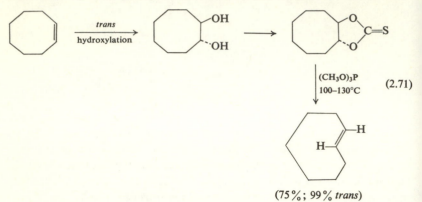

(2.71)

(75%; 99% *trans*)

Wright (1968). The benzylidene derivative of *trans*-cyclooctan-1,2-diol on treatment with butyllithium in hexane at 20°C and then with water gave *trans*-cyclooctene in 75 per cent yield.

2.9. Claisen rearrangement of allyl vinyl ethers. Finally, attention is drawn to the Claisen rearrangement of allyl vinyl ethers (Rhoads, 1963) as a route to $\gamma\delta$-unsaturated aldehydes and ketones (2.72). This reaction

$$\text{R}^1 \overset{\text{R}^2}{\underset{\text{O}}{\bigsqcup}} \text{R}^3 \overset{\Delta}{\longrightarrow} \text{R}^1 \overset{\text{R}^2}{\underset{\text{O}}{\bigsqcup}} \text{R}^3 \qquad (2.72)$$

has formed an important step in the synthesis of a number of natural products, as in the example (2.73). Where structurally possible a mixture of the *cis* and *trans* forms of the new double bond is produced in which

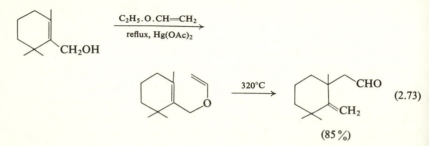

(2.73)

(85%)

the *trans* isomer predominates. It has been shown recently, however, that
when R^1 is not hydrogen the reaction becomes much more stereoselective,
and Claisen rearrangement of such allyl vinyl ethers provides another
convenient method for the stereoselective synthesis of *trans* trisubstituted

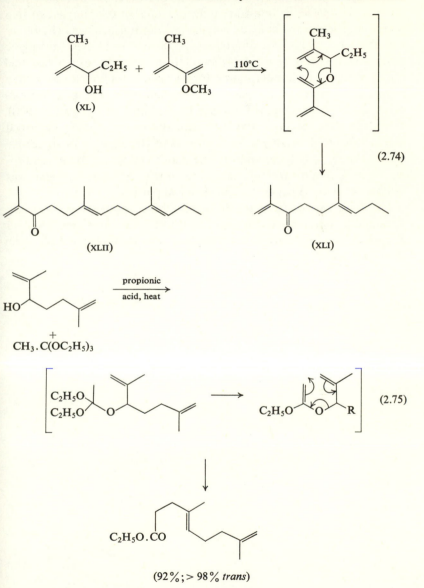

(XLII)   (XLI)

(2.74)

(2.75)

(92%; > 98% *trans*)

ethylenes. Thus, reaction of the allylic alcohol (XL) and 2-methoxy-3-methylbutadiene (2.74) gave the $\gamma\delta$-unsaturated ketone (XLI) which was exclusively the *trans* isomer. Reduction of the carbonyl group with sodium borohydride and repetition of the process then gave the ketone (XLII) in good yield, demonstrating the potential of this simple two-step procedure for the synthesis of polyolefinic natural products (Faulkner and Petersen, 1969). An alternative procedure is to heat an allylic alcohol with excess of ethyl orthoacetate in presence of a weak acid (2.75). A mixed orthoester is first formed and loses ethanol to form a ketene acetal which rearranges to a $\gamma\delta$-olefinic ester (Johnson, Werthemann, Bartlett, Brocksom, Tsung-tee Li, Faulkner and Petersen, 1970). This orthoester reaction, like the methoxyisoprene method described above, lends itself readily to the synthesis of structures with successive isoprene units containing *trans* double bonds and joined in the head-to-tail manner found in many natural products. All-*trans*-squalene was synthesised with 95 per cent stereochemical purity.

The high stereoselectivity shown in these reactions is probably attributable to non-bonded interaction between the substituents R^1 and R^3, which develops only in the transition state leading to the *cis* olefin (Faulkner and Petersen, 1969).

3　The Diels–Alder reaction

3.1. General. The Diels–Alder reaction, one of the most useful syn-thetic reactions in organic chemistry, is one of a general class of cyclo-addition reactions (Huisgen, Grashey and Sauer, 1964). In it a 1,3-diene reacts with an olefinic or acetylenic dienophile to form an adduct with a six-membered hydroaromatic ring (3.1). In the reaction two new σ bonds are formed at the expense of two π bonds in the starting materials (Onischenko, 1964; Sauer, 1966, 1967).

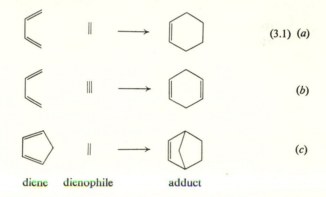

| diene | dienophile | adduct |

In general the reaction takes place easily, simply by mixing the components at room temperature or by gentle warming in a suitable solvent, although in some cases with unreactive dienes or dienophiles more vigorous conditions may be necessary. The Diels–Alder reaction is reversible, and many adducts dissociate into their components at quite low temperatures, particularly those formed from cyclic dienes such as cyclopentadiene, fulvene or furan. In these cases heating is disadvantageous and better yields are obtained by using an excess of one of the components, or a solvent from which the adduct separates readily. It has recently been found that many Diels–Alder reactions are accelerated by Lewis acid catalysts (see p. 146).

The usefulness of the Diels–Alder reaction in synthesis arises from its versatility and from its remarkable stereoselectivity. By varying the nature of the diene and the dienophile many different types of structures can be built up. In the majority of cases all six atoms involved in forming the new ring are carbon atoms, but ring closure may also take place at atoms other than carbon, giving rise to heterocyclic compounds. It is very frequently found, moreover, that although reaction could conceivably give rise to a number of structurally- or stereo-isomeric products, one isomer is formed exclusively or at least in preponderant amount.

Many dienes can exist in a *cisoid* and a *transoid* conformation, and it is only the *cisoid* form which can undergo addition. If the diene does not have or cannot adopt a *cisoid* conformation no reaction occurs.

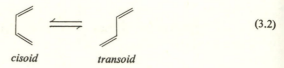

$$\text{(3.2)}$$

cisoid   transoid

3.2. The dienophile.

Ethylenic and acetylenic dienophiles; quinones. Many different kinds of dienophile can take part in the Diels–Alder reaction. They may be derivatives of ethylene or acetylene (the majority of cases) or reagents in which one or both of the reacting atoms is a heteroatom. All dienophiles do not undergo the reaction with equal ease; the reactivity depends greatly on the structure. In general, the greater the polarisation caused by electron-attracting substituents on the double or triple bond the more reactive is the dienophile. Thus, ethylene reacts only slowly with butadiene at 20°C and 90 atmospheres pressure (3.1*a*), whereas maleic anhydride affords a quantitative yield of adduct in boiling benzene or, more slowly, at room temperature (3.3). Tetracyanoethylene, with four

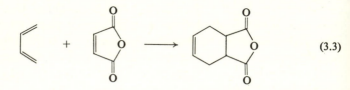

$$\text{(3.3)}$$

electron-attracting substituents, reacts extremely rapidly even at 0°C. Similarly, acetylene reacts with electron rich dienes only under severe conditions, but propiolic acid, phenylpropiolic acid and acetylene dicarboxylic acid react readily, and have been frequently used as dienophiles in the Diels–Alder reaction (Holmes, 1948). Table 3.1 gives some

TABLE 3.1   *Reaction of dienophiles with cyclopentadiene and 9,10-dimethylanthracene*

Dienophile	Cyclopentadiene $10^5 k_1$ (l mol^{-1} s^{-1})	9,10-Dimethylanthracene $10^5 k_1$ (l mol^{-1} s^{-1})
Tetracyanoethylene	*c.* 43 000 000	*c.* 1 300 000 000
Tricyanoethylene	*c.* 480 000	590 000
1,1-Dicyanoethylene	45 500	12 700
Acrylonitrile	1.04	0.089
Dimethyl fumarate	74	215
Dimethyl acetylene dicarboxylate	31	140

values for the rates of addition of a number of dienophiles to cyclopentadiene and 9,10-dimethylanthracene in dioxan at 20°.

It should be noted, however, that there are a number of Diels–Alder reactions in which the above generalisation does not hold and in which reaction takes place between electron *rich* dienophiles and electron *poor* dienes such as o-quinones, thiophen-1,1-dioxides and perchlorocyclopentadiene (see Sauer, 1967). The essential feature is that the two components should have complementary electronic character.

The most commonly encountered activating substituents for the 'normal' Diels–Alder reaction are —CO—, —COOR, —C≡N and —NO$_2$, and dienophiles containing one or more of these groups in conjugation with the double or triple bond react readily with dienes (Kloetzel, 1948; Holmes, 1948).

$\alpha\beta$-Unsaturated carbonyl compounds are very reactive dienophiles and probably represent the most valuable components for synthetic processes. Typical examples of this class are acrolein, acrylic acid and its esters, maleic acid and its anhydride and acetylenedicarboxylic acid. Thus, acrolein reacts rapidly with butadiene in benzene solution at 0°C to give tetrahydrobenzaldehyde in quantitative yield (3.4). Acrylonitrile also reacts readily with butadiene or with 2,3-dimethylbutadiene to form cyclic nitriles and acetylenedicarboxylic acid and butadiene give 3,4-dihydrophthalic acid.

Substituents exert a pronounced steric effect on the reactivity of dienophiles. Comparative experiments show that the yields of adducts obtained in the condensation of butadiene and 2,3-dimethylbutadiene with acrylic dienophiles decrease with the introduction of substituents into the α- position of the dienophile molecule, and $\alpha\beta$-unsaturated

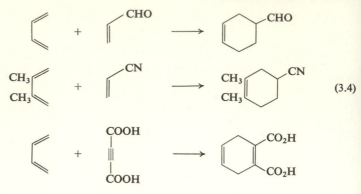

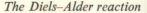(3.4)

ketones with two alkyl substituents in the β- position are almost unable to take part in diene condensations. Similarly the reactions of butadiene and 2,3-dimethylbutadiene with citraconic anhydride (methylmaleic anhydride) require more drastic conditions than the reactions with maleic anhydride.

It is of interest that esters of maleic and citraconic acid condense with dienes far less readily than their *trans* isomers. Diene condensations with *cis* dienophiles occur readily only in the case of anhydrides.

Another important group of dienophiles of the $\alpha\beta$-unsaturated carbonyl class are quinones (Butz and Rytina, 1949). *p*-Benzoquinone itself reacts readily with butadiene at room temperature to form a high yield of the mono-adduct, tetrahydronaphthaquinone (3.5). Under more

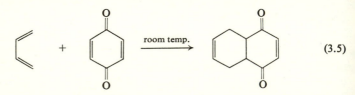

(3.5)

vigorous conditions a bis-adduct is obtained which can be converted into anthraquinone by oxidation of an alkaline solution with atmospheric oxygen. 1,4-Naphthaquinone behaves similarly, readily furnishing adducts with acyclic dienes which can be oxidised to anthraquinones (3.6). If the reaction is conducted in nitrobenzene solution the anthraquinone e.g. (I) is often obtained directly by dehydrogenation of the initial adduct by the solvent.

As with other dienophiles, substitution on the double bond leads to a weakening of the dienophilic properties of quinones. It is found in

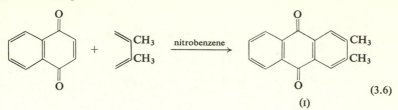

(3.6)

(I)

general that monosubstituted *p*-benzoquinones add dienes at the unsubstituted double bond only.

A double bond of a conjugated diene can sometimes itself act as a dienophile, being activated by the neighbouring double bond. This is seen in the dimerisation of dienes, for example of cyclopentadiene and of isoprene (3.7). Self-condensation of simple conjugated dienes some-

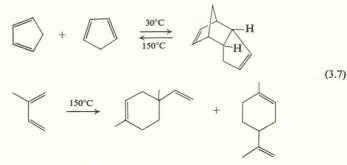

(3.7)

times has an adverse effect on the yields obtained in Diels–Alder reactions, particularly in the case of reactions which proceed slowly or require vigorous conditions.

In contrast to the reactive dienophiles discussed above, in which the double or triple bond is activated by conjugation with unsaturated electron-attracting groups, ethylenic compounds such as allyl alcohol and its esters, allyl halides, vinyl compounds and styrene are relatively unreactive, although they can frequently be induced to react with dienes under forcing conditions. Thus vinyl acetate and butadiene at 180°C for 12 h give only a 6 per cent yield of 4-acetoxycyclohexene (3.8), although with the more reactive diene, cyclopentadiene, dihydro-norbornyl acetate is formed in almost quantitative yield under the same conditions.

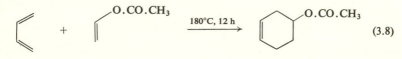

(3.8)

Vinyl ethers and esters have been widely used in the synthesis of dihydropyrans and chromans by reaction with $\alpha\beta$-unsaturated carbonyl compounds. 2-Alkoxydihydropyrans are obtained in good yields at temperatures between 150 and 200°C with an excess of vinyl ether present (Colonge and Descotes, 1967), (3.9). The high polarity of the vinyl

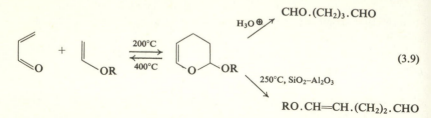

$$CHO.(CH_2)_3.CHO$$

$$RO.CH{=}CH.(CH_2)_2.CHO$$

(3.9)

ethers appears to facilitate the addition at the expense of the dimerisation or polymerisation of the unsaturated carbonyl compound. The dihydropyrans are useful intermediates for the preparation of glutaraldehydes and alkoxypentenals.

Enamines and vinyl carbamates also react readily as dienophiles with $\alpha\beta$-unsaturated carbonyl compounds. The products are again easily hydrolysed in acid solution to glutaraldehydes (3.10).

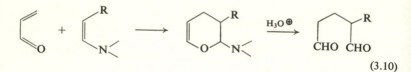

(3.10)

Cycloalkenes and cycloalkynes. A number of cyclic olefins and acetylenes with pronounced angular strain are reactive dienophiles. The driving force for these reactions is thought to be the reduction in angular strain associated with the transition state for the addition. Thus, cyclopropene reacts rapidly and stereospecifically with cyclopentadiene at 0°C to form the *endo* adduct (II) in 96 per cent yield, and methylcyclopropene behaves similarly (3.11). The less reactive butadiene forms norcarene (III) in 37 per cent yield.

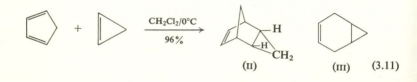

(3.11)

In cyclobutene derivatives as well the angular strain has a reaction-promoting effect. 3,3,4,4-Tetrafluorocyclobutene reacts readily with a number of dienes (3.12) and so do unsaturated four-membered cyclic sulphones. The existence of the transient species benzocyclobutadiene (ɪᴠ)

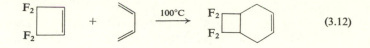

$$(3.12)$$

has been shown by trapping experiments with the reactive dienophile diphenylisobenzofuran (3.13).

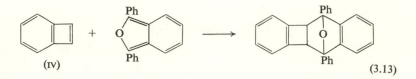

$$(3.13)$$

Some cyclic acetylenic compounds are also powerful dienophiles. Because of its linear structure a carbon–carbon triple bond can only be incorporated without strain into a ring with nine or more members. The increasing strain with decreasing ring size in the sequence cyclo-octyne to cyclopentyne is shown in an increasing tendency to take part in 1,4-addition reactions. Cyclooctyne has been prepared as a stable liquid with pronounced dienophilic properties. It reacts readily with di-phenylisobenzofuran to give an adduct in 91 per cent yield (3.14). The

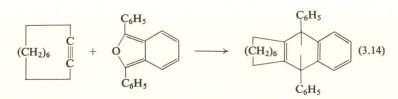

$$(3.14)$$

lower cycloalkynes have not been isolated but their existence has been shown by trapping them with diphenylisobenzofuran (Wittig, 1962).

Arynes, such as dehydrobenzene, also readily undergo Diels–Alder addition reactions. Cyclopentadiene, cyclohexadiene and even benzene and naphthalene add to the highly reactive species C_6H_4 (Wittig, 1962) (3.15). Analogous addition reactions are shown by dehydroaromatics in the pyridine and thiophen series.

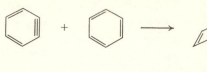

(3.15)

+ other products

Ketenes and allenes. It is evident that, under the appropriate conditions, most olefinic and acetylenic compounds can function as dienophiles. A notable exception is ketene. The C=C linkage in the system C=C=O does indeed react with dienes but the additions are not diene syntheses; the products are four-membered ring compounds formed by 1,2-addition. For example, cyclopentadiene and dimethylketene form the cyclobutanone derivative (v), (3.16).

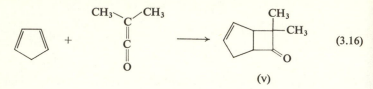

(3.16)

(v)

On the other hand, a number of diene additions involving allene derivatives have been recorded. Allene itself only reacts with electron-deficient dienes, for example hexachlorocyclopentadiene, but allene carboxylic acid, in which a double bond is activated by conjugation with the carboxyl group, reacts readily with cyclopentadiene to give a 1:1 adduct in 84 per cent yield (3.17).

(3.17)

Heterodienophiles. One or both of the carbon atoms of the dienophile multiple bond may frequently be replaced by a heteroatom without loss of activity. But in general Diels–Alder reactions involving heterodienophiles have not been so widely employed in synthesis as those using olefines and acetylenes (Needleman and Chang Kuo, 1962).

Carbonyl groups in aldehydes and ketones add to dienes, and the reaction has been used to prepare derivatives of 5,6-dihydropyran. Formaldehyde reacts only slowly, but reactivity increases in compounds such as chloral in which the carbonyl group is deprived of electrons by suitable electronegative substituents (3.18).

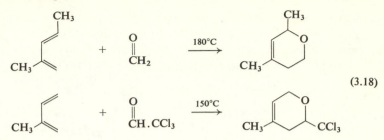

(3.18)

Nitriles also react with dienes affording pyridine derivatives, but the method is of limited preparative value since most of the reactions require very high temperatures and are often conducted in the gas phase. Dihydropyridines are obtained by reaction of iminochlorides (available from amides and phosphorus oxychloride) with aliphatic dienes. For example 2,3-dimethylbutadiene and acetamide in presence of phosphorus oxychloride readily afford 3,4,6-trimethyl-1,2-dihydropyridine (3.19).

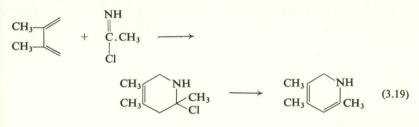

(3.19)

With styrenes, derivatives of 3,4-dihydroisoquinoline are obtained.

Nitroso compounds also react with dienes, to form oxazine derivatives (Hamer and Ahmad, 1967). Aromatic nitroso compounds are the most reactive. Nitrosobenzenes add readily to butadiene and a variety of butadiene derivatives, as well as to some cyclic dienes. Thus, with butadiene nitrosobenzene gives a high yield of N-phenyl-3,6-dihydro-oxazine (3.20). Cyclopentadiene reacts at $-40°C$ to form a related

$$C_6H_5.NO \ + \ \text{(diene)} \xrightarrow[0°C]{\text{ethanol}} C_6H_5—N \text{(ring)} \quad (95\%) \quad (3.20)$$

unstable adduct. Electron-releasing substituents in the *ortho* and *para* positions of the nitrosobenzene retard its activity considerably through mesomeric effects.

With aliphatic nitroso compounds only derivatives with an electron-withdrawing group on the α-carbon atom undergo the reaction. Thus, α-chloro or α-cyano nitroso compounds react readily with butadienes. The adducts obtained from the α-chloro compounds are of interest for in ethanolic solution they are hydrolysed directly, and good yields of 3,6-dihydro-1,2-oxazines are obtained (3.21).

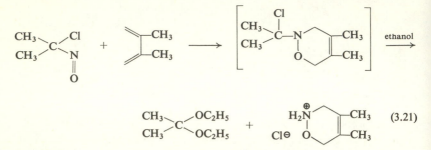

(3.21)

Some azo compounds with electron-attracting groups attached to the nitrogen atoms are very reactive dienophiles (Gillis, 1967), and react with a variety of dienes leading to derivatives of tetrahydropyridazine. The most commonly used reagent of this type is ethyl azodicarboxylate and the reaction of this dienophile with cyclopentadiene formed one of the earliest known examples of Diels–Alder addition. Most azodicarb-oxylic esters have the *trans* configuration; recent work has shown that the *cis* compounds are even more reactive, partly owing to the decrease in steric hindrance. This contrasts with the reactivity of isomeric pairs of olefinic dienophiles (see p. 118).

The Diels–Alder reaction of azo dienophiles is of interest because it provides a convenient route to the pyridazine ring system. Thus, 2,3-dimethylbutadiene and ethyl azodicarboxylate react readily at room

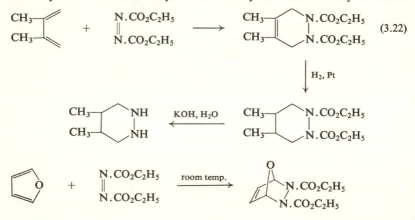

(3.22)

temperature giving an almost quantitative yield of the tetrahydro-pyridazine, which is readily converted into the cyclic hydrazine by the steps shown (3.22).

Ethyl azodicarboxylate forms adducts with a wide variety of other dienophiles, including anthracene and furan, both of which react readily.

Other azo dienophiles which have been used are azodiaroyls (VI), 3,6-pyridazinediones (VII) and 1,4-phthalazinedione (VIII), (3.23). It is observed that as the azo group becomes more electron deficient by induction or resonance its reactivity increases. Thus, dimethyldiazenium bromide (IX) is extremely reactive.

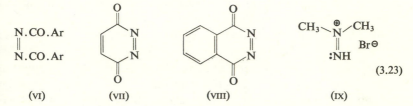

$$(3.23)$$

(VI) (VII) (VIII) (IX)

Oxygen as a dienophile. The reaction of oxygen with 1,4-dienes to form endoperoxides, which is formally a 1,4-cycloaddition, is of outstanding interest (Gollnick and Schlenck, 1967) (3.24). Generally,

$$\text{diene} + O_2 \longrightarrow \text{endoperoxide} \qquad (3.24)$$

addition of oxygen to dienes must be effected under the influence of light, either directly or in presence of a photosensitiser, and it is doubtful whether these reactions are strictly comparable to true thermal Diels–Alder additions. It has been found recently that similar conversions of dienes into endoperoxides can be effected with sodium hypochlorite and hydrogen peroxide and by other means, and it has been suggested that singlet oxygen is the reactive species in each case (Foote and Wexler, 1964; Corey and Taylor, 1964). In most cases the required *cisoid* diene forms part of a carbocyclic or heterocyclic system and there appear to be very few recorded examples of addition of oxygen to dienes in which the double bonds are not in one ring.

The light-induced 1,4-addition of oxygen to dienes was discovered independently by Clar and by Windaus. Clar found in 1930 that when a solution of the linear pentacyclic aromatic hydrocarbon pentacene (X) in benzene was irradiated with ultraviolet light in presence of oxygen, the transannular peroxide (XI) was obtained (3.25). A similar product

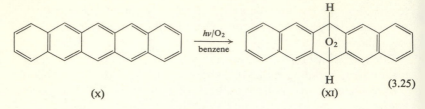

was obtained from anthracene in carbon disulphide solution. More than a hundred photoperoxides of this type have now been prepared in the anthracene and naphthacene series. In general, photosensitisers are not used in these reactions, but the rate of formation of the oxide is often strongly dependent on the solvent.

The photosensitised addition of oxygen to conjugated alicyclic dienes was discovered by Windaus in 1928 in the course of his classical studies of the conversion of ergosterol into vitamin D. Irradiation of a solution of ergosterol in alcohol in presence of oxygen and a sensitiser led to the formation of a peroxide which was subsequently shown to have been formed by 1,4-addition of a molecule of oxygen to the conjugated diene system.

Numerous other endoperoxides of the same type have since been obtained by sensitised photo-oxidation of other steroidal dienes with *cisoid* 1,3-diene systems. For many years it was believed that photo-sensitised addition of oxygen to conjugated dienes was restricted to steroids, but it is now known that this is not the case and endoperoxides have been obtained from many mono-, bi- and tri-cyclic dienes. Thus, irradiation of cyclohexadiene in presence of oxygen with chlorophyll as a sensitiser leads to the endoperoxide norascaridole, the structure of which was established by reduction with hydrogen and platinum to *cis*-1,4-dihydroxycyclohexane (3.26). Similar unstable peroxides have been obtained from cyclopentadiene and from cycloheptadiene; reduction of the product in each case led to the *cis*-1,4-diol.

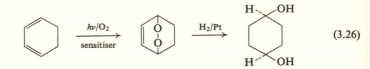

Photosensitised oxidation of 1,3-dienes has been used with conspicuous success as a key step in the synthesis of a number of natural products. The longest known, and so far the only endoperoxide found in nature, is the terpene peroxide ascaridole (XII) the principal component of chenopodium oil (3.27). Its structure was elucidated by hydrogenation to *p*-methane-*cis*-1,4-diol. The compound was readily synthesised by Schenck and Ziegler by photosensitised oxidation of α-terpinene. The

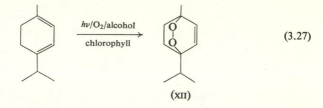

(3.27)

(XII)

product from chenopodium oil is optically inactive and it has been suggested that, in the plant, ascaridole is formed by chlorophyll-sensitised photo-oxidation of α-terpinene rather than by an enzymic process which would be expected to lead to optically active ascaridole. In the synthesis of the growth inhibitory substance abscissic acid (XIV) the unsaturated ketol grouping was introduced *via* photosensitised addition of oxygen to the conjugated diene (XIII) (Cornforth, Milborrow and Ryback, 1965) (3.28).

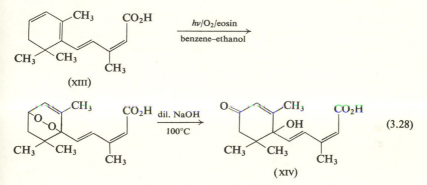

(3.28)

(XIV)

3.3. The diene. It has already been pointed out (p. 116) that the diene component must have or must be able to adopt the *cisoid* conformation before it can take part in Diels–Alder reactions with dienophiles. The majority of dienes which satisfy this condition undergo the reaction more or less easily depending on their structure.

Acyclic dienes. Acyclic conjugated dienes react readily often forming the adducts in almost quantitative yield. For example, butadiene itself reacts quantitatively with maleic anhydride in benzene at 100°C in

5 h, or more slowly at room temperature, to form *cis*-1,2,5,6-tetra-hydrophthalic anhydride.

Substituents in the butadiene molecule influence the rate of cyclo-addition both through their electronic nature and by a steric effect on the conformational equilibrium. Alder found that the rate of the re-action is often increased by electron-donating substituents (e.g. —NMe$_2$, —OMe, —Me) in the diene as well as by electron-attracting substituents in the dienophile. Bulky substituents which discourage the diene from adopting the *cisoid* conformation hinder the reaction. Thus, whereas 2-methyl-, 2,3-dimethyl- and 2-t-butylbutadiene react normally with maleic anhydride the 2,3-diphenyl compound is less reactive and 2,3-di-t-butylbutadiene is completely unreactive. Apparently the molecule of the 2,3-di-t-butyl compound is prevented from attaining the necessary planar *cisoid* conformation by the steric effect of the two bulky t-butyl substituents. In contrast, 1,3-di-t-butylbutadiene, in which the sub-stituents do not interfere with each other even in the *cisoid* form, reacts readily with maleic anhydride.

cis Alkyl or aryl substituents in the 1-position of the diene reduce its reactivity by sterically hindering formation of the *cisoid* conformation through non-bonded interaction with a hydrogen atom at position 4. Accordingly, a *trans* substituted 1,3-butadiene reacts with dienophiles much more readily than the *cis* isomer. Thus, *cis*-piperylene gave only a 4 per cent yield of adduct when heated with maleic anhydride at 100°C for 8 h, whereas the *trans* isomer formed an adduct in almost quanti-tative yield in benzene at 0°C, (3.29).

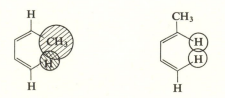

(3.29)

Similarly, *trans,trans*-1,4-dimethylbutadiene reacts readily with many dienophiles, but the *cis,trans* isomer yields an adduct only when the components are heated in benzene at 150°C. *cis* Substituents in both the 1- and 4- positions prevent reaction.

1,1-Disubstituted butadienes also react with difficulty, and with such compounds addition may be preceded by isomerisation of the diene to a more reactive species. Thus, in the reaction of 1,1-dimethylbutadiene with acrylonitrile the diene first isomerises to 1,3-dimethylbutadiene which then reacts in the normal way.

This difference in reactivity towards maleic anhydride of *cis-* and *trans-*1-substituted dienes has been exploited as a method for the determination of the stereochemistry of diene systems in a number of naturally occurring dienes and polyenes (see p. 131; also Alder and Schumacher, 1953).

Very many Diels–Alder reactions with alkyl- and aryl-substituted butadienes have been effected (cf. Onischenko, 1964). Derivatives of hydroxybutadiene also react with dienophiles, providing a route to synthetically useful intermediates. Thus, 2-alkoxybutadienes react easily with maleic anhydride and other dienophiles to form adducts which, as enol ethers, are readily hydrolysed to cyclohexanone derivatives (3.30).

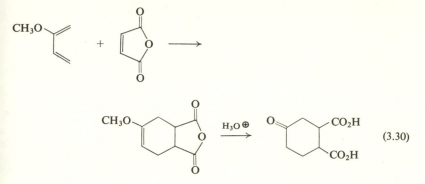

(3.30)

Again, the reaction of *trans,trans-*1,4-diacetoxybutadiene with methyl acrylate formed the first step in a stereospecific total synthesis of the important biosynthetic intermediate shikimic acid (xv) (Smissman, Suh, Oxman and Daniels, 1962) (3.31). Another route to this compound begins with the Diels–Alder reaction between butadiene and propiolic acid (Grewe and Hinrich, 1964).

Reaction of diacetoxybutadiene with olefinic and acetylenic dienophiles provides a useful synthetic route to benzene derivatives, for the initial adducts readily eliminate acetic acid to give the aromatic compound. In many cases it is unnecessary to isolate the intermediate adduct and the benzene derivative is obtained directly by heating the components together at 100–120°C (Hill and Carlson, 1965), (3.32).

Conjugated polyenes react normally with dienophiles by 1,4-addition to form six-membered rings. The position of attack on the polyene chain is governed by the same principles as operate in the case of butadienes.

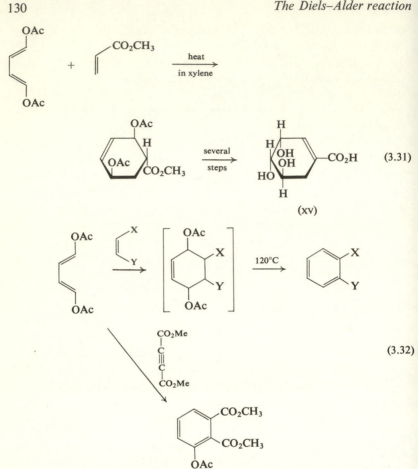

(3.31)

(xv)

(3.32)

Thus, *trans,trans*-1,3,5-heptatriene reacts readily with maleic anhydride in ether at room temperature to give a 90 per cent yield of a mixture of the two possible adducts (3.33). In contrast, the isomeric *cis,trans*-

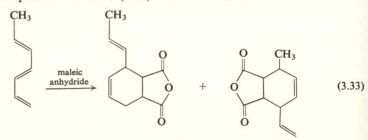

(3.33)

heptatriene yields only one adduct by reaction at the *trans*-substituted butadiene grouping.

If the *cisoid* conformations are not hindered a tetraene can add two molecules of maleic anhydride. Thus the naturally occurring β-parinaric acid (XVI) has been shown to be the all-*trans*-compound by the formation of a bis-adduct with maleic anhydride.

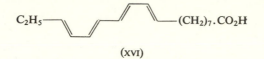

(XVI)

Compounds containing more than four conjugated double bonds, such as the natural dyes of the carotene and bixin series and naturally occurring polyacetylenes, also react readily with dienophiles.

Trienes with an allene arrangement of the double bonds condense with dienophiles according to the general scheme of the diene synthesis. 1,2,4-Pentatriene reacts readily with a variety of dienophiles forming adducts which can be easily aromatised. With naphthaquinone an adduct is obtained which is converted into 1-methylanthraquinone with palladium in boiling ethanol, and acetylenedicarboxylic acid gives 3-methylphthalic acid directly at 90°C, (3.34).

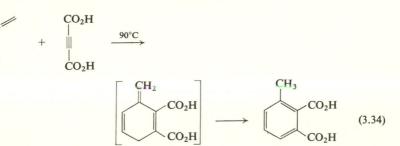

(3.34)

Eneynes and dienynes. Eneynes and dienynes, with the groupings

$$-\overset{|}{C}=\overset{|}{C}-C\equiv C- \quad \text{and} \quad -\overset{|}{C}=\overset{|}{C}-C\equiv C-\overset{|}{C}=\overset{|}{C}-,$$

react readily with dienophiles to form 1,4-addition compounds. The reaction of eneynes with dienophiles differs from that of dienes in that migration of a hydrogen atom takes place during the reaction. It has been suggested that addition of the dienophile is preceded by tautomerism of the eneyne to a zwitterionic diene, as illustrated (3.35) for the dimerisation of 3-methylpent-3-en-1-yne (XVII) to give the benzene derivative (XVIII).

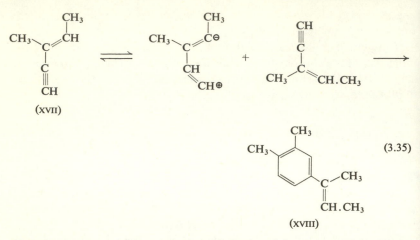

(3.35)

(xviii)

Many examples of diene condensations involving enynes have been recorded. With acetylenic dienophiles aromatic compounds are formed directly, as in the example cited above.

Dienynes form bis-adducts. For example, 2,5-dimethylhexa-1,5-dien-3-yne condenses with two molecules of maleic anhydride at 130°C to form a bis-anhydride which is assigned the structure (xix), because on decarboxylation and dehydrogenation it is converted into 1,5-dimethyl-naphthalene (3.36).

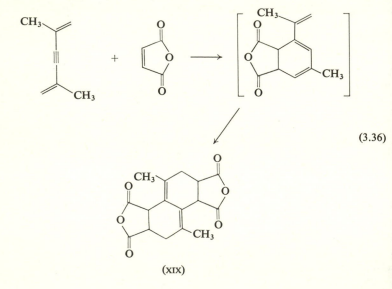

(3.36)

(xix)

Heterodienes. Heterodienes, in which one or more of the atoms of the conjugated diene is a heteroatom, are not so numerous as hetero-dienophiles (see p. 122) and only the 1,4-addition reactions of αβ-unsaturated carbonyl compounds have been used synthetically to any extent so far. (Colonge and Descotes, 1967) Diene condensations of heterodienes containing nitrogen atoms in the diene system have been much less studied (see Needleman and Chang Kuo, 1962).

αβ-Unsaturated carbonyl compounds react most readily as dienes with electron-rich dienophiles such as enol ethers and enamines. With less reactive dienophiles dimerisation of the αβ-unsaturated carbonyl compound is a competing reaction. For example, acrolein is converted into a dimer, 2-formyl-2,3-dihydropyran, in 40 per cent yield by heating in benzene solution, one molecule acting as the diene component and the other as dienophile. The reaction with *N*-vinyl compounds, such as enamines and *N*-vinyl carbamates, and with enol ethers, proceeds readily to yield derivatives of dihydropyran which can be converted into syn-thetically useful 1,5-dicarbonyl compounds, (3.37).

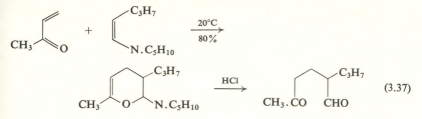

1,2-Dimethylenecycloalkanes. In 1,2-dimethylenecycloalkanes the *cisoid* conformation of the double bonds necessary for 1,4-addition is fixed and dienes of this type are excellent reagents for dienophiles often forming the adducts in almost quantitative yield. This and the ready availability of the compounds make 1,4-additions to dimethylene-cycloalkanes a convenient route to polycyclic compounds.

Thus, 1,2-dimethylenecyclohexane reacts exothermally with maleic anhydride, and with benzoquinone a bis-adduct (xx) is obtained which can be readily converted into the polycyclic aromatic hydrocarbon pentacene (x) (3.28). A variety of alkyl derivatives of this and other linearly condensed polycyclic aromatic compounds has been conveniently obtained by this general route.

The ready reaction of 1,2-dimethylenecycloalkanes with dienophiles has been used to identify this structural feature in unstable intermediates.

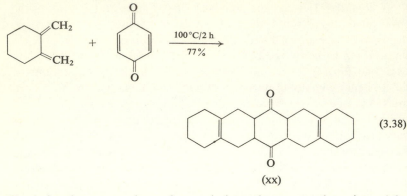

(3.38)

(xx)

Thus, in the conversion of $\alpha,\alpha,\alpha',\alpha'$-tetrabromo-*o*-xylene into 1,2-dibromobenzocyclobutene the dimethylenecyclohexadiene (xxi) was shown to be formed transiently by trapping it with maleic anhydride (3.39).

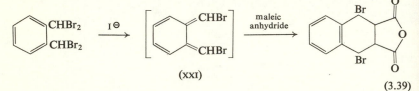

(xxi)

(3.39)

Vinylcycloalkenes and vinylarenes. Many vinylcycloalkenes react readily with dienophiles. In contrast, dienes of the types

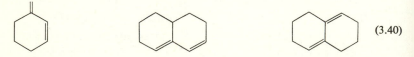

(3.40)

in which the double bonds are constrained in the *transoid* conformation do not react. 1-Vinylcyclohexene itself reacts exothermally with maleic anhydride to form a high yield of the adduct (xxii), (3.41). 1-Vinyl-3,4-

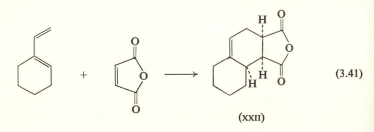

(3.41)

(xxii)

dihydronaphthalene similarly forms a chrysene derivative with benzo-quinone. Reactions of this type using cyclopentenones and cyclo-hexenones as dienophiles have been extensively used to build up tetracyclic structures related to steroid systems (cf. Onischenko, 1964).

1,1'-Dicyclohexenyls and related compounds also react readily with dienophiles provided the molecule is not prevented by steric factors from assuming a planar *cisoid* conformation. Since the dienes are readily available from the cyclic ketones *via* a pinacol condensation, the sequence constitutes another very convenient route to polycyclic compounds. Thus 1,1'-bicyclohexenyl reacts readily with benzoquinone to form an adduct which is readily converted into the aromatic hydrocarbon tri-phenylene and 3,4,5,6-dibenzophenanthrene can be prepared from tetrahydrobinaphthyl by way of its adduct with maleic anhydride, as illustrated in (3.42).

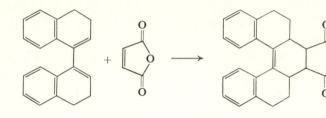

Pd–C, 300°C (3.42)

Many vinylaromatic compounds react normally with dienophiles, even though the initial reaction results in partial loss of aromatic con-jugation. Styrene itself gives only a poor yield of adduct because it is consumed by polymerisation, but *p*-alkoxy groups facilitate the reaction and α-substituted styrenes, which polymerise less readily than styrene, form adducts in good yield. *p*-Methylisopropenylbenzene, for example, reacts with maleic anhydride at 180°C to form the adduct (xxɪɪɪ) (3.43). 1-Vinylnaphthalene reacts similarly.

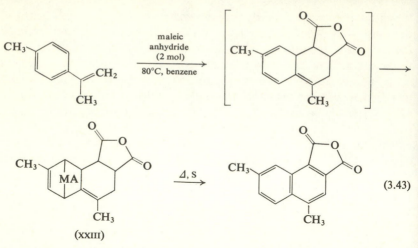

(3.43)

(xxiii)

Aromatic hydrocarbons. A number of types of polycyclic aromatic hydrocarbons react with dienophiles by 1,4-addition, but the reaction is particularly characteristic of anthracene and the higher linear acenes (Badger, 1954). Since the reaction results in loss of aromatic resonance, it is not surprising to find that benzene does not normally undergo thermal Diels–Alder reaction, although a bis adduct has been obtained under the influence of ultraviolet light (see p. 167). Interestingly, however, reaction of benzene with hexafluorobut-2-yne affords a small yield of *o*-bis(trifluoromethyl)benzene (3.44). This substance is thought to arise by initial 1,4-addition of the dienophile to benzene followed by

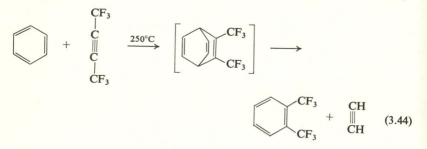

(3.44)

loss of acetylene from the adduct. With durene the initial adduct can be isolated (Krespan, McKusick and Cairns, 1961). More recently dicyanoacetylene has been found to react with benzene at 180°C to form the crystalline adduct (xxiv), and in presence of $AlCl_3$ as catalyst (see p. 146) the adduct was obtained in 63 per cent yield at room temperature (Ciganek, 1967). *p*-Xylene gave a mixture of two isomeric adducts and with hexamethylbenzene the activating effect of methyl groups is shown by the fact that the adduct was isolated in 83 per cent yield from an uncatalysed reaction at 130°C.

(xxiv)

Naphthalene reacts very slowly with maleic anhydride but the poly-methylnaphthalenes are more reactive, and in presence of an excess of maleic anhydride in boiling benzene almost quantitative yields of 1,4-adducts have been obtained, such as (xxv) from 1,2,3,4-tetra-methylnaphthalene.

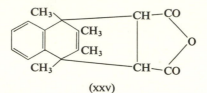

(xxv)

Anthracene reacts with an equimolecular amount of maleic anhydride in boiling xylene to form the 9,10-addition product (xxvi) in quantitative yield. Many anthracene derivatives and anthracene benzologues react

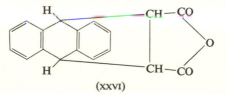

(xxvi)

similarly, although the ease of the reaction varies with the structure of the hydrocarbon (cf. Badger, 1954). Naphthacene, pentacene and the higher acenes react with dienophiles even more readily than anthracene. These addition reactions are reversible, and in many cases the hydro-carbon can be recovered by sublimation or distillation of the adduct.

Cyclic dienes. Cyclopentadiene, in which the double bonds are con-strained in a planar *cisoid* conformation, reacts easily with a variety of dienophiles. 1,3-Cyclohexadiene is also reactive, but with increase in the size of the ring the reactivity rapidly decreases because the double bonds can no longer adopt the necessary coplanar configuration . *cis,cis-* and *cis,trans*-1,3-Cyclooctadienes form only copolymers when treated with maleic anhydride and *cis,cis-* and *cis,trans*-1,3-cyclodecadienes similarly do not form adducts with maleic anhydride. Dienes with fourteen- and fifteen-membered rings again react with dienophiles but only under relatively severe conditions.

Cyclopentadiene is a very reactive diene and reacts easily with acetylenic and olefinic dienophiles to form bridged compounds of the bicyclo-[2,2,1]heptane series (3.1c). With maleic anhydride the adduct (xxvii) is obtained in a stereospecific exothermic reaction and tetrolic acid forms the bicycloheptadiene (xxviii).

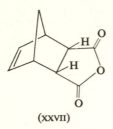

(xxvii)

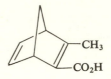

(xxviii)

Less reactive dienophiles such as allyl alcohol, vinyl chloride and vinyl acetate react with cyclopentadiene at elevated temperatures, and even ethylene and propylene form adducts under forcing conditions.

Cyclopentadiene itself can act both as a diene and a dienophile, and readily forms a dimer (xxix) which dissociates into its components on

(xxix)

moderate heating. At higher temperatures the dimer itself can act as a dienophile and trimers, tetramers, etc. are formed which do not readily dissociate.

The reaction of cyclopentadiene with mono- and *cis*-di-substituted ethylenes could apparently give rise to two stereochemically distinct products, the *endo-* and the *exo*-bicyclo[2,2,1]heptene derivatives (3.45).

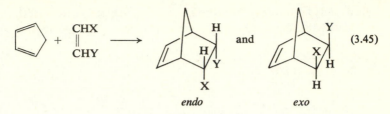

$$ (3.45) $$

endo   *exo*

It is found in practice, however, that the *endo* isomer always predominates, except under conditions where isomerisation of the original adduct occurs.

Compounds of the bicyclo[2,2,1]heptane type are widely distributed in nature among the bicyclic terpenes, and the Diels–Alder reaction provides a very convenient method for the synthesis of these compounds. Thus, cyclopentadiene and vinyl acetate react smoothly when warmed together to form the acetate (xxx) which is easily transformed into norcamphor (xxxi) and related compounds (3.46).

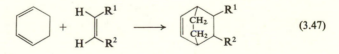

$$ (3.46) $$

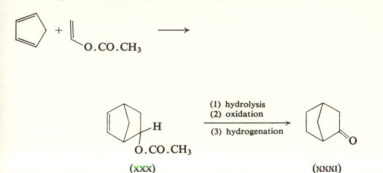

(xxx)   (xxxi)

1,3-Cyclohexadienes react with ethylenic dienophiles to form derivatives of bicyclo[2,2,2]octene (3.47). In general, the additions proceed more slowly than the corresponding reactions with cyclopentadiene.

$$ (3.47) $$

With acetylenic dienophiles derivatives of bicyclo[2,2,2]octadiene are formed initially, but these often undergo a retro Diels–Alder reaction (see p. 144) with elimination of an ethylene and formation of a benzene derivative. Thus, reaction of cyclohexadiene and propargyl aldehyde yields only benzaldehyde formed by thermal decomposition of the initial adduct (3.48). This reaction sequence has been used to locate the position

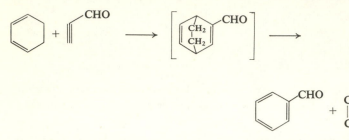

(3.48)

of the double bonds in naturally-occurring cyclohexadienes such as α-phellandrene, which, when heated with acetylenedicarboxylic ester gives diethyl 4-methylphthalate, whilst the isomeric α-terpinene gives diethyl 3-methyl-6-isopropylphthalate.

Cyclopentadienones and ortho-*quinones.* Derivatives of cyclopentadien-one and of *o*-quinones also form adducts with ethylenic and acetylenic dienophiles. Many cyclopentadienones can act both as dienes and dienophiles and this is particularly true of cyclopentadienone itself which dimerises spontaneously. Its transient existence during dehydro-bromination of the bromoketone (xxxii) was shown by a trapping experiment with cyclopentadiene (3.49). The adducts obtained from

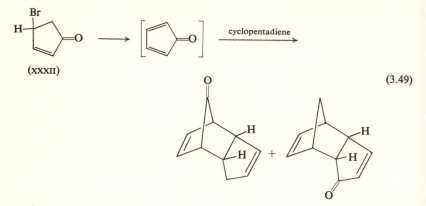

(3.49)

cyclopentadienone lose carbon monoxide easily on heating, with formation of dihydrobenzene or benzene derivatives. With olefinic dienophiles dehydrogenation may follow cleavage of carbon monoxide to form a compound of aromatic type. With acetylenic dienophiles the primary adducts formed are generally not stable and lose carbon monoxide spontaneously to give aromatic compounds. Thus, tetra-phenylcyclopentadienone and diphenylacetylene form hexaphenyl-benzene when heated together at 250°C (3.50).

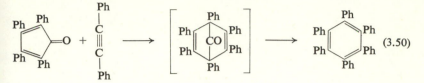

ortho-Quinones can also react both as dienes and dienophiles. *o*-Benzoquinone spontaneously forms the stable crystalline dimer (xxxiii) in acetone solution at room temperature; 4,5-dimethylbenzo-quinone behaves similarly. But it appears that in general their capacity

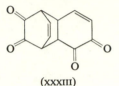

(xxxiii)

for acting as dienes predominates. Thus, tetramethylbenzoquinone and cyclopentadiene in boiling ethanol form the adduct (xxxiv) in 63 per cent yield; in this reaction the cyclopentadiene behaves as the dienophile (but see Ansell *et al.*, 1971).

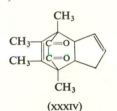

(xxxiv)

Furans. Many furan derivatives react with ethylenic and acetylenic dienophiles to form bicyclo compounds with an oxygen bridge (3.51) (Alder, 1948) (cf. p. 121). With simple furans a powerful dienophile is needed. Furan itself reacts with maleic anhydride or maleimide at room temperature to form derivatives of oxabicycloheptene, for example (xxxv), in high yield. Acetylenedicarboxylic ester also reacts readily to form the adduct (xxxvi) which may react further with excess of furan (3.52). On the other hand, no reaction takes place with dimethylmaleic anhydride or with tetrolic acid.

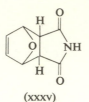

(3.51)

(xxxv)

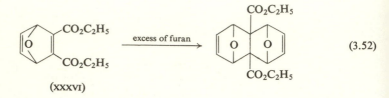

(3.52)

(xxxvi)

The oxabicycloheptenes obtained from furans and maleic acid derivatives are readily converted into benzene derivatives on treatment with
acid. These transformations are valuable as a means of identifying the
furan adducts, and also as a route to some benzene derivatives. For
example, 3,6-dimethylphthalic anhydride is conveniently obtained by
this route from 2,5-dimethylfuran.

Most of the adducts obtained from furan derivatives are thermally
labile and readily dissociate into their components on warming. For this
reason, of course, heat cannot be used to speed up their rate of formation.
But this limitation is not serious for addition generally proceeds rapidly
at room temperature and by using a solvent from which the adduct
crystallises out, high yields can be obtained.

The adduct from furan and maleic anhydride has been shown to have
the *exo* structure (xxxvii) apparently violating the rule (p. 149) that the
endo isomer usually predominates. The reason for this is found in the related observation that the normal *endo* adduct formed from maleimide
and furan at 20°C dissociates at temperatures only slightly above room
temperature and more rapidly on warming, allowing conversion of the
endo adduct formed in the kinetically controlled reaction into the
thermodynamically more stable *exo* isomer (3.53).

Pyrrole and its derivatives are unsuitable as dienes in the Diels–Alder
reaction because the susceptibility of the nucleus to electrophilic sub-

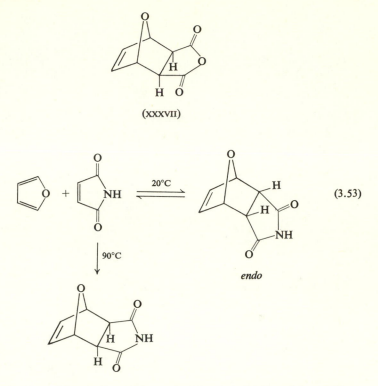

(XXXVII)

(3.53)

endo

stitution leads to side reactions. Normal adducts at the 2- and 5-positions have been obtained from some *N*-substituted pyrroles and acetylenic dienophiles under vigorous conditions. It has recently been found, however (Barlow, Haszeldine and Hubbard, 1969), that the very reactive dienophile (XXXVIII) reacts with pyrrole itself in ether to form the 1:1 adduct (XXXIX) in 58 per cent yield (3.54).

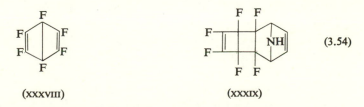

(XXXVIII)

(XXXIX)

(3.54)

Thiophen and its simple derivatives show no inclination to add dienophiles, but the 1,1-dioxide behaves both as a diene and dienophile; all attempts to prepare it have led to the sulphone (XL) through loss of

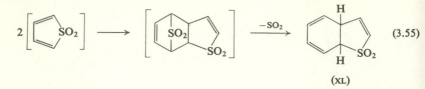

(XL)

sulphur dioxide from the intermediate dimer. The more stable 3,4-dichlorothiophen dioxide reacts readily with cyclopentadiene to give the adducts (XLI) and (XLII), (3.56).

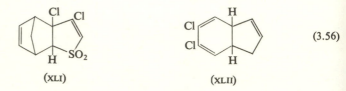

(XLI)   (XLII)

3.4. The retro Diels–Alder reaction. Diels–Alder reactions are reversible, and on heating many adducts dissociate into their components, sometimes under quite mild conditions. This can be made use of as in, for example, the separation of anthracene derivatives from mixtures with other hydrocarbons through their adducts with maleic anhydride, and in the preparation of pure D vitamins from the mixtures obtained by irradiation of the provitamins (cf. Alder, 1948).

More interesting are reactions in which the original adduct is modified chemically and subsequently dissociated to yield a new diene or dienophile. Thus, catalytic hydrogenation of the adduct obtained from 4-vinylcyclohexene and anthracene followed by thermal dissociation affords 4-vinylcyclohexane, whereas direct hydrogenation of vinylcyclohexene itself results in reduction of both double bonds. In this case adduct formation has been used to protect one of the double bonds of the vinylcyclohexene.

A similar technique is used to obtain 22,23-dihydroergosterol from ergosterol by protection of the cyclic 1,3-diene system through the adduct with maleic anhydride, followed by hydrogenation of the side-chain double bond and regeneration of the diene system.

Again, benzoquinone and its derivatives cannot be oxidised directly to epoxides with alkaline hydrogen peroxide because of the sensitivity of the quinone ring to alkaline conditions. This difficulty can be circumvented by oxidation of the 1:1 adduct with cyclopentadiene (XLIII),

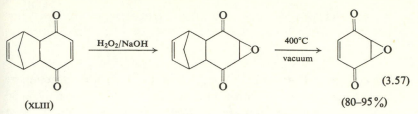

(XLIII)   (3.57)
                                                                    (80–95%)

(3.57). The quinone epoxide is then obtained in high yield by thermal decomposition of the oxidised adduct.

It is not always the bonds formed in the original diene addition which are broken in the retro reaction. Examples have already been noted for the adducts formed from cyclohexadiene and acetylenic dienophiles (p. 139). A reaction of this kind was used (Vogel, Grimme and Korte, 1963) in an ingenious synthesis of benzocyclopropene (XLVI) (3.58). Addition of methyl acetylenedicarboxylate to 1,6-methanocyclodeca-pentaene (XLIV) afforded the adduct (XLV) from which on pyrolysis at 400°C benzocyclopropene was obtained in 45 per cent yield, with elimination of dimethyl-phthalate. In another example furan 2,3-

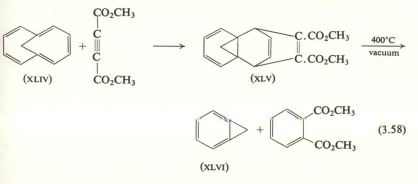

(XLIV)   (XLV)   (3.58)

(XLVI)

dicarboxylic acid was readily prepared by decomposition of the hydrogenated adduct from furan and acetylenedicarboxylic ester (3.59).

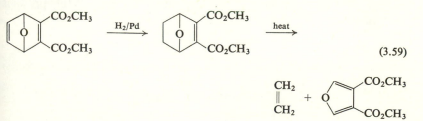

(3.59)

Most retro Diels–Alder reactions are brought about by heat, but some photo-induced reactions have been observed (Nozake, Kato and Nayori, 1969).

3.5. Catalysis by Lewis acids.   Catalysis of the Diels–Alder reaction by acids has been known for some time, but the influence of catalysts on the rate has generally been small (Wassermann, 1965). It has been found recently however that some Diels–Alder condensations are accelerated remarkably by aluminium chloride and other Lewis acids such as boron trifluoride and stannic chloride (Yates and Eaton, 1960; Sauer, 1967). Thus equimolecular amounts of anthracene, maleic anhydride and aluminium chloride in methylene chloride solution gave a quantitative yield of adduct in 90 s at room temperature. It is estimated that reaction in absence of the catalyst would require 4800 h for 95 per cent completion. Similarly, butadiene and methyl vinyl ketone react in 1 h at room temperature in presence of stannic chloride to give a 75 per cent yield of acetylcyclohexene. In absence of a catalyst no reaction takes place.

The catalytic action is thought to be due to complex formation between the Lewis acid and the polar groups of the dienophile leading to further reduction of the electron density at the double bond and increased dienophilic character. Interestingly, however, the addition of ethylene, which has no polar groups, to tetraphenylcyclopentadienone is also accelerated by aluminium chloride.

Pure *cis* addition is observed in reactions carried out in presence of Lewis acids just as in the uncatalysed reaction (see p. 147), that is the relative orientation of substituents in the diene and dienophile is preserved in the adduct. But there are indications that the ratio of structurally isomeric or stereoisomeric 1:1 adducts may be different in the catalysed reactions. Thus, on the addition of methyl vinyl ketone to isoprene two structural isomers are formed of which the predominant one is that with the greatest separation between the two groups (3.60). Lutz and Bailey (1964) have recently shown that the proportion of this product is even greater in the catalysed reaction. This is believed to be due to a

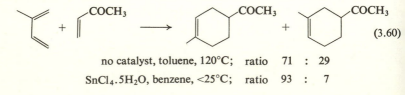

$$\text{no catalyst, toluene, } 120°C; \quad \text{ratio} \quad 71 : 29$$
$$\text{SnCl}_4.5\text{H}_2\text{O, benzene, } <25°C; \quad \text{ratio} \quad 93 : 7$$

steric effect caused by co-ordination of the large Lewis acid with the carbonyl group of the ketone, favouring a transition state in which the two substituents are as far removed from each other as possible. A similar effect has been noted for the reaction of methyl acrylate with isoprene.

Similarly, in the addition of acrylic acid to cyclopentadiene the proportion of *endo* adduct was found to increase noticeably in presence of aluminium chloride etherate (Sauer and Kredel, 1966) (3.61). Other

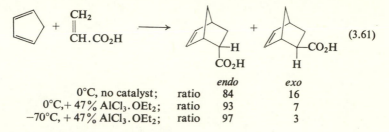

		endo	*exo*
0°C, no catalyst;	ratio	84	16
0°C, + 47% AlCl₃.OEt₂;	ratio	93	7
−70°C, + 47% AlCl₃.OEt₂;	ratio	97	3

Lewis acid catalysts had the same effect (see also Inukai and Kojima, 1966).

These catalysed reactions are of great preparative usefulness, but unfortunately limitations are imposed by the instability of some dienes and dienophiles in presence of Lewis acids.

3.6. Stereochemistry of the Diels–Alder reaction. The great synthetic usefulness of the Diels–Alder reaction depends not only on the fact that it provides easy access to a variety of six-membered ring compounds, but also on its remarkable stereoselectivity. This factor more than any other has contributed to its successful application in the synthesis of a number of complex natural products (cf. Alder and Schumacher, 1953; Martin and Hill, 1961). It should be noted, however, that the high stereoselectivity applies only to the kinetically-controlled reaction and may be lost by epimerisation of the product or starting materials, or by easy dissociation of the adduct allowing thermodynamic control of the reaction. These factors are fully discussed by Martin and Hill (1961).

The cis *principle*. The stereochemistry of the adduct obtained in many Diels–Alder reactions can be selected on the basis of two empirical rules formulated by Alder and Stein in 1937. According to the '*cis* principle', which is very widely followed, the relative stereochemistry of substituents in both the dienophile and the diene is retained in the adduct. That is, a dienophile with *trans* substituents will give an adduct in which the *trans* configuration of the substituents is retained, while a *cis* disubstituted dienophile will form an adduct in which the substituents are *cis* to each other. For example, in the reaction of cyclopentadiene with dimethyl maleate the *cis* adducts (XLVII) and (XLVIII) are formed while in the reaction with dimethyl fumarate the *trans* configuration of the ester groups is retained in the adduct (XLIX), (3.62).

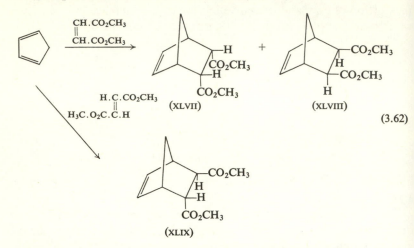

(3.62)

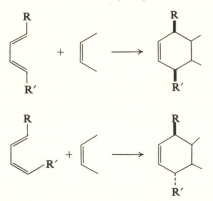

(XLIX)

Similarly with the diene component, the relative configuration of the substituents in the 1- and 4- positions is retained in the adduct; a *trans, trans*-1,4-disubstituted diene gives rise to adducts in which the 1- and 4- substituents are *cis* to each other, and a *cis,trans*-disubstituted diene gives adducts with *trans* substitutents (3.63).

(3.63)

The almost universal application of the *cis* principle provides strong evidence for a mechanism for the Diels–Alder reaction in which both the new bonds between the diene and the dienophile are formed at the same time. But a two-step mechanism is not completely excluded, for the same stereochemical result would obtain if the rate of formation of the second bond in the (diradical or zwitterionic) intermediate (L), (3.64), were faster than the rate of rotation about a carbon–carbon bond.

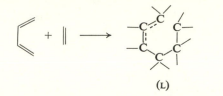

$$(3.64)$$

(L)

The endo *addition rule*. In the addition of maleic anhydride to cyclo-pentadiene two different products, the *endo* and the *exo*, might con-ceivably be formed depending on the manner in which the diene and the dienophile are disposed in the transition state. According to Alder's *endo* addition rule, in a diene addition reaction the two components arrange themselves in parallel planes, and the most stable transition state arises from the orientation in which there is 'maximum accumu-lation of double bonds'. Not only the double bonds which actually take part in the addition are taken into account, but also the π bonds of the activating groups in the dienophile. The rule appears to be strictly applicable only to the addition of cyclic dienophiles to cyclic dienes, but it is a useful guide in many other additions as well.

Thus, in the addition of maleic anhydride to cyclopentadiene the *endo* product, formed from the orientation with maximum accumulation of double bonds, is produced almost exclusively (3.65). The thermo-dynamically more stable *exo* compound is formed in yields of less than 1.5 per cent.

From benzoquinone and cyclopentadiene again only the *endo* adduct (LI) was isolated, the configuration of the product being shown by its conversion into the 'caged' compound (LII) with ultraviolet light (3.66).

The products obtained from the cyclic diene furan and maleic an-hydride and from diene addition reactions of fulvene do not obey the *endo* rule. The reason is that the initial *endo* adducts easily dissociate at moderate temperatures, allowing conversion of the kinetic *endo* adduct into the thermodynamically more stable *exo* isomer. In other cases prolonged reaction times may lead to the formation of some *exo* isomer at the expense of the *endo*.

In the addition of open-chain dienophiles to cyclic dienes, the *endo*

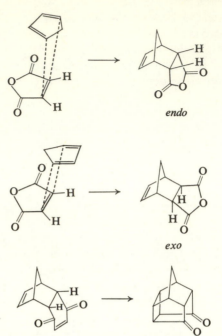

(3.65)

(3.66)

rule is not always obeyed and the composition of the mixture obtained may depend on the precise structure of the dienophile and on the reaction conditions. Thus, in the addition of acrylic acid to cyclopentadiene

TABLE 3.2   *Proportion of* endo *and* exo *acids formed in addition of α–substituted acrylic acids to cyclopentadiene*

X	endo CO$_2$H	exo CO$_2$H
H	75	25
CH$_3$	35	65
C$_2$H$_5$	—	100
C$_6$H$_5$	60	40
Br	30	70

the *endo* and *exo* products were obtained in the ratio 75:25 but in the α- substituted acrylic acids (LIII) the product ratio varied depending on the nature of the group X (Martin and Hill, 1961, p. 550) (see Table 3.2). Equally variable ratios are observed in reactions with β-substituted acrylic acids. With acrylic acid itself, the proportion of *endo*-adduct formed was noticeably increased by the presence of Lewis acid catalysts (p. 147) (see also Kobuke, Fueno and Furukawa, 1970).

Solvent and temperature may also affect the product ratio. Thus, in the kinetically controlled addition of cyclopentadiene to methyl acrylate, methyl methacrylate and methyl *trans*-crotonate in different solvents, the proportion of *endo* product increased with the polarity of the solvent, and the product ratio was also slightly affected by the temperature of the

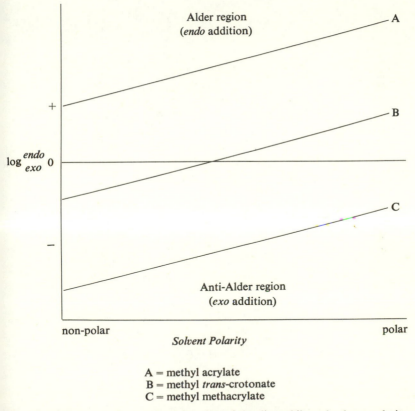

A = methyl acrylate
B = methyl *trans*-crotonate
C = methyl methacrylate

Fig. 3.1. *Endo/exo* product ratio as function of the dienophile and solvent polarity (schematic). (From Berson, Hamlet and Mueller, 1962.)

reaction. In all cases mixtures of the *endo* and *exo* products were obtained and with methyl methacrylate the *exo* isomer was the predominant product under all experimental conditions. With methyl crotonate the *exo* adduct was predominant in some solvents (e.g. trimethylamine at 30°C) and the *endo* in others (ethanol, acetic acid) (see Fig. 3.1) (Berson, Hamlet and Mueller, 1962).

The adducts obtained from acyclic dienes and cyclic dienophiles are frequently formed in accordance with the *endo* rule. Thus, in the addition of maleic anhydride to *trans,trans*-1,4-diphenylbutadiene the *cis* adduct (LIV) is formed almost exclusively (3.67) through the orientation of diene

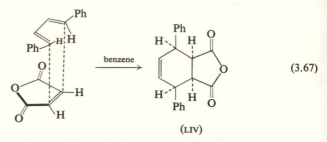

(3.67)

(LIV)

and dienophile with 'maximum accumulation of double bonds'. Similarly, maleic anhydride and dimethyl *trans*-muconate afford the adduct (LV), (3.68). Again in accordance with the *endo* rule *cis*-1-ethyl-1,3-butadiene and maleic anhydride afford the adduct (LVI) in which in this case the ethyl substituent is *trans* to the anhydride group.

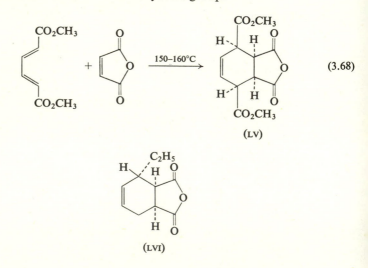

(3.68)

(LV)

(LVI)

The stereoselectivity of addition of cyclic dienophiles to acyclic dienes was exploited by Criegee and Becher in their synthesis of the naturally-occurring tetraol, conduritol D (LVII) (3.69). The all-*cis* arrangement of the hydroxy groups was shown by oxidation to allo-mucic acid. In Woodward's synthesis of reserpine three of the five

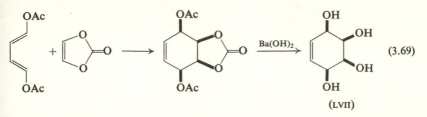

(3.69)

(LVII)

asymmetric centres in ring *E* (LVIII) were set up correctly, in one step, by reaction of vinylacrylic acid with benzoquinone to give the adduct (LIX) (3.70).

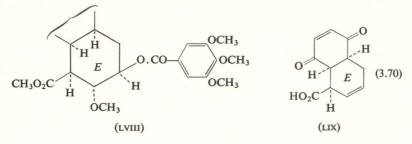

(3.70)

(LVIII)　　　　　(LIX)

In the addition reactions of open-chain dienes and open-chain dienophiles the *endo* adduct is the main product at moderate temperatures, but in a number of cases it has been found that the proportion of the *exo* isomer increases with rise in temperature. Thus in the addition of acrylonitrile to *trans*-1-phenylbutadiene at 100°C the *cis* (*endo*) isomer (LX) was obtained, whereas at reflux temperature the main product was the *trans* (*exo*) isomer (LXI) (3.71). Table 3.3 shows the effect of temperature on the ratio of *cis* and *trans* adducts obtained by reaction of *trans*-butadiene-1-carboxylic acid and acrylic acid.

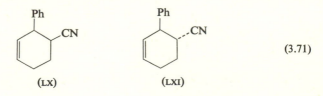

(3.71)

(LX)　　　　　(LXI)

TABLE 3.3 *Effect of temperature on ratio of* cis- *and* trans-*adducts formed in reaction of* trans-*butadiene-1-carboxylic acid with acrylic acid* (from Sauer, 1967).

Temperature (°C)	75	90	100	110	130
Ratio cis:trans	cis only	7:1	4·5:1	2:1	1:1

The factors which determine the steric course of diene additions are still not completely clear. It appears that a number of different forces operate in the transition state and the precise steric composition of the product depends on the balance among these (see Martin and Hill, 1961; Berson, Hamlet and Mueller, 1962; Wassermann, 1965). The predominance of *endo* addition has been ascribed to the tendency of dienophile substituents to be so oriented in the favoured transition state of the reaction that they lie directly above the residual unsaturation of the diene, either for reasons of spatial orbital overlap or for reasons of steric accommodation. Non-bonded charge-transfer type interactions may also play a part in the stabilisation of the transition state in some cases. Alder and Stein originally summarised the steric nature of the transition state by their principle of 'maximum accumulation of double bonds' leading to spatial orbital overlap of unsaturated centres of the diene and dienophile, but this is clearly not the whole story and recent work outlined on pp. 150–151 appears to show that these attractive forces are easily overweighed by steric factors caused by structural changes in the dienophile and, in some cases, by changes in the experimental conditions. It is difficult to understand also how spatial orbital overlap alone can account for the high preponderance of *endo* products formed in the addition of cyclopentene (98 per cent *endo*) and other cycloalkenes to cyclopentadiene. It appears that the transition state which is best stabilised by spatial orbital overlap and by non-bonded interactions and simultaneously least destabilised by unfavourable steric repulsions has the lowest free energy and consequently predominates in the kinetically controlled product. For reactions with cyclic dienophiles this is usually the *endo* adduct, but with open-chain dienophiles the interplay of the different factors makes it more difficult to predict precisely the steric course of the additions.

Further insight into the origins of stereoselectivity in Diels–Alder addition reactions has been gained recently by Hoffmann and Woodward (1965) by consideration of orbital symmetry relationships. They have pointed out that the Diels–Alder dimerisation of butadiene, regarded by them as a concerted cycloaddition reaction, may take place through one or other of two transition states which are distinguished from each other mainly by the proximity of a β- and a β'-orbital in the *endo* approach (3.72). The significant orbital interactions in the transition state will

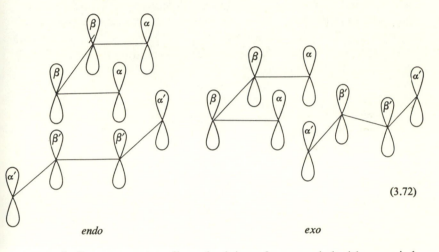

$$(3.72)$$

endo   *exo*

come only from symmetry-allowed mixing of unoccupied with occupied levels and it can be shown (compare diagrams in (3.73)) that in the *endo* approach both possibilities of mixing lead to bonding, that is energy-lowering, interaction of the neighbouring β and β' orbitals.

Thus the *endo* transition state for this reaction is stabilised in comparison with the *exo* alternative by symmetry-controlled secondary orbital interactions. Similar circumstances are said to hold for other $(4 + 2)$ concerted cycloadditions which were considered. This approach has been criticised by Herndon and Hall (1967) who point out that it would not seem to be able to explain the high *endo* stereoselectivity observed in the reaction of cyclopentadiene with cyclopentene and other cyclic olefins, where the possibility of secondary orbital interactions does not exist. They conclude as a result of calculations of the difference in activation energies for the *endo* and *exo* reactions that secondary interactions may be unimportant and that the stabilities of the *endo* and *exo* transition states are governed largely by the geometrical overlap of the π bonds at the primary centres where the new bonds are actually developing. This geometric factor is thought to be energetically sufficient to account for preferred *endo* addition. This hypothesis has the merit of accounting for the preferred *endo* addition of cyclopentadiene with cyclic olefins, but the general validity of Hoffmann and Woodward's approach is strongly supported by the observation that reaction of the cyclopentadienone (LXII) with cycloheptatriene and of cyclopentadiene with tropone led to the two *exo* addition products (LXIII) and (LXIV), in line with their prediction, based on consideration of secondary orbital interactions in the transition state, that $(6 + 4)$ concerted cycloaddition reactions should lead preferentially to the *exo* product and not the *endo* as in Diels–Alder and other $(4 + 2)$ cycloaddition reactions (Woodward, 1967) (3.74).

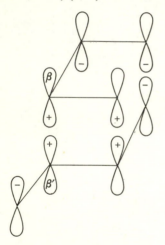

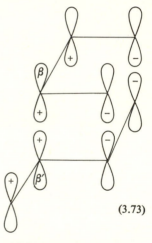

(3.73)

Mixing of highest occupied 'diene' orbital (top) with lowest unoccupied 'olefin' orbital (bottom).

Mixing of lowest unoccupied 'diene' orbital (top) with highest occupied 'olefin' orbital (bottom).

(From Hoffmann and Woodward, 1965.)

Orientation of addition of unsymmetrical components. Addition of an unsymmetrical diene to an unsymmetrical dienophile could take place in two ways to give two structurally isomeric products. It is found in practice, however, that formation of one of the isomers is strongly

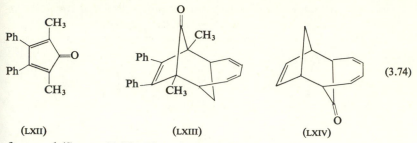

(LXII)　　　　　　　　　(LXIII)　　　　　　　　(LXIV)

favoured (Sauer, 1967). Thus, in the addition of acrylic acid derivatives to 1-substituted butadienes the '*ortho*' (1,2-) adduct is favoured, irrespective of the electronic nature of the substituent (see Table 3.4). The orientating forces are relatively weak and with greater steric demand of the substituents the proportion of the '*meta*' (1,3-) isomer approaches the statistical value. The very strong '*ortho*' orientation observed with acrylic acid and butadiene-1-carboxylic acid is lost when the anions are used, presumably because of the Coulomb repulsion of the two charged groups. It has been thought that increase in temperature led to formation of an increasing proportion of '*meta*' adduct, but recent work in which the product distribution was determined by gas–liquid chromatography indicates that this is not the case.

Similarly, in the addition of methyl acrylate to 2-substituted butadienes the '*para*' (1,4-) adduct is formed predominantly, irrespective of the electronic nature of the substituent (see Table 3.5).

TABLE 3.4　*Proportions of structural isomers formed in addition of acrylic acid derivatives to 1-substituted butadienes.*

R^1	R^2	R	T (°C)	Ratio of products 1,2-	1,3-
$N(C_2H_5)_2$	H	C_2H_5	20	1,2- only	
CH_3	H	CH_3	20	18	1
CO_2H	H	H	70	1,2- only	
CO_2Na	H	Na	220	1	1
$C(CH_3)_3$	$C(CH_3)_3$	CH_3	200	0·9	1

TABLE 3.5 *Proportions of structural isomers formed in addition of methyl acrylate to 2-substituted butadienes.*

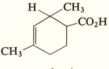

R_1	$T (°C)$	Ratio 1,4- to 1,3-
OC_2H_5	160	1,4- only
C_6H_5	150	4·5 to 1
CN	95	1,4- only

Addition of acetylenic dienophiles to 1- and 2- substituted butadienes also results in preferential formation of the '*ortho*' and '*para*' adducts as with olefinic dienophiles.

As would be expected, in 1,3-disubstituted butadienes the directive influence of the substituents is additive. Thus, in the addition of 1,3-dimethylbutadiene and acrylic acid the adduct (LXV) is formed almost exclusively.

(LXV)

From work on the addition of dienophiles to 1,4-disubstituted butadienes Alder has concluded that the orienting effect of substituents decreases in the order $C_6H_5 > CH_3 > CO_2H$.

No satisfactory mechanistic interpretation of these orientation effects is yet available.

It has been found recently that the relative amounts of the structurally isomeric products is strongly influenced by the presence of Lewis acid catalysts, and when these conditions are applicable very high yields of a single isomer can be obtained. Thus, for the addition of methyl vinyl ketone or acrolein to isoprene the proportion of the '*para*' adduct was increased in presence of stannic chloride, so that it became almost the exclusive product of the reaction (Lutz and Bailey, 1964) (3.75). Similar effects were noted for the addition of isoprene to methyl acrylate (Inukai and Kojima, 1966).

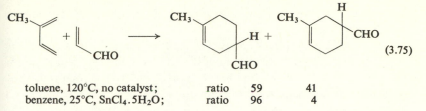

(3.75)

toluene, 120°C, no catalyst;	ratio	59	41	
benzene, 25°C, SnCl$_4$.5H$_2$O;	ratio	96	4	

It is thought that complexing with the catalyst increases the steric requirements of the dienophile, resulting in preference for the transition state for '*para*' addition in which the two substituents are as far removed from each other as possible.

Asymmetric induction. In reactions of disymmetric dienes and dienophiles in which the plane of the double bonds is not a plane of symmetry it is found, as might have been expected, that the diene and dienophile approach each other from the less hindered side of each. Thus, in the addition of maleic anhydride to cyclo-octatetraene, which reacts as if in the form (LXVI), the product is exclusively (LXVII) (3.76).

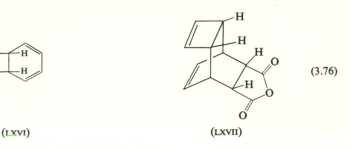

(3.76)

(LXVI)   (LXVII)

The interesting possibility arises that in the addition of a planar diene or dienophile to an asymmetric dienophile or diene, approach of the reagents might take place preferentially from one direction, and if reaction gives rise to a new asymmetric centre this might be formed in non-statistical ratio, resulting in an asymmetric synthesis. This possibility has been realised in a number of instances recently. In the normal thermal reaction optical yields are low, but in reactions catalysed by Lewis acid catalysts, particularly when carried out at low temperatures, very high optical yields have been obtained. Furthermore, it appears that the absolute stereochemistry of the predominant isomer can often be predicted, and the catalysed sequence offers a convenient new method for the asymmetric synthesis of a variety of optically active compounds of known absolute configuration (Sauer and Kredel, 1966b; Farmer and Hamer, 1966).

Thus, reaction of optically active (−)-menthyl fumarate with butadiene or isoprene followed by reduction of the adduct with lithium aluminium hydride with removal of the auxiliary asymmetric (−)-menthol, gave optically active specimens of the compounds (LXVIII), (R = H and CH$_3$) in optical yields of 1–9 per cent (3.77). Similarly, addition of R-(−)-menthyl acrylate to cyclopentadiene and reduction with lithium aluminium hydride gave the optically active product (LXIX) in 1–9 per cent optical yield, but when the addition was effected at −70°C in presence of boron trifluoride etherate the (+)-isomer of (LXIX) was obtained in nearly 90 per cent optical yield.

(LXVIII)　　　　　　　　　　(LXIX)

(3.77)

3.7. Mechanism of the Diels–Alder reaction. The precise mechanism of the Diels–Alder reaction has been the subject of much debate. There is general agreement that the rate-determining step in adduct formation is bimolecular and that the two components approach each other in parallel planes roughly orthogonal to the direction of the new bonds about to be formed. Formation of the two new σ bonds takes place by overlap of molecular π orbitals in a direction corresponding to endwise overlap of atomic *p*-orbitals. But there is still uncertainty about the nature of the transition state and, in particular, about the timing of the changes in covalency that result in the formation of the new bonds (see Woodward and Katz, 1959; Sauer, 1967; Wassermann, 1965).

Two main views have been considered. One, (*A*), that the reaction is a concerted four-centre addition in which both of the new single bonds are formed at the same time (3.78). The other, (*B*), that reaction takes place in two steps the first of which, the formation of a single bond between atoms of the reactants, is rate controlling; the addition is then completed by formation of the second bond in a fast reaction. The intermediate in the second alternative may have either zwitterionic or diradical character.

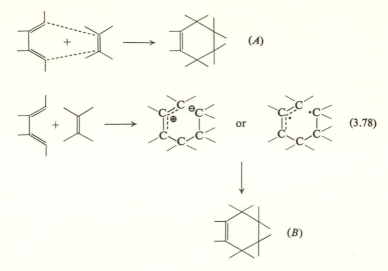

$$(A)$$

$$(3.78)$$

$$(B)$$

Evidence has been adduced in support of both mechanisms, but it is now generally believed that most thermal Diels–Alder additions are concerted. The Woodward–Hoffmann ideas discussed above (p. 155) strongly support a one-step process. A major factor bringing about acceptance of this view has been the high stereoselectivity of the reaction, although, as pointed out on page 148, a two-step mechanism is not entirely ruled out by this evidence if we assume that rotation about carbon–carbon single bonds in the intermediate is slow compared with the rate of formation of the second bond. In this connection it is noteworthy that the Diels–Alder reactions of *cis-* and *trans-*1,2-dichloro-ethylene with cyclopentadiene, which, if the two-step mechanism is correct, could involve biradical intermediates (LXX) and (LXXI) gave strict *cis* addition; there was no interconversion of possible intermediates

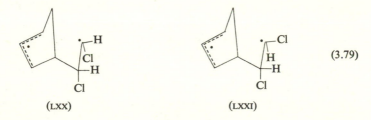

$$(3.79)$$

(LXX)   (LXXI)

(LXX) and (LXXI) (3.79). On the other hand, in the addition of dichloro-difluoroethylene to the isomeric hexadienes to form cyclobutane derivatives, which certainly proceeds by a two-step mechanism with a biradical intermediate, the step leading to formation of the ring is not stereoselective, because the rate of rotation about a carbon–carbon single bond in the intermediate is comparable with the rate of formation of the ring (3.80).

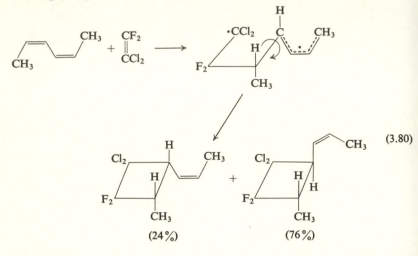

(24%)   (76%)

Attempts to show the involvement of free radicals in the Diels–Alder reaction have been negative. No biradical intermediates have been detected, and compounds that catalyse singlet–triplet transitions have no effect on the reaction. Similarly, the kinetic effects of *para* substituents in 1-phenylbutadiene, although large in absolute terms, are considered much too small for a rate-determining transition state corresponding to a zwitterion intermediate, and indicate a synchronous mechanism. Thus, replacement of methoxyl by nitro in the reaction of *para* substituted phenylbutadienes with maleic anhydride results in a decrease in the rate constant by only 10, whereas the rate of solvolysis of *para* substituted α,α-dimethylbenzyl chlorides to give the ion pairs decreases by a factor of 7×10^9 on replacement of a *para* methoxyl by a nitro substituent.

The two-step mechanism similarly receives no support from studies of the isomerisation of Diels–Alder adducts. Many Diels–Alder adducts can be isomerised by heat. For example, the adduct (LXXII) from cyclopentadiene and methyl methacrylate is isomerised to the isomer (LXXIII) on heating in decalin solution (3.81). Strong support for the possibility of a two-step Diels–Alder mechanism would be provided if a biradical (or zwitterion) such as (LXXIV) could be shown to be an intermediate in

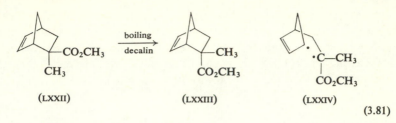

(3.81)

this isomerisation. But in fact it is known that reaction takes place by complete dissociation of the adduct into its components followed by recombination, for on heating the optically active *exo*-carboxylate (LXXII) completely racemised *endo*-carboxylate (LXXIII) was obtained. In other cases where this mechanism is precluded, as in the stereospecific rearrangement of the deuterated methacrolein dimer (LXXV), the conversion is thought to take place by a Cope rearrangement and has no bearing on the mechanism of the Diels–Alder reaction (3.82).

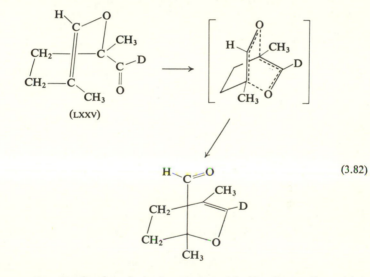

(3.82)

The one-step mechanism is in harmony with kinetic studies which have been interpreted as requiring a highly special orientation of the reactants and a geometry for the activated complex similar to that of the adduct, and also with recent work on kinetic isotope effects.

Whether or not both of the new bonds in the concerted mechanism are formed to the same extent at the transition state is an open question. A current view is that although they both begin to be formed at the same time the process may take place at a different rate in the two bonds so that the transition state is 'lop-sided', with one bond formed to a greater extent than the other. Streitwieser represents the transition state as shown in (3.83). Berson and Ramenick (1961) have put forward the view that there may be a gradation of mechanisms for different

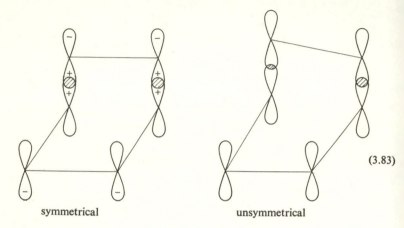

symmetrical   unsymmetrical

(3.83)

Diels–Alder reactions, extending from a completely concerted four centre mechanism with a symmetrical transition state at one extreme to something approaching a two-step process at the other.

It has always been difficult to understand why addition of olefins to conjugated systems to form six-membered rings takes place so easily, while thermal addition to form cyclobutanes is so uncommon, although it is easily effected photochemically. From a consideration of molecular orbital symmetries Hoffmann and Woodward (1965) have recently been able to explain why this is so. Using extended Hückel calculations, they have shown that the hypothetical concerted reaction of two ethylene molecules to form cyclobutane is a highly unfavourable ground state process, because of symmetry restrictions, and requires an activation energy comparable to that of electronic excitation of ethylene, an energy unavailable in ordinary thermal reactions. On the other hand, concerted formation of a six-membered ring by addition of an olefin to a conjugated system involves ground state molecules only and is a thermally allowed reaction. Hoffmann and Woodward have embodied their results in a set of selection rules, according to which intermolecular cycloadditions involving ethylenic bonds and conjugated systems take place easily in a concerted fashion when six-membered rings are being formed, but not when the products are four- or eight-membered rings (see also Woodward and Hoffmann, 1969).

These selection rules describe the probable course of a reaction, but they do not have to be followed; they are permissive but not obligatory. Any process which is 'forbidden' by the selection rules will have a high activation energy by a *concerted* mechanism but it may occur easily by a stepwise process involving biradicals or an intermediate dipolar ion. Bartlett (1968) and others have recently described a number of thermal reactions involving 1,2-addition of olefins to conjugated dienes. 1,1-Dichloro-2,2-difluoroethylene, for example, reacts with butadiene at 79°C to form 99 per cent of the 1,2-addition product (3.84), and

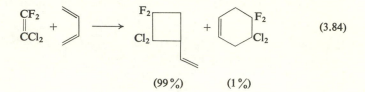

$$(99\%) \qquad (1\%) \tag{3.84}$$

α-acetoxyacrylonitrile gives a mixture of products formed by 1,2- and 1,4-addition (3.87).

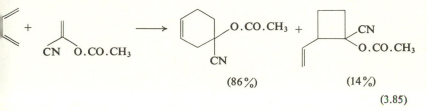

$$(86\%) \qquad (14\%)$$

$$(3.85)$$

In these reactions it is highly probable that the formation of the four-membered rings takes place by a stepwise process involving biradicals, and not by a concerted mechanism.

3.8. Photosensitised Diels–Alder reactions. A number of light-induced Diels–Alder reactions have been described recently. It is probable that the mechanism of these photo-reactions differs from that of thermal Diels–Alder additions. Concerted (4 + 2) photocycloadditions are forbidden by the Hoffmann–Woodward selection rules and a different mechanism is indicated also by the fact that the products of the photo and thermal reactions are not the same. Thus, in the photodimerisation of cyclopentadiene the two dimers (LXXVII) and (LXXVIII) are obtained in equal amount with the normal thermal adduct (LXXVI) (3.86), and irradiation of butadiene in presence of ketonic sensitisers affords a

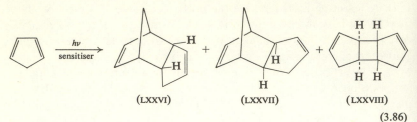

(3.86)

mixture of *cis*- and *trans*-divinylcyclobutane along with the normal Diels–Alder adduct vinylcyclohexene (LXXIX) (Hammond, Turro and

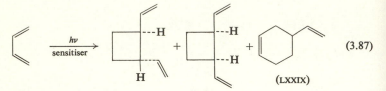

(3.87)

Liu, 1963) (3.87). Cyclohexadiene also affords a mixture of three products on irradiation. None of the thermal *endo* adduct (LXXX) is produced in this case (3.88).

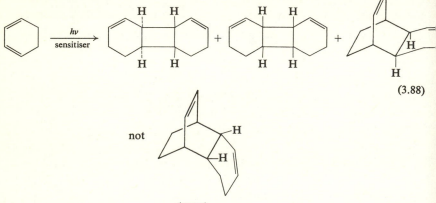

(3.88)

Hammond, Turro and Liu (1963) have suggested that these photosensitised Diels–Alder products are formed through diallylic biradicals which themselves arise by addition of triplet diene to another molecule of diene.

Perhaps the most spectacular photochemical Diels–Alder type reaction is the formation of the stable 2:1 adduct (LXXXI) by ultraviolet irradiation of a solution of maleic anhydride in benzene. Similar adducts are formed by reaction of maleic anhydride with toluene and with *ortho*- and *para*-xylene (3.89). It was originally thought that the mono-

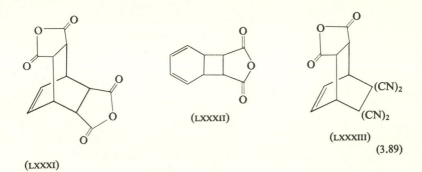

(LXXXII)

(LXXXIII)

(3.89)

(LXXXI)

adduct (LXXXII), formed by photochemical addition of maleic anhydride to benzene, was an intermediate in the reaction and that formation of the bis adduct followed by a rapid thermal addition of a second molecule of maleic anhydride to the diene (LXXXII), but it now appears that this is not the case. When reaction was effected in presence of tetracyanoethylene, which does not itself form an adduct with benzene but which would be expected to react rapidly with (LXXXII), only the normal adduct (LXXXI) was formed; none of the mixed product (LXXXIII) was obtained (Angus and Bryce-Smith, 1960; Bryce-Smith, 1968).

3.9. The homo Diels–Alder reaction. In the homo Diels–Alder reaction of a 1,4-diene with a dienophile, *three* π bonds are converted into three σ bonds with formation of two new rings (3.90). Most recorded reactions

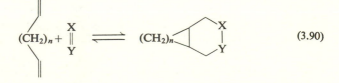

(3.90)

of this type involve additions to bicycloheptadiene, and the dienophiles which have been used include tetracyanoethylene, dicyanoacetylene, 4-phenyl-1,2,4-triazolin-3,5-dione and (less readily) acrylonitrile. Thus,

tetracyanoethylene and bicyclo[2,2,1]heptadiene form the adduct
(LXXXIV) in quantitative yield in boiling benzene (3.91) and methyl

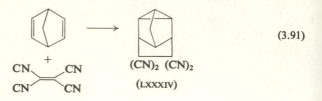

(3.91)

azodicarboxylate readily gives the adduct (LXXXV) (3.92). Bicycloheptadiene itself, when heated with rhodium on charcoal, forms a
mixture of two stereoisomers of the homoadduct (LXXXVI) (3.93).

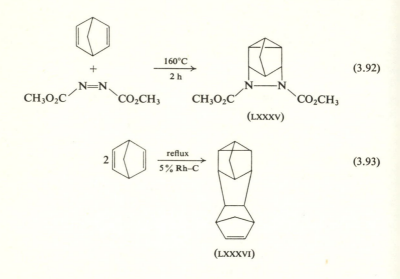

There is evidence (Berson and Olin, 1969) that the homo Diels–Alder
reaction, like the Diels–Alder reaction itself, is a concerted process.
A concerted thermal addition of this type, since it involves $(4 + 2)\ \pi$
electrons is 'allowed' by Woodward and Hoffmann's selection rules.

3.10. The 'ene' synthesis. Formally related to the Diels–Alder reaction
is the so-called 'ene synthesis' in which an olefin reacts thermally with
a dienophile with formation of a new carbon–carbon σ bond, migration
of a hydrogen atom and change in the position of the double bond in
the olefin (see Alder and von Brachel, 1962; Hoffmann, 1969) (3.94).

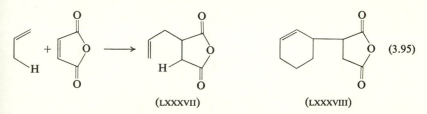

(3.94)

Thus, maleic anhydride and propylene react at 200°C to give the product (LXXXVII), and cyclohexene forms the derivative (LXXXVIII), (3.95). Particularly interesting synthetically is the reaction of allyl

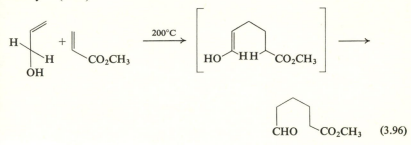

(3.95)

(LXXXVII)   (LXXXVIII)

alcohol with dienophiles; the initial adduct is the enolic form of an aldehyde (3.96).

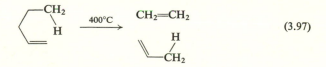

(3.96)

Like the Diels–Alder reaction itself the ene synthesis is reversible. This is shown for example by the decomposition of n-pentene at 400°C to give ethylene and propylene (3.97). A synthetically useful reaction

(3.97)

of this type is the thermal decomposition of $\beta\gamma$-unsaturated alcohols to olefins and formaldehyde (3.98).

The ene synthesis resembles the Diels–Alder reaction also in its stereoselectivity, showing *cis* addition and a preference for the formation of *'endo'* products. *Cis* addition is illustrated by the reaction of n-heptene

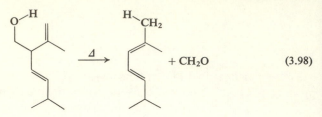

(3.98)

with dimethyl acetylenedicarboxylate to form the adduct (LXXXIX) (3.99). None of the fumaric acid derivative was obtained. Similarly, in

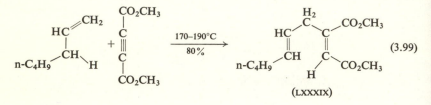

(3.99)

(LXXXIX)

the reaction with propiolic ester a *trans* $\alpha\beta$-unsaturated ester was formed (3.100).

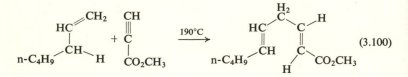

(3.100)

In the reaction of *trans*-but-2-ene with maleic anhydride the major product was the *erythro* adduct, whereas with *cis*-but-2-ene the *threo* product was obtained. Assuming a concerted reaction, this result indicates a strong preference for an '*endo*' type transition state, illustrated in (3.101) for the case of *cis*-but-2-ene. 'Endo' addition is said to be

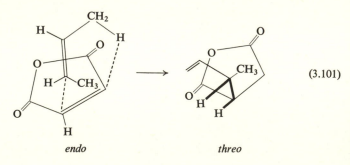

(3.101)

endo   threo

favoured by orbital symmetry relationships, although the preference is not so marked as in the Diels–Alder reaction (Berson, Wall and Perlmutter, 1966).

It is believed, although it has not yet been conclusively proved, that the ene synthesis proceeds by a concerted mechanism. Early attempts to ascribe to the reaction a free radical or ionic character have been abandoned in favour of the concerted process. Strong evidence against a stepwise diradical mechanism (carbon–carbon bond formation followed by intramolecular hydrogen transfer) is provided by the absence of cyclo-adducts in the ene synthesis and of 'ene products' in these cyclo-additions which are known to proceed by way of diradicals (p. 162).

4 Reactions at unactivated C–H bonds

Most synthetically useful ionic organic reactions at saturated carbon atoms take place either by displacement of a suitable leaving group, or by replacement of a hydrogen atom at a carbon atom which is 'activated' by the presence of a neighbouring activating group (cf. Ch. 1). Ionic attack at unactivated C–H bonds is uncommon. Free radicals, on the other hand, can often be obtained with enough energy to break un-activated C–H bonds (see Walling, 1957), but intermolecular reactions of this type are of limited value synthetically, because the reagents are unselective and mixtures of products generally result. It has recently been found, however, that in many molecules which meet certain structural and geometrical requirements *intramolecular* free radical attack at unactivated C–H bonds can become quite specific, leading to the intro-duction of functional groups at the site of specific C–H bonds. A number of reactions of this type have become of synthetic importance.

The key step in these reactions is an intramolecular abstraction of hydrogen from a carbon atom, resulting in the transfer of a hydrogen atom from carbon to the attacking free radical in the same molecule. Because of the geometrical requirements of the transition state for this step, the most frequently observed intramolecular hydrogen transfers of this type are 1,5-shifts, corresponding to specific attack on a hydrogen atom attached to a δ-carbon. The reactions can be generally represented as in (4.1). Homolytic cleavage of the Y–X bond is followed by hydrogen

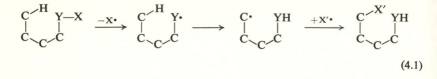

$$(4.1)$$

transfer from the δ-carbon atom to Y·, and the resulting carbon radical finally reacts with a free radical X′· which may or may not be identical with X·.

A number of important reactions of this type in which $Y = O$ are discussed below. Another useful reaction in which $Y = N$ is the well-known Hofmann–Loeffler–Freytag reaction.

4.1. The Hofmann–Loeffler–Freytag reaction. This reaction provides a convenient and useful method for the synthesis of pyrrolidine derivatives from *N*-halogenated amines (Wolff, 1963) (4.2). The reaction is effected

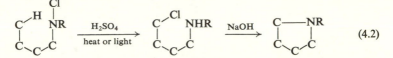

$$(4.2)$$

by warming a solution of the halogenated amine in strong acid (concentrated sulphuric acid or trifluoroacetic acid are often used), or by irradiation of the acid solution with ultraviolet light. The immediate product of the reaction is the δ-halogenated amine, but this is not generally isolated, and by basification of the reaction mixture it is converted directly into the pyrrolidine. Both *N*-bromo- and *N*-chloro-amines have been used as starting material, but the *N*-chloro-amines are said to give better yields. They are easily obtained from the amines by action of sodium hypochlorite or *N*-chloro-succinimide.

The first example of this reaction was reported by A. W. Hofmann in 1883. In the course of a study of the reactions of *N*-bromoamides and *N*-bromoamines, he treated *N*-bromoconiine (I) with hot sulphuric acid and obtained, after basification, a tertiary base which was later identified as δ-coneceine (II) (4.3). Later, further examples of the reaction were

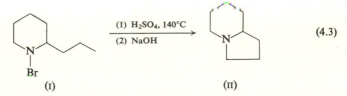

$$(4.3)$$

reported by Loeffler, including a neat synthesis of the alkaloid nicotine (III) (4.4). Numerous other cyclisations leading to both simple pyrroli-

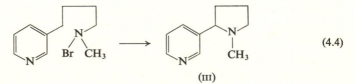

$$(4.4)$$

dines and to more complex polycyclic structures have since been recorded. Thus, *N*-methyl-*N*-chloro-octylamine affords 1-methyl-2-butylpyrrolidine in high yield (4.5), and the cyclopentyl compound (IV) is converted into the bridged pyrrolidine (V) (4.6).

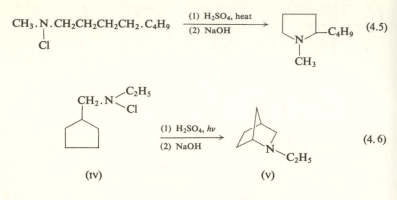

(4.5)

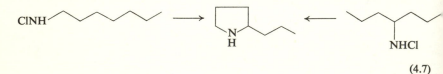

(4.6)

It was long believed that only *N*-halogenated derivatives of secondary amines would undergo the reaction, but it is now known that pyrrolidines lacking an *N*-alkyl substituent can be obtained from *N*-chloro- primary amines in strongly acid solution by using ferrous ions as initiators (Schmitz and Murawski, 1966). Thus, *N*-chlorobutylamine is converted into pyrrolidine in 72 per cent yield, and 2-n-propylpyrrolidine is readily obtained from both *N*-chloro-1-aminoheptane and *N*-chloro-4-aminoheptane (4.7).

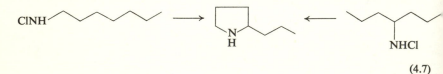

(4.7)

Although the majority of applications of the reaction have employed *N*-haloamines as starting materials, *N*-haloamides have been used occasionally and these also give rise to secondary pyrrolidines by loss of the acyl group. *N*-Butyl-*N*-chloroacetamide, for example, when heated with sulphuric acid and subsequently treated with alkali, is converted into pyrrolidine in 50 per cent yield.

With *N*-halocycloalkylamines cyclisation leads to bridged-ring structures, but in these cases the products are not exclusively pyrrolidine derivatives. Thus, whereas *N*-bromo-*N*-methylcycloheptylamine was converted into tropane (VI) in 40 per cent yield when warmed with sulphuric acid, *N*-chloro-4-ethylpiperidine gave a mixture of the azabicycloheptane (VII) and quinuclidine (VIII), and *N*-chloro-*N*-methylcyclooctylamine is reported to yield *N*-methylgranatamine (IX) exclusively (4.8). The reasons for the preferential formation of a six-membered

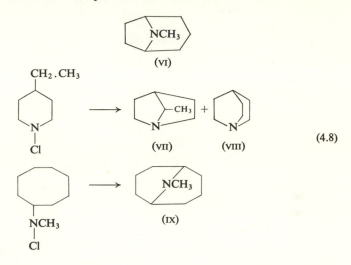

(4.8)

ring in this case are not clearly understood (Wawzonek and Thelen, 1950).

In general, yields obtained in these 'bicyclo' reactions, and the ease of reaction, are noticeably less than with open chain amines. This is because it may not be so easy in the cycloalkylamines for the molecule to adopt the necessary conformation in the transition state to allow the nitrogen radical to abstract a hydrogen atom from the δ-carbon. An extreme example of this is shown in *N*-chloro-*N*-methylcyclohexylamine in which irradiation in sulphuric acid gave only a very low yield of cyclised product. In this compound the cyclohexane ring has to adopt the unfavourable boat conformation to allow 1,5-intramolecular hydrogen transfer from C-4 to N to take place.

On the other hand, with *N*-chlorocamphidine (X), in which the reacting groups are already suitably disposed in space for ready formation of the six-membered transition state, irradiation in sulphuric acid led to the tertiary amine cyclocamphidine (XI) in nearly 70 per cent yield (4.9).

The reaction is believed to proceed by a free radical chain process, and the reaction path shown in (4.10) has been proposed (Corey and Hertler, 1960).

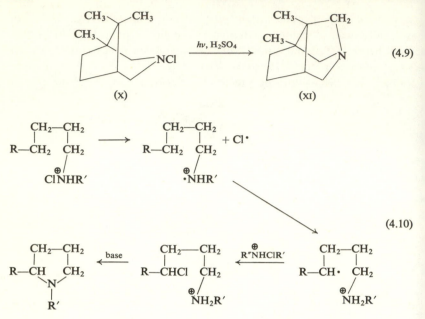

The reacting species may be either the free chloramine or the chlor-ammonium ion. This dissociates under the reaction conditions to form the amminium radical which then abstracts a suitably situated hydrogen atom on the δ-carbon to give the corresponding carbon radical. This in turn abstracts a chlorine atom from another molecule of chloramine, thus propagating the chain and at the same time forming the δ-chloro-amine, from which the cyclic amine is subsequently obtained.

The free radical nature of the reaction is suggested by the fact that it does not proceed in the dark at 25°C, and that it is initiated by heat, light or ferrous ions and inhibited by oxygen. The hydrogen abstraction step must be intramolecular for only thus can the specificity of reaction at the δ-carbon be understood. Strong evidence that the decomposition of *N*-chloroamines in acid involves an intermediate in which the δ-carbon atom is trigonal is provided by the observation that the optically active chloroamine (XII) on decomposition in sulphuric acid at 95°C gave a 43 per cent yield of pure 1,2-dimethylpyrrolidine (XIII) which was optically inactive (4.11). The intermediacy of δ-chloroamines in the reaction has been confirmed by their isolation in a number of cases.

Where alternative modes of cyclisation are possible it is found that, as in other free radical reactions, secondary hydrogen atoms react more

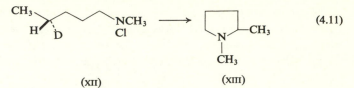

(4.11)

(xii)   (xiii)

readily than primary. In nearly all cases reaction at the δ-carbon atom
with formation of a pyrrolidine is favoured. Thus, in the reaction with
N-chlorobutylamylamine, attack by the nitrogen radical on the δ-methyl
group would lead subsequently to 1-amylpyrrolidine, whereas attack on
the δ-methylene would result in formation of 1-*n*-butyl-2-methylpyrroli-
dine. Only the latter compound was formed (4.12). Tertiary hydrogens

$$C_4H_9N.CH_2.CH_2.CH_2.CH_2.CH_3 \longrightarrow$$

(4.12)

react very readily, but the resulting tertiary chlorides are rapidly sol-
volysed under the conditions of the reaction, and no cyclisation products
are formed.

An outstanding application of the Hofmann–Loeffler–Freytag
reaction is found in the synthesis of the steroidal alkaloid derivative
dihydroconessine (xiv) illustrated in (4.13). In this synthesis the five-
membered nitrogen ring is constructed by attack on the unactivated.

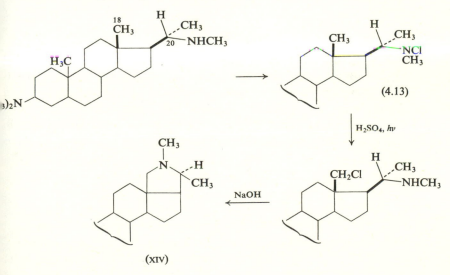

(4.13)

(xiv)

C-18 angular methyl group of the precursor by a suitably placed nitrogen radical at C-20. (Corey and Hertler, 1959). In a similar series of reactions the *N*-chloro derivative (xv) was converted into the conanine (xvi) (4.14). The ease of these reactions is undoubtedly due to the fact that

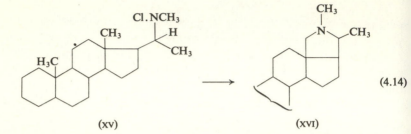

(xv)   (xvi)

(4.14)

in the rigid steroid framework the β-C-18 angular methyl group and the β-C-20 side chain carrying the nitrogen radical are suitably disposed in space to allow easy formation of the chair-like six-membered transition state necessary for 1,5-hydrogen transfer from the methyl group to nitrogen (4.15).

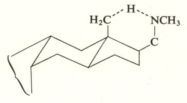

(4.15)

4.2. Cyclisation reactions of nitrenes. Intramolecular reactions leading to the formation of five-membered rings containing nitrogen atoms also take place on thermal decomposition of azides (Abramovitch and Davis, 1964; Smolinsky and Feuer, 1964). Thus, the azide (xvii) on heating in boiling diphenyl ether is converted into the indoline (xviii) in 45–50 per cent yield (4.16), and the cyclohexyl derivative shown is

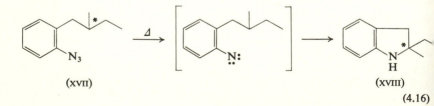

(xvii)   (xviii)

(4.16)

similarly converted into hexahydrocarbazole (4.17). Carbazoles themselves are formed by thermal or photolytic decomposition of 2-azido-diphenyls.

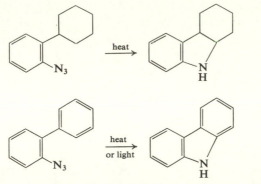

(4.17)

There is much evidence to suggest that these reactions involve the intermediate formation of nitrenes (electron deficient species in which nitrogen has only six electrons in its outer shell; they may have either a singlet R—N̈: or a triplet diradical structure R—N̈˙). The cyclisation step is thought to take place by direct insertion of singlet nitrene into the C–H bond, and not through a diradical intermediate, for reaction with the optically active azide (XVII) gave the optically active indoline (XVIII). A diradical intermediate would have been expected to give an optically inactive product.

It has been reported that photolysis of alkyl azides also gives rise to pyrrolidine derivatives. Thus, n-butyl azide on irradiation in cyclo-hexane solution is stated to give pyrrolidine in 22 per cent yield, and n-octyl azide to form 2-n-butylpyrrolidine in 35 per cent yield. It has been suggested that in these reactions cyclisation does involve a diradical intermediate formed by 1,5-hydrogen abstraction in the initially formed nitrene. However, the experimental conditions for cyclisation appear to be very critical and have not yet been clearly defined. Subsequent attempts to repeat some of this work have led only to the formation of imines, and the status of the earlier experiments is at present uncertain (Barton and Morgan, 1962; Barton and Starratt, 1965).

Acylnitrenes can also be obtained, most readily by photolysis of acyl azides (Lwowski, 1967). In general they rearrange readily to isocyanates, but in suitable cases where the geometry of the molecule brings a C–H bond into close proximity to the nitrogen atom, cyclisation may take place. Thus, the decalin derivative (XIX) on irradiation in cyclohexane solution affords the two lactams (xx) and (xxi), as well as the isocyanate (4.18). This reaction has been used to prepare compounds related to the

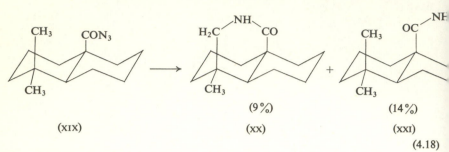

(9%)   (14%)

(xix)   (xx)   (xxi)

(4.18)

diterpene alkaloids. Thus, the azide of podocarpic acid methyl ether (xxii) on photolysis in cyclohexane afforded the lactam (xxiii) by attack on the unactivated angular methyl group (4.19), a synthesis which completed the structural proof of the atisine family of alkaloids (ApSimon and Edwards, 1962). A small amount of γ-lactam was also produced,

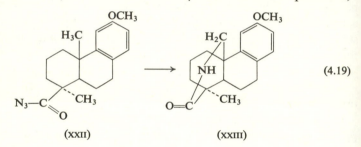

(4.19)

(xxii)   (xxiii)

but the δ-lactam was the main product. The factors affecting the relative proportions of γ- and δ- lactams produced on photolysis of acyl azides are not clearly understood.

Nitrenes are also thought to be intermediates in the reductive cyclisation of aromatic nitro compounds with triethyl phosphite, a reaction which provides a convenient route to carbazole derivatives and a variety of other aromatic heterocyclic systems (Cadogan, 1968). 2-Nitrobiphenyl, on heating with triethyl phosphite, affords carbazole in high yield, and 2-*o*-nitrophenylpyridine is similarly converted into pyrid[1,2-*b*] indazole (4.20). *o*-Nitrostyrenes or -stilbenes give rise to indoles, and, analogously, *o*-nitrobenzylideneanilines are cyclised to 2-arylindazoles in high yield (4.21). These reactions are believed to take place by reduction of the nitro group to a nitrene, followed by direct insertion of the singlet nitrene

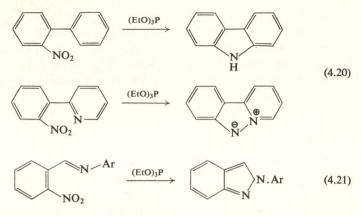

(4.20)

(4.21)

into a C–H bond. This is supported by the finding that the optically active nitro compound (XXIV) affords the indoline shown which retains optical activity. In nearly every case studied cyclisation resulted in the

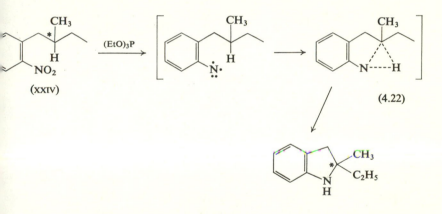

(4.22)

formation of a five-membered ring, even when cyclisation to a six-membered ring seemed favourable.

4.3. The Barton reaction and related processes. In the general equation for intramolecular free radical hydrogen transfer reactions (4.1) Y may be an oxygen atom, and a number of synthetically useful reactions of this type, leading to attack on unactivated C–H bonds, have recently been discovered in which Y = O and X = NO, Cl, I or OH.

Photolysis of organic nitrites. It has been known for many years that vapour phase pyrolysis or photolysis of organic nitrites ($Y = O$, $X = NO$) affords alkoxyl radicals and nitric oxide. But in many cases the radicals produced are consumed in synthetically useless reactions such as fragmentation, disproportionation and non-selective intermolecular hydrogen abstraction. It has been found recently, however, that when the structure of the molecule is such as to bring the $-\overset{|}{\underset{|}{C}}-O-NO$ group and a C–H bond into close proximity, or potentially close proximity, the alkoxyl radicals produced by photolysis of the nitrites in solution have sufficient energy to bring about selective *intramolecular* hydrogen abstraction according to the general scheme (4.1), with subsequent capture of nitric oxide by the carbon radical and formation of a nitroso-alcohol, which may be isolated as the dimer or, where the structure permits, rearranged to an oxime (Barton, Beaton, Geller and Pechet, 1961) (4.23). The nitroso (oximino) compounds produced may be

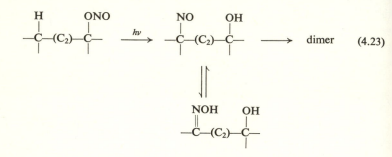

$$\hspace{20em}(4.23)$$

further transformed into other functional derivatives such as carbonyl compounds, amines or cyano derivatives. The photolytic conversion of organic nitrites into nitroso alcohols has become known as the Barton reaction, after its inventor, and the sequence has been widely used (see Nussbaum and Robinson, 1962), particularly in the synthesis of biologically important steroid derivatives.

The reaction is effected by irradiation under nitrogen of a solution of the nitrite in a suitable non-hydroxylic solvent with light from a high pressure mercury arc lamp. A pyrex filter is usually employed to limit the radiation to wavelengths greater than about 300 nm, thus avoiding deleterious side reactions induced by more energetic lower wavelength radiation. This is possible because of the multiplicity of weak absorption

bands of organic nitrites in the region 320–380 nm. It is the absorption
of the light due to these weak bands which brings about the dissociation
of the nitrite.

Detailed examination of a range of examples leads to the conclusion
that the reaction proceeds in discrete steps by an intramolecular radical
mechanism, and is strongly favoured by the possibility of a cyclic
six-membered transition state. Mechanistic studies using ^{15}N have
shown that photolysis of a nitrite gives, in a reversible step, an alkoxyl
radical and nitric oxide which are completely dissociated from each
other. The alkoxyl radical rearranges rapidly to a carbon radical which
can be captured by deuterium atom transfer or by radical trapping
reagents to give 'transfer products'. The normal fate of the carbon
radical in the absence of trapping reagents is to react relatively slowly
with nitric oxide to give the nitroso alcohol (Akhtar and Pechet, 1964).
The sequence is pictured in (4.24).

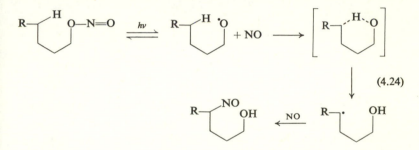

$$(4.24)$$

There is ample evidence to support the view that the hydrogen transfer
step takes place through a six-membered cyclic transition state. In
practice, reaction occurs almost exclusively by abstraction of a hydrogen
atom from the δ-carbon atom. Only a very few examples in which this
is not the case have been found. Thus, photolysis of n-octyl nitrite in
benzene solution gave a 45 per cent yield of the dimer of 4-nitroso-1-
octanol, by way of the transition state (4.25). No evidence for the form-

$$(4.25)$$

$$CH_3(CH_2)_3 . \overset{\displaystyle H}{\underset{\displaystyle CH}{|}} \quad \overset{\displaystyle O}{\underset{\displaystyle CH_2}{\diagup}} \overset{\displaystyle CH_2}{\underset{\displaystyle \diagdown CH_2 \diagup}{|}}$$

ation of any other nitroso-octanol was found. Similarly, 4-phenyl-1-butyl
nitrite and 5-phenyl-1-pentyl nitrite are readily converted into nitroso

dimers whose structures can be rationalised by assuming a six-membered transition state in the hydrogen transfer step (4.26). In the latter example,

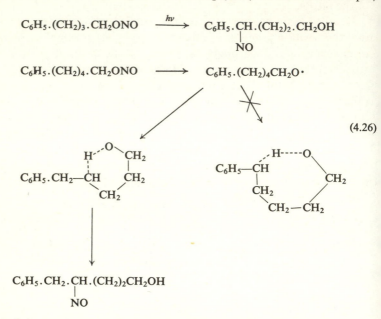

(4.26)

none of the product formed by abstraction of a benzylic hydrogen atom, which would have required a seven-membered transition state, was obtained. In the same way, photolysis of 3-phenylpropyl nitrite did not yield any product corresponding to abstraction of a benzylic hydrogen atom through a five-membered transition state, even although abstraction of the benzylic hydrogen should be highly favourable thermodynamically.

The exact spatial arrangement of the six atomic nuclei forming the transition state is not known with certainty and may not be the same in every case. The chair, boat and semi-chair forms have all been invoked, as well as the form in which the C, H and O atoms are approximately linear (4.27), but reaction in many cases appears to be favoured by the availability of a chair-form transition state (Heusler and Kalvoda, 1964). But even in geometrically favourable cases reaction only takes place provided the distance between the reacting centres is not too great. This may become the deciding factor in rigid molecules. For example, in the radical (xxv) the distance between the δ-methyl group and the oxygen atom is too large for hydrogen abstraction to take place and the alkoxyl radical undergoes fragmentation instead (4.28). On the other

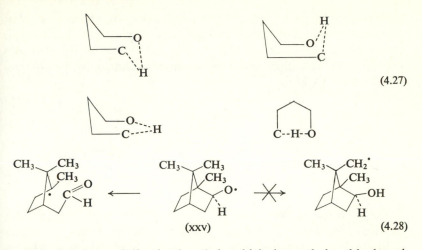

(4.27)

(xxv) (4.28)

hand, in the pinane derivative (xxvi), in which the methyl and hydroxyl groups are nearer each other, attack of the oxy radical on the methyl hydrogen atoms takes place readily, (4.29), thus providing a convenient method for functionalising the unactivated bridge methyl groups of the pinane structure (Gibson and Erman, 1967). Other similar cases are

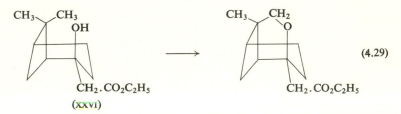

(4.29)

(xxvi)

found in the steroid series. It has been estimated that the activation energy for the intramolecular abstraction of hydrogen reaches a minimum in fixed systems with an O–C distance of 2·5 to 2·7 Å in the starting material. For distances exceeding 2·8 Å the rate of this reaction falls below that of intermolecular hydrogen abstraction or fragmentation reactions.

Interesting results which further support the idea of a six-membered cyclic transition state in the Barton reaction have been obtained in the photolysis of certain cyclohexyl nitrites. Nitroso dimers or monomers were obtained in only small amounts when cyclohexyl nitrite or *cis*- or *trans*-3-methylcyclohexyl nitrite were irradiated. In these compounds formation of a six-membered cyclic intermediate is unfavourable on conformation grounds. In contrast, rearrangement of both *cis*- and *trans*-2-ethylcyclohexylnitrites proceeded readily and in good yield, clearly facilitated by the fact that the transition states for the hydrogen transfer steps can adopt stable conformations similar to the *cis*- and *trans*-decalins (4.30).

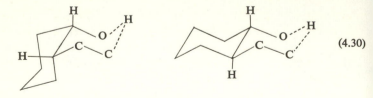

(4.30)

The products obtained from other simple alicyclic nitrites depend on the size of the ring. With cyclobutyl and cyclopentyl nitrites, photolysis results in ring cleavage to give the corresponding linear nitroso-aldehydes (4.31). With cycloheptyl and cyclooctyl nitrites, however, 4-nitroso-

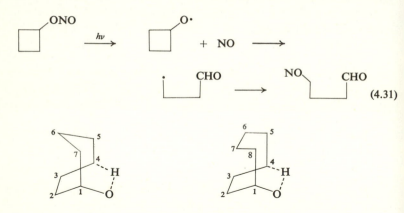

(4.31)

cycloheptan-1-ol and 4-nitrosocyclooctan-1-ol are obtained by trans-annular attack of the oxy radical through a six-membered transition state.

Other factors being equal, it is found that in general, abstraction of a secondary hydrogen atom by an alkoxy radical is much easier than abstraction of a primary hydrogen, although in certain cases attack at the primary hydrogens may be facilitated by favourable geometrical factors. Thus 2-hexyl nitrite afforded a 30 per cent yield of nitroso dimer on photolysis, but with 2-pentyl nitrite, where formation of a six-membered cyclic transition state requires abstraction of a primary hydrogen atom, the yield fell to 6 per cent. Presumably a tertiary hydrogen would react even more easily than a secondary, but no comparative studies have apparently been made (Kabasakalian, Townley and Yudis, 1962).

Intramolecular hydrogen abstraction by alkoxy radicals is always accompanied to a greater or lesser extent by disproportionation, radical decomposition and intermolecular reactions. In the case of oxy radicals derived from tertiary nitrites reaction follows the normal course if tertiary or secondary hydrogen atoms are available for abstraction. But if only primary hydrogens are available the Barton reaction is superseded by alkoxy radical decomposition (4.32). With primary and secondary

$$\underset{\underset{ONO}{\overset{CH_3}{|}}}{CH_3\!-\!C\!-\!(CH_2)_2.CH_3} \xrightarrow{h\nu} \underset{CH_3}{\overset{CH_3}{>}}C\!=\!O + ON.(CH_2)_2.CH_3$$

$$(4.32)$$

alkoxy radicals, disproportionation to form an alcohol and a carbonyl compound is often favoured over decomposition and always takes place to an appreciable extent, and particularly when the geometric and structural requirements of the Barton reaction are not met.

The most important synthetic applications of the Barton reaction have been in the steroid series, particularly in the functionalisation of the two non-activated C-18 and C-19 angular methyl groups by photolysis of the nitrites of suitably disposed hydroxyl groups. In principle C-18 can be attacked by an alkoxy radical at C-8, C-11, C-15 or C-20, and C-19 by an alkoxyl radical at C-2, C-4, C-6 and C-11 (4.33). Most of these approaches have been realised in practice, either through the Barton reaction or by one of the related reactions described below (see Nussbaum and Robinson, 1962; Heusler and Kalvoda, 1964; Akhtar, 1962). The reactions are facilitated by the conformational rigidity of

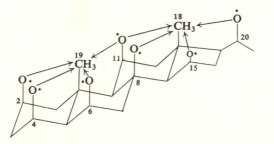

$$(4.33)$$

the steroid framework and by the 1,3-diaxial relationship of the inter-
acting groups, which allows easy formation of conformationally favourable six-membered cyclic transition states. Because of this, attack on the
primary hydrogen atoms of the methyl groups is much easier than in
the aliphatic series, and good yields of nitroso monomers or dimers are
often obtained.

Thus, the nitrite of the steroidal alcohol 3β-acetoxy-5-α-pregnan-20β-
ol (xxvIII) on photolysis in benzene solution is converted into the oxime
(xxIX) in 34 per cent yield, (4.34).

Similarly, the nitrite of 3β-acetoxycholestan-6β-ol (xxx) in which the
methyl and nitrite groups are in the favourable 1,3-diaxial relation, on

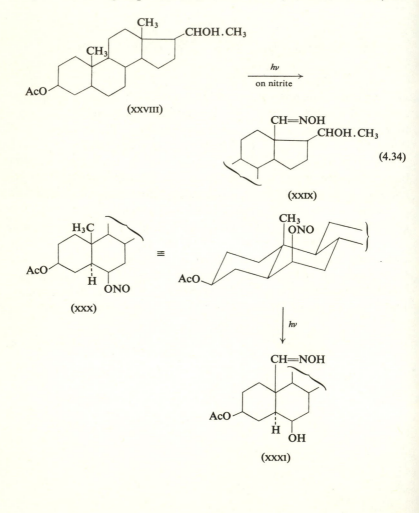

photolysis in toluene solution gave a 67 per cent yield of the rearranged nitroso dimer which was converted into the oxime (XXXI) by refluxing in propanol solution. The reaction was used by Barton and his co-workers to effect the key step in their elegant synthesis of aldosterone (XXXIV), (R = H), (4.35), a biologically important hormone of the adrenal cortex (Barton, Beaton, Geller and Pechet, 1961), by photolysis of the 11β-nitrite (XXXII) in toluene solution. The oxime (XXXIII) separated from the solution in 21 per cent yield and on hydrolysis with nitrous acid afforded aldosterone-21-acetate (XXXIV), (R = Ac) directly.

It may happen that more than one site in a molecule is favourably situated for attack by the newly formed alkoxy radical, and in such cases

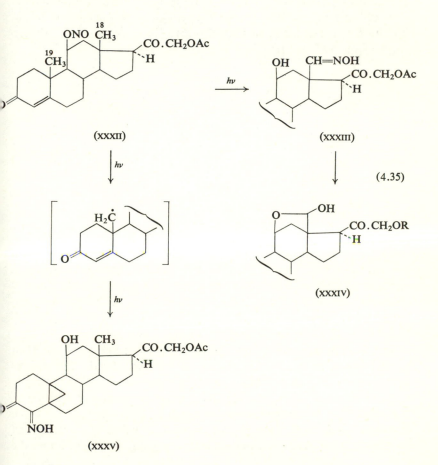

a mixture of products may result. Thus, in the reaction (4.35), attack at C-19 instead of C-18 led to (xxxv).

Photolysis of hypohalites. The generation of alkoxy radicals which can undergo intramolecular hydrogen abstraction can also be achieved by photolysis of hypochlorites. The initial products of the reaction are 1,4-chlorohydrins, formed again by preferential abstraction of hydrogen attached to the δ-carbon atom. In some cases small amounts of 1,5-chloro alcohols are formed, but in no case has the product of a 1,2-, 1,3-, 1,4- or 1,7-hydrogen shift been observed. The 1,4-chlorohydrins produced are easily converted into tetrahydrofurans by treatment with alkali and the sequence provides a convenient route to these compounds. The reactions are thought to involve long chains, and, as in the photolysis of nitrites, probably proceed through a six-membered cyclic transition state (Walling and Padwa, 1961; 1963. Jenner, 1962; Akhtar and Barton, 1961; 1964) (4.36).

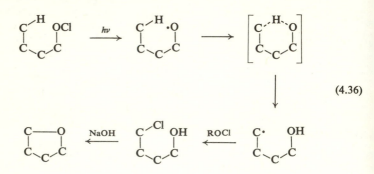

(4.36)

The hypochlorites are prepared by the action of chlorine dioxide on the corresponding alcohols. They show two sets of ultraviolet absorption bands at 250–260 nm and 300–320 nm and since reaction takes place by irradiation through Pyrex it is the longer wavelength absorption which brings about the photolysis.

In the aliphatic series, a number of alkyl hypochlorites have been shown to undergo photo-decomposition to give δ-chloro alcohols. As in the photolysis of nitrites, abstraction of hydrogen by the oxy radical takes place most readily from a tertiary carbon atom and least readily from a primary. Thus, the dimethylpentyl hypochlorite (xxxvi) is converted into the chloro alcohol (xxxvii) in 70 per cent yield (4.37), but with the isomeric hypochlorite (xxxviii) in which a six-membered cyclic transition state requires attack on a primary hydrogen atom, the yield of chloride (XXXIX) fell to 29 per cent. In the latter example, no product from attack on the tertiary hydrogen, by 1,4-hydrogen shift, was observed.

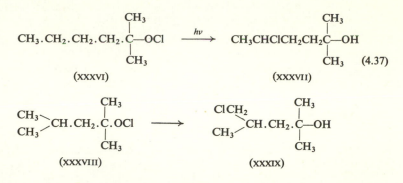

$$(4.37)$$

(XXXVI)   (XXXVII)

(XXXVIII)   (XXXIX)

With 1-methylcyclooctyl hypochlorite transannular abstraction of hydrogen by the oxy radical took place, and a mixture of *cis*- and *trans*-4- and 5-chloro-1-methylcyclooctanol was obtained. The reactions are most easily effected with tertiary hypochlorites, but with suitable precautions 1,4-chlorohydrins can be obtained from primary and secondary hypochlorites as well.

A competing reaction is β-cleavage of the oxy radical to form, in the case of a tertiary hypochlorite, a ketone and an alkyl chloride (4.38).

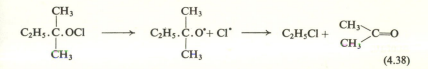

$$(4.38)$$

The extent of this reaction varies with the structure of the hypochlorite. It becomes the predominant reaction if the carbon chain is too short to permit 1,5-abstraction of hydrogen, or if the geometrical requirements for a six-membered transition state are not met.

A number of reactions in the steroid series involving attack on the non-activated angular methyl groups have been reported. Thus, the hypochlorite (XL) on irradiation in benzene solution gave the 1,4-chlorohydrin (XLI) in 25 per cent yield, smoothly cyclised to the ether (XLII) with methanolic potassium hydroxide (4.39).

Hypobromites have also been employed in a limited number of cases. Thus, the trimethylcyclohexyl hypobromite (XLIII), on irradiation and

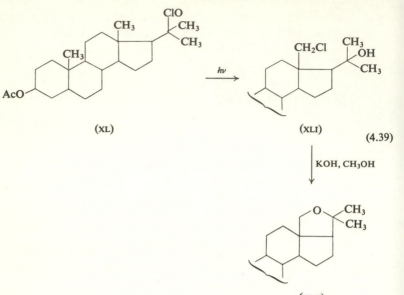

(XL)   (XLI)

(4.39)

KOH, CH₃OH

(XLII)

treatment with alkali, was converted into the cyclic ether (XLIV) in 60
per cent yield, presumably by way of the 1,4-bromohydrin (4.40).

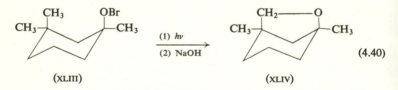

(XLIII)   (XLIV)   (4.40)

Interestingly, it has been reported that the same change can be effected
in 75 per cent yield, in absence of light, when the alcohol is treated with
bromine and silver oxide at room temperature. An ionic mechanism is
suggested (Sneen and Matheny, 1964) (4.41).

In practice, a more convenient method for generating oxy radicals for
intramolecular hydrogen abstractions is by cleavage of hypoiodites
prepared *in situ* from the corresponding alcohol. A number of methods
have been employed, including treatment of the alcohol with lead tetra-
acetate and iodine, with or without irradiation, and irradiation of a
solution of the alcohol in presence of iodine and mercuric oxide. The
products of the reaction may be either five-membered ring oxides or
acetals (Akhtar and Barton, 1964; Heusler and Kalvoda, 1964).

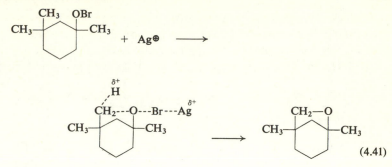

$$+ \quad Ag^{\oplus} \quad \longrightarrow$$

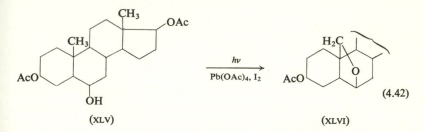

(4.41)

Most of the applications of the hypoiodite method have been in the steroid series. The diacetate (XLV), for example, on irradiation in carbon tetrachloride solution in presence of lead tetra-acetate and iodine, gave the cyclic ether (XLVI) in 90 per cent yield (4.42).

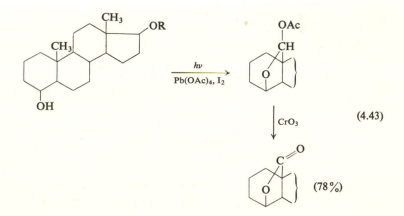

Under certain conditions, especially with lead tetra-acetate and iodine, a second substitution of the group being attacked can take place, and hemi-acetals are produced. These can be oxidised to lactones (4.43).

All these reactions are thought to take place by initial formation of the unstable hypoiodite which is converted into the 1,4-iodohydrin in a Barton type transformation through a six-membered cyclic transition state (4.44). In subsequent steps which are influenced by steric factors,

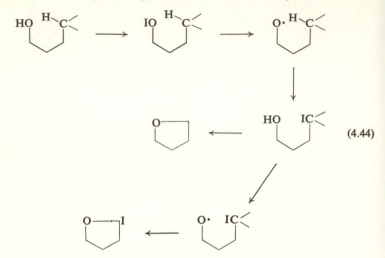

(4.44)

the iodohydrin may be converted into the tetrahydrofuran, or, by further reaction with the reagent, into an iodo derivative from which the hemiacetal is derived.

A novel and useful extension of the hypoiodite reaction, again involving attack on an unactivated C–H bond, is the formation of γ-lactones from carboxylic acids by photolysis of the corresponding N-iodoamides. The iodoamides are prepared *in situ* from the amides with lead tetraacetate and iodine or t-butylhypochlorite and iodine, and lactonisation is conveniently effected by photolysis of the amide in presence of the iodinating agent, followed by hydrolysis of the product (Barton, Beckwith and Goosen, 1965). Under these conditions stearamide forms γ-stearolactone together with a smaller amount of the δ-lactone, and 4-phenylbutyramide is converted into phenylbutyrolactone (4.45).

$$C_6H_5 . CH_2 . CH_2 . CH_2 . CONH_2 \xrightarrow[\text{Pb(OAc)}_4, \text{I}_2]{h\nu} \begin{array}{c} \overset{H_2}{\underset{}{C}} \\ C_6H_5 . \overset{}{C}H \quad CH_2 \\ \underset{O\text{———}CO}{|} \quad | \end{array}$$

(4.45)

The reaction is thought to proceed by homolysis of the N–I bond and intramolecular hydrogen abstraction by the nitrogen radical through a

six-membered cyclic transition state, followed by formation of the
γ-iodoamide. Intramolecular substitution to an imino-lactone and
hydrolysis then leads to the γ-lactone (4.46). In agreement with the

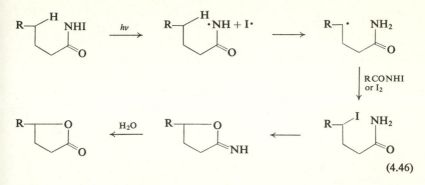

(4.46)

proposed mechanism, the isolation of a derivative of the postulated
imino-lactone was achieved in the reaction with γ-phenylbutyramide.

Photolysis of the amide of (+)-4-methylhexanamide in presence of
t-butylhypochlorite and iodine, and hydrolysis of the resulting iodine
chloride complex, gave racemic 4-methyl-4-hexanolactone, in line with
the stepwise radical mechanism and against a direct insertion mechanism
(4.47). The yields obtained in the lactonisations parallel the strengths

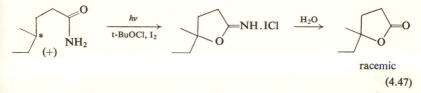

racemic

(4.47)

of the C–H bonds broken in the hydrogen abstraction step. Abstraction
of benzylic or tertiary hydrogen gives high yields. Abstraction of secondary hydrogen affords satisfactory yields but, as in the photolysis of
nitrites, abstraction of primary hydrogen is difficult and gives poor yields.

In sterically favourable cases, disubstitution at the site of reaction
may occur, and the product is then an anhydride. Thus, the amide
(XLVII) when treated under the usual conditions, was converted into the
anhydride (XLVIII) instead of the expected γ-lactone (Baldwin, Barton,
Dainis and Pereira, 1968) (4.48).

Photochemical lactonisation of iodo-amides has been used in the
synthesis of a number of lactones related to diterpenes. The octahydro-

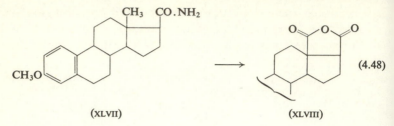

(XLVII)   (XLVIII)

(4.48)

phenanthrene derivative (XLIX) on irradiation in presence of lead tetra-acetate and iodine gave a mixture of two lactones. In this reaction the γ-lactone was the main product (4.49). In the kaurane series, in contrast,

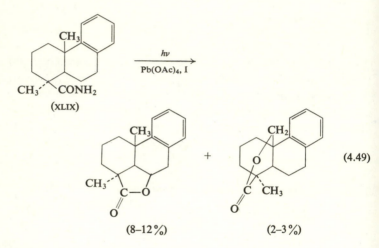

(XLIX)

(8–12%)   (2–3%)

(4.49)

the amide (L) gave mainly the δ-lactone (4.50). This is presumably due to steric congestion at the C-6 methylene group in the kauranes, favouring attack at the C-10 methyl group through a seven-membered cyclic transition state.

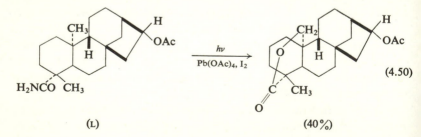

(L)   (40%)

(4.50)

γ-Lactones can also be conveniently obtained from carboxylic acids by photolysis of the *N*-chloro-amides, and this method offers certain advantages, particularly with the lower aliphatic acids where the iodo-amide method gives poor yields (Beckwith and Goodrich, 1965). Chloro-amides are readily obtained from the amides by action of t-butyl hypo-chlorite, and irradiation of a range of examples, followed by hydrolysis, afforded γ-lactones in yields of 20–60 per cent. δ-Lactones were also produced, but always in minor amounts. It is presumed that γ-chloro-amides are intermediates and that the reaction path is similar to that suggested for the iodo-amide reaction. Interestingly, similar transforma-tions could be effected, in comparable yields, by reaction of the *N*-chloro-amides with catalytic amounts of cuprous chloride.

In a number of open-chain secondary amides where there is a choice in the derived radical of intramolecular hydrogen abstraction from the δ-carbon atom of the alkyl or acyl side chain it was found by Chow and Joseph (1969) that reaction always occurred in the alkyl chain, as in (4.51). None of the alternative product was detected. But when attack

$$\overset{\text{Br}}{\underset{|}{(CH_3)_3C.CH_2.CO.N.C(CH_3)_2.CH_2.C(CH_3)_2.CH_3}}$$

$$\downarrow h\nu$$

(4.51)

$$(CH_3)_3C.CH_2.CO.N \overset{\displaystyle CH_3}{\underset{\displaystyle CH_3}{\overset{\displaystyle |}{\underset{\displaystyle |}{\bigvee}}}} \begin{array}{c} CH_3 \\ -CH_3 \end{array} \quad (93\,\%)$$

on the alkyl chain was not possible reaction occurred in the acyl group. In no case was any product observed corresponding to hydrogen abstraction by the oxy radical $\overset{\text{O}^{\cdot}}{\underset{|}{-N{=}C-}}$.

4.4. Reaction of monohydric alcohols with lead tetra-acetate. Another method which has been extensively used for the preparation of tetra-hydrofurans, particularly in the steroid series, is oxidation of monohydric alcohols with lead tetra-acetate (see Heusler and Kalvoda, 1964). Like the hypoiodite reaction, this method has the advantage over the photolysis of nitrites and hypochlorites that the unstable reactive intermediate

does not have to be prepared and isolated in a separate step. It is produced from the alcohol *in situ* and converted directly into the alkoxy radical.

It was originally suggested that the lead tetra-acetate reaction proceeded *via* an electron-deficient oxonium ion intermediate, but it is now believed that it represents yet another example of an alkoxy radical rearrangement. The initial reaction is formation of the triacetoxy lead alkoxide

$$\text{R—OH} + \text{Pb(OAc)}_4 \quad \rightleftharpoons \quad \text{RO—Pb(OAc)}_3 + \text{HOAc} \tag{4.52}$$

which is easily split thermally or photolytically to give the corresponding alkoxy radical. Formation of the lead alkoxide is dependent on steric effects in the alcohol, and tertiary or strongly hindered secondary alcohols react only slowly.

The steric requirements for the formation of tetrahydrofurans from alcohols with lead tetra-acetate are substantially the same as for the abstraction of hydrogen in the homolysis of nitrites, hypochlorites and hypoiodites. Cleavage of the alkoxide leads through the oxy radical to a favoured six-membered cyclic transition state which in certain cases may be oxidised directly to the tetrahydrofuran with the aid of a lead tri-acetate radical. In other cases, 1,5-hydrogen transfer occurs, but direct formation of an ether from the resulting carbon radical is energetically unfavourable, and conversion into the tetrahydrofuran is thought to proceed by oxidation to a carbonium ion, which subsequently cyclises (4.53).

This reaction has been extensively applied in the steroid series, and it also provides a useful method for the preparation of simple alkyltetrahydrofurans from primary and secondary aliphatic alcohols. Thus, n-heptanol, treated with lead tetra-acetate in boiling cyclohexane gives 2-n-propyltetrahydrofuran in 50 per cent yield, accompanied by a small amount of 2-ethyltetrahydropyran and heptan-2-ol is converted into a mixture of *cis*- and *trans*-2-ethyl-5-methyltetrahydrofuran. The *cis* and *trans* forms were obtained in nearly equal amounts, showing that the

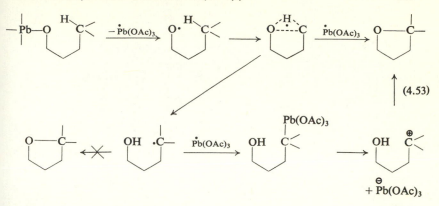

$$(4.53)$$

oxidation of aliphatic alcohols with lead tetra-acetate is not stereo-selective.

Similar conversions of primary and secondary alcohols into tetra-hydrofuran derivatives can be readily achieved at room temperature by irradiation of a mixture of the alcohol and lead tetra-acetate in benzene solution, or by irradiation in presence of silver oxide and bromine or mercuric oxide and iodine (Mihailović, Čeković and Stanković, 1969).

In the steroid series the products in general are similar to those obtained in the hypoiodite reaction (4.54) (see p. 193). However, in some cases

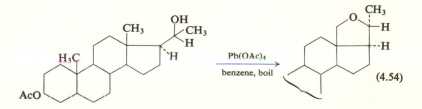

$$(4.54)$$

the yields are less good. For example, the 6β-alcohol (LI) with lead tetra-acetate and iodine is converted into the ether (LII) in high yield but with lead tetra-acetate alone the ketone (LIII) is the main product (4.55).

Products of the type of (LIII) are produced to some greater or lesser extent in all the reactions involving oxy radicals which have been discussed. Carbonyl-forming fragmentation reactions may also take place (4.56), the extent of which is influenced by the structure of the compound. In particular, fragmentations which give rise to allyl or benzyl radicals, or to a radical adjacent to an oxygen function are

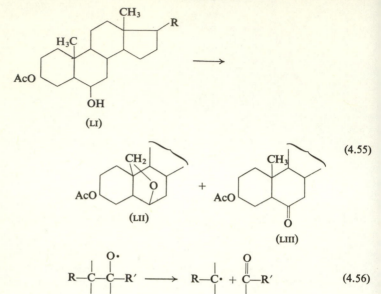

(4.55)

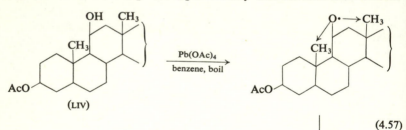

(4.56)

strongly favoured, and in such cases hydrogen abstraction may be completely suppressed.

In certain steroid compounds, particularly with the lead tetra-acetate

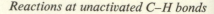

(4.57)

$11\beta,18$- and $11\beta,19$- ether (~5%)

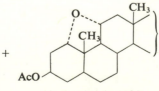

$11\alpha,1$- ether (~15%)

reaction, the fragmentation may be reversible. In such cases recyclisation can regenerate oxy radicals which may be epimeric at one or both of the α- or β-carbon atoms and which may then cyclise to ethers epimeric with the starting alcohol at one or two centres. Thus, the 11β- hydroxy steroid (LIV) on treatment with lead tetra-acetate, affords a mixture of 11β,18- and 11β,19- ethers, and the epimeric 11α,1- ether (4.57).

4.5. Miscellaneous reactions

Olefinic alcohols from hydroperoxides. Introduction of an olefinic bond into a saturated aliphatic hydrocarbon chain, although it occurs in nature, is not a reaction which is easily effected in the laboratory with ionic reagents. It has recently been found, however, that t-alkyl hydroperoxides are conveniently converted into olefinic alcohols by reduction with ferrous ion in presence of cupric salts (Acott and Beckwith, 1964). The key step in this reaction is once again a 1,5-intramolecular hydrogen transfer to the oxy radical produced by reduction of the hydroperoxide; subsequent oxidation of the resulting carbon radical by cupric ion with simultaneous expulsion of a proton leads to formation of the olefin. Thus, 2-methyl-2-hexyloxy radical generated by ferrous ion reduction of the hydroperoxide (LV) undergoes intramolecular hydrogen transfer to give the carbon radical (LVI) which in presence of cupric ion is oxidised to a mixture of 2-methylhex-4-en-2-ol and (mainly) 2-methylhex-5-en-2-ol (4.58). Similarly, the dimethylhexyl hydroperoxide (LVII) affords the two olefins (LVIII) and (LIX) (4.59).

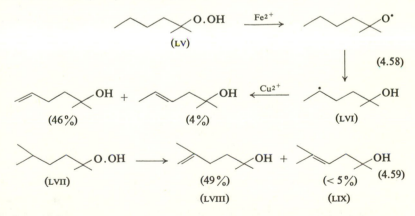

It is of interest that in all cases studied so far, the olefin with the double bond in the 4,5-position to the hydroxyl group, and not necessarily the thermodynamically more stable olefin, is the main product. This is thought to be due to a directive effect of the hydroxyl group exerted through a cyclic transition state of the form (LX) which models show is much less strained than the form (LXI) leading to the γ-olefin, the minor product of the reactions (4.60).

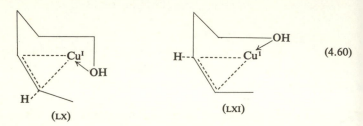

(4.60)

(LX)   (LXI)

Cyclobutanols by photolysis of ketones. Intramolecular abstraction of hydrogen by an oxy radical is also involved in the formation of cyclo-butanol derivatives by photolysis of aldehydes and ketones, and in the photolytic fragmentation of ketones to give olefins and ketones of smaller carbon number by cleavage between the α- and β- carbon atoms. Thus, irradiation of methyl neopentyl ketone gives acetone and α,α-dimethylethylene (4.61). A structure in which there is a hydrogen atom

$$(CH_3)_3C.CH_2.CO.CH_3 \xrightarrow{h\nu} (CH_3)_2.C{=}CH_2 + CH_3COCH_3$$

(4.61)

on the γ-carbon atom is necessary for those reactions to take place, and a mechanism involving hydrogen transfer through a diradical six-membered cyclic transition state has been proposed (4.62). In many

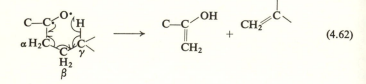

(4.62)

cases these fragmentation reactions are accompanied by the formation of cyclobutanol derivatives, and although the yields obtained are small the reaction provides a practicable route to otherwise difficultly accessible compounds (Yang and Yang, 1958). Thus, pentan-2-one on irradiation in iso-octane solution affords acetone, ethylene and methylcyclobutanol

(12 per cent) and octan-2-one gives 1-methyl-2-n-propylcyclobutanol in addition to acetone and pent-1-ene (4.63).

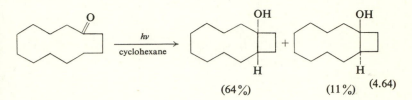

$$CH_3 . CO . (CH_2)_5 . CH_3 \xrightarrow[\text{iso-octane}]{h\nu} \quad (17\%) \quad (4.63)$$

With 1,2-diketones, 2-hydroxycyclobutanones are obtained, and in these cases the yields are much higher owing to the stabilisation of the intermediate excited state by the α-carbonyl group.

The reaction has also been effected with cyclic ketones (4.64), and

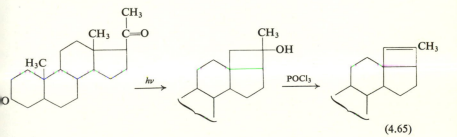

has been applied in the steroid series for reaction at the unactivated C-18 methyl group through attack by a suitably situated carbonyl group at C-20 (see Schaffner, Arigoni and Jeger, 1960) (4.65).

These reactions are thought to take place by initial photo-excitation of the carbonyl group to a diradical, followed by 1,5-intramolecular hydrogen transfer to the oxy radical and intramolecular combination of the two carbon radicals with formation of the four-membered ring (4.66). There is evidence however that the reaction may not be entirely a stepwise process involving discrete radical intermediates. Experiments using optically active ketones with an asymmetric carbon atom at the γ-position gave cyclobutanol derivatives which retained some optical

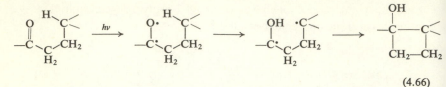

(4.66)

activity, showing partial retention of configuration during cyclisation (Orban, Schaffner and Jeger, 1963; Schutte-Elte and Ohloff, 1964). For example the aldehyde (LXII) on irradiation under the usual conditions gave the cyclobutanol (LXIII) which had at least 24 per cent of the expected optical activity (4.67). This can be explained by competitive

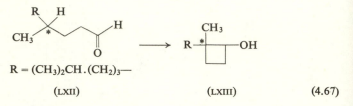

(LXII)   (LXIII)   (4.67)

participation of stepwise and concerted mechanisms, or by intervention of a short-lived diradical whose rates of racemisation and cyclisation are of the same order of magnitude.

In an interesting extension of this reaction Breslow and Winnik

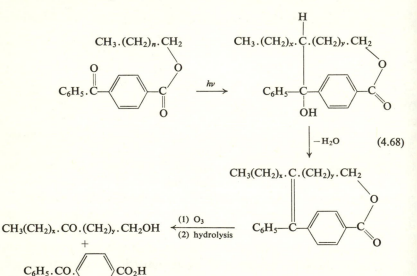

(1969) have shown how photochemical cyclisation of ketones can be used to effect oxidation of remote unactivated methylene groups in a series of esters derived from *p*-benzophenone carboxylic acid and long-chain alcohols. The general scheme is shown in (4.68). It was found that hydrogen abstraction can occur over very large distances, much further than the six atoms common in the Barton reaction and in the other reactions discussed above. With the ester derived from the C-16 alcohol there was considerable preference for reaction at C-14.

Intramolecular hydrogen abstraction by carbon radicals. There are only a few authenticated examples of intramolecular hydrogen abstraction by carbon radicals, and reactions incorporating such a step are not yet of practical synthetic importance. One example is seen in the reaction of 2-(4'-methylbenzoyl)benzenediazonium salts (LXIV) with carbon tetrachloride and alkali, which leads to the formation of small amounts of the products (LXV) and (LXVI) (4.69). Formation of the

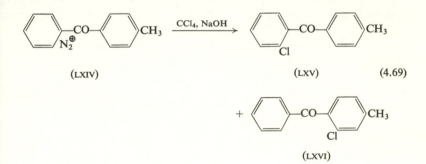

(LXIV)   (LXV)   (4.69)

(LXVI)

latter is attributed to an intramolecular hydrogen transfer involving an aromatic C–H bond (De Los de Tar and Relyea, 1956) (4.70). Again,

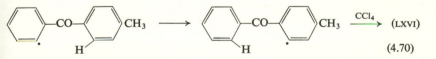

(4.70)

decomposition of ε-phenylcaproyl peroxide in boiling cumene gives, in addition to the expected 1,10-diphenyldecane, a 15 per cent yield of 5,6-diphenyldecane formed by recombination of rearranged radicals (Grob and Kammüller, 1957) (4.71). A six-membered cyclic transition state for the hydrogen transfer step is suggested by the observation that the next lower homologue, δ-phenylvaleryl peroxide, on decomposition

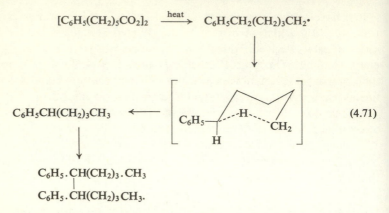

$$[C_6H_5(CH_2)_5CO_2]_2 \xrightarrow{\text{heat}} C_6H_5CH_2(CH_2)_3CH_2 \cdot$$

$$C_6H_5CH(CH_2)_3CH_3 \quad\longleftarrow\quad \left[\begin{array}{c} \\ C_6H_5 \end{array} \right] \qquad (4.71)$$

$$C_6H_5 . \underset{|}{C}H(CH_2)_3 . CH_3$$
$$C_6H_5 . CH(CH_2)_3 CH_3.$$

in boiling benzene, gives tetralin, but no product derivable by an intra-molecular hydrogen transfer reaction.

Formally similar to these radical reactions are the intramolecular insertion reactions of alkyl carbenes on neighbouring C–H bonds, to form cyclic compounds (see p. 55). There is strong evidence, however, that these reactions are one-step concerted 'three centre' processes, and do not involve radical intermediates.

5 Synthetic applications of organoboranes

One of the most rapidly expanding areas of organic chemistry in recent years has been in the field of organoboranes. These versatile reagents, which are readily obtained by addition of borane, BH_3 (which exists as the gaseous dimer diborane B_2H_6), to olefins or acetylenes, undergo a wide variety of reactions, many of which are of great value in synthesis. In addition, diborane itself is a powerful reducing agent and attacks a variety of unsaturated groups besides carbon–carbon multiple bonds (Brown and Subba Rao, 1960) (see Table 5.1).

5.1. Reductions with diborane and dialkylboranes. Diborane for reductions, or for the hydroboration reactions described later, is conveniently prepared (5.1) by reaction of boron trifluoride etherate with sodium borohydride in diglyme solution (diglyme is the dimethyl ether of diethylene glycol). In many cases the reagent is generated in the

$$3NaBH_4 + 4BF_3 \longrightarrow 3NaBF_4 + 2B_2H_6 \qquad (5.1)$$

presence of the compound being reduced or hydroborated, and reaction takes place directly. In other cases where especially pure material is required, or where it is desired to avoid reduction of the substrate by the sodium borohydride the diborane may be prepared separately and distilled into tetrahydrofuran. The resulting solution of diborane, which probably contains an equilibrium mixture of the free dimer and the tetrahydrofuran–borane complex, can be standardised by a titration procedure (House, 1965).

Reaction of diborane with unsaturated groups takes place readily at room temperature, and the products shown in Table 5.1 are isolated in high yield after hydrolysis of the intermediate boron compound. Diborane reacts rapidly with water, and reactions must be effected under anhydrous conditions, best under nitrogen since diborane itself, and the lower alkylboranes, may ignite in air.

A valuable feature of reductions with diborane is that they do not

TABLE 5.1   *Reduction of functional groups with diborane*

Reactant	Product
—CO$_2$H	—CH$_2$OH
—CH=CH—	—CH$_2$—CH— (before hydrolysis) 　　　　｜ 　　　　B 　　　／＼
—CHO	—CH$_2$OH
—CO 　｜	—CHOH 　　｜
—C≡N	—CH$_2$NH$_2$
／—CO＼ (C)$_n$　　O ＼—CH$_2$／	／CH$_2$OH (C)$_n$ ＼CH$_2$OH
＞C—C＜ 　　＼O／	＞CH—COH 　　　｜
—CO$_2$R	—CH$_2$OH + ROH
—COCl	no reaction
—NO$_2$	no reaction

H.O. House, *Modern synthetic reactions*, copyright 1965, W. A. Benjamin, Inc., Menlo Park, California.

simply parallel those with sodium borohydride. This is because sodium borohydride is a nucleophile and reacts by addition of hydride ion to the more positive end of a polarised multiple bond (see p. 353), whereas diborane is a Lewis acid and attacks electron-rich centres. For example, while sodium borohydride very rapidly reduces acid chlorides to primary alcohols, the reaction being facilitated by the electron-withdrawing effect of the halogen, diborane does not react with acid chlorides under the usual mild conditions. Reduction of carbonyl groups by diborane is believed to take place by addition of the electron-deficient borane to the oxygen atom, followed by the irreversible transfer of hydride ion from boron to carbon (5.2). The inertness of acid chlorides can be

$$>\!C\!=\!O + BH_3 \rightleftharpoons >\!C\!=\!\overset{\oplus}{O}\!-\!\overset{\ominus}{BH_3} \longrightarrow \underset{\underset{H}{|}}{-\!C}\!-\!OBH_2 \quad (5.2)$$

ascribed to the decreased basic properties of the carbonyl oxygen resulting from the electron-withdrawing effect of the halogen. For a similar reason esters are reduced only slowly by diborane.

Remarkable is the ready reduction of carboxylic acids to primary alcohols with diborane, which can often be selectively effected in presence of other unsaturated groups. *p*-Nitrobenzoic acid, for example, is reduced to *p*-nitrobenzyl alcohol in 79 per cent yield. Reduction of carboxylic acids is believed to proceed by way of a triacylborane, the carbonyl groups of which are rapidly reduced by further reaction with diborane.

The reduction of epoxides with diborane is also noteworthy since it gives rise to the *least* substituted carbinol in preponderant amount, in contrast to reduction with complex hydrides (see p. 360). The reaction is catalysed by small amounts of sodium or lithium borohydride and high yields of the alcohol are obtained (Brown and Nung Min Yoon, 1968*a*) (5.3). With 1-alkylcycloalkene epoxides the 2-alkylcycloalkanols

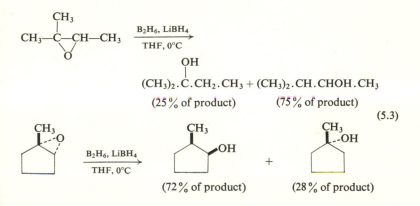

produced are entirely *cis*, and this reaction thus complements the hydroboration–oxidation of cycloalkenes described on p. 224, which leads to *trans*-2-alkylcycloalkanols. Another method, which is applicable only to arylethylene epoxides, is reduction with diborane in presence of boron trifluoride; styrene epoxide, for example, is converted into 2-phenylethanol in 98 per cent yield (Brown and Nung Min Yoon, 1968*b*).

A more selective reducing agent than diborane itself is the sterically hindered di(3-methylbut-2-yl)borane (disiamylborane), which is easily obtained by action of diborane on 2-methylbut-2-ene (see p. 213) (Brown and Bigley, 1961) (5.4). Disiamylborane is a much milder reducing

$$B_2H_6 + (CH_3)_2C{=}CH.CH_3 \longrightarrow (CH_3{-}\underset{\underset{CH_3}{|}}{CH}{-}\underset{\underset{CH_3}{|}}{CH}{-})_2BH \qquad (5.4)$$

agent than diborane, and because of the large steric requirements of the alkyl groups the rate of reaction is strongly influenced by the structure of the compound being reduced. Aldehydes and ketones are converted into the corresponding alcohols, but the reactivity of ketones varies widely with their structure, and selective reduction of an aldehyde in presence of a ketone seems feasible in favourable cases (Brown, 1962). Acid chlorides, acid anhydrides and esters do not react, and epoxides are reduced only very slowly. In contrast to their rapid reaction with diborane, carboxylic acids are not reduced; they simply form the dialkyl-boron carboxylate which on hydrolysis regenerates the acid. Presumably the large size of the dialkylboron group in the carboxylate prevents further attack by the reagent on the carbonyl group.

Two useful reactions of disiamylborane, which are not paralleled by diborane itself, are the reductions of lactones to hydroxyaldehydes and of dimethylamides to aldehydes (5.5). High yields of aldehyde are

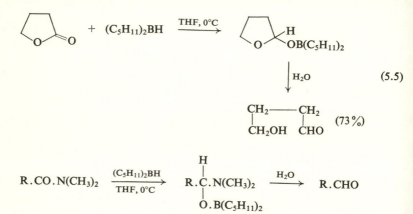

obtained, and reduction with disiamylborane supplements the reactions with hydridoalkoxy-aluminates described on p. 363 for the conversion of carboxylic acid derivatives into aldehydes.

Some interesting stereochemical consequences of substituting alkyl-boranes for diborane in the reduction of cyclic ketones have been observed. 2-Methylcyclopentanone and 2-methylcyclohexanone when reduced with diborane gave mainly the more stable *trans*-2-methylcyclo-alkanol, but with disiamylborane the preponderant product was the *cis* isomer, and with the bulky optically active terpenoid derivative diisopinocamphenylborane (p. 219), the yield of *cis* isomer rose to 94 per cent. The product of the last reaction was optically active. Very high stereoselectivity is shown also by the new reagent lithium perhydro-9*b*-boraphenalylhydride, which is readily obtained from *cis,cis,trans*-perhydro-9*b*-boraphenalene with lithium hydride in tetrahydrofuran. With this reagent 2-methylcyclopentanone gave 94 per cent *cis*-2-methylcyclopentanol and norcamphor gave 99 per cent *endo*-norborneol in rapid reaction at 0° (Brown and Dickason, 1970) (5.6).

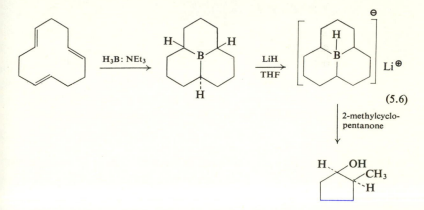

(5.6)

2-methylcyclo-pentanone

5.2. Hydroboration. Undoubtedly the most important synthetic application of diborane in organic chemistry is in the preparation of alkyl- and vinyl-boranes by addition to olefins and acetylenes, a process known as hydroboration (Brown, 1962) (5.7). This reaction, which is catalysed

$$>C=C< \; + \; H-B< \; \longrightarrow \; H-\overset{|}{\underset{|}{C}}-\overset{|}{\underset{|}{C}}-B<$$

$$-C\equiv C- \; + \; H-B< \; \longrightarrow \; H-\overset{|}{C}=\overset{|}{C}-B<$$

(5.7)

by ethers, has been applied to a large number of olefins of widely different structures. In nearly all cases the addition proceeds rapidly and simply at room temperature and only the most hindered olefins do not react, although the nature of the product may vary with the structure of the olefin. With simple olefins (mono- and disubstituted ethylenes) all three hydrogen atoms of the borane are used, to form a trialkylborane (5.8).

$$CH_3.CH=CH.CH_3 \xrightarrow[25°C]{B_2H_6, \text{ diglyme}} (CH_3.CH_2.\overset{\overset{\displaystyle CH_3}{\displaystyle |}}{C}H)_3B \qquad (5.8)$$

But if the double bond carries a bulky substituent rapid reaction may proceed only to the dialkylborane stage. Similarly, trisubstituted ethylenes normally give the dialkylborane under ordinary mild conditions, and tetrasubstituted olefins form only monoalkylboranes (5.9). This

$$(CH_3)_3C.CH=CH.CH_3 \xrightarrow[25°C]{B_2H_6, \text{ diglyme}} [(CH_3)_3C.CH_2.\overset{\overset{\displaystyle CH_3}{\displaystyle |}}{C}H]_2BH$$

$$(CH_3)_2C=CH.CH_3 \xrightarrow[25°C]{B_2H_6, \text{ diglyme}} [(CH_3)_2.CH.\overset{\overset{\displaystyle CH_3}{\displaystyle |}}{C}H]_2BH \qquad (5.9)$$

$$(CH_3)_2C=C(CH_3)_2 \xrightarrow[25°C]{B_2H_6, \text{ diglyme}} (CH_3)_2.CH.\overset{\overset{\displaystyle CH_3}{\displaystyle |}}{\underset{\underset{\displaystyle CH_3}{\displaystyle |}}{C}}-BH_2$$

order of reactivity has been exploited in the preparation of a number of mono- and dialkylboranes which are less reactive and more selective than diborane itself (see p. 213).

Other methods which have been used for preparing organoboranes from olefins include reaction with the trimethylamine addition compound of t-butylborane (Hawthorne, 1961), and reaction with the readily available tri-isobutylborane (Köster, 1958).

Directive effects. Addition of diborane to an unsymmetrical olefin could, of course, give rise to two different products, by addition of the boron atom at either end of the double bond. It is found in practice, however, that, in the absence of strongly polar neighbouring substituents, the reactions are highly selective and give predominantly the isomer in which boron is bound to the less highly substituted carbon atom (5.10). In disubstituted internal olefins there is some preference for addition of boron to the end of the double bond adjacent to the less branched alkyl substituent.

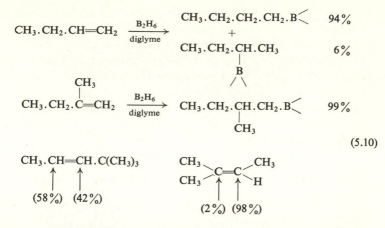

$$CH_3.CH_2.CH=CH_2 \xrightarrow[\text{diglyme}]{B_2H_6}$$

$$CH_3.CH_2.CH_2.CH_2.B\diagdown \quad 94\%$$
$$+$$
$$CH_3.CH_2.\underset{\overset{|}{B}}{CH}.CH_3 \quad 6\%$$
$$\diagup\diagdown$$

$$\underset{\overset{|}{CH_3}}{CH_3.CH_2.C}=CH_2 \xrightarrow[\text{diglyme}]{B_2H_6} CH_3.CH_2.\underset{\overset{|}{CH_3}}{CH}.CH_2.B\diagdown \quad 99\%$$

(5.10)

Even greater selectivity can be achieved by hydroboration with sterically hindered less reactive organoboranes such as di(3-methyl-but-2-yl)borane (disiamylborane) and 2,3-dimethylbut-2-ylborane (thexylborane), which are readily obtained by hydroboration of 2-methyl-but-2-ene and 2,3-dimethylbut-2-ene. Another very useful reagent is 9-borabicyclo[3,3,1]nonane, formed by reaction of diborane with cyclo-octa-1,5-diene in tetrahydrofuran. This is a crystalline compound which, unlike the two reagents above can be isolated and stored. On reaction with olefins it readily affords *B*-alkyl derivatives which undergo all the reactions of other trialkylboranes. Boron-alkyl and boron-aryl derivatives of 9-borabicyclo[3,3,1]nonane can also be obtained by reaction with the corresponding organolithium compounds, thus making available

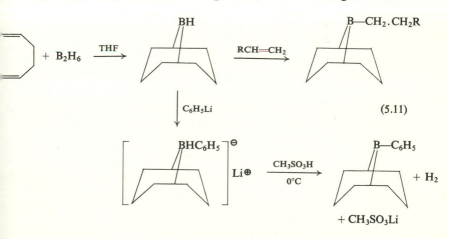

(5.11)

derivatives, such as the *B*-phenyl compound, which cannot be prepared by way of an olefin (Brown and Rogic, 1969) (5.11). Hydroboration of hex-1-ene with diborane itself gives a product containing 94 per cent of the primary and 6 per cent of the secondary borane, but with disiamyl-borane the primary borane is formed almost exclusively. Similarly, diborane shows little discrimination between the two carbon atoms of the double bond in *trans*-4-methylpent-2-ene, but with disiamylborane or 9-borabicyclo[3,3,1]nonane reaction occurs predominantly at C-2 (5.12).

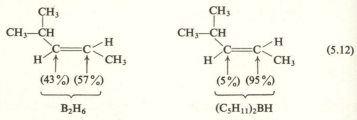

$$(5.12)$$

Because of the large steric requirements of disiamylborane it reacts at different rates with different kinds of olefinic double bonds, and this allows some remarkably selective reactions. Thus, a mixture of 18 per cent *cis*- and 82 per cent *trans*-pent-2-ene, by a single treatment with disiamylborane gave *trans*-pent-2-ene of 98 per cent purity through preferential reaction of the *cis* isomer. Another striking example is the conversion of limonene into the monohydroborated product by pre-ferential reaction at the disubstituted double bond (5.13).

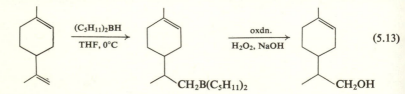

$$(5.13)$$

Mechanism of hydroboration. All the available evidence suggests that hydroboration is a concerted process and takes place through a cyclic four-membered transition state formed by addition to the double bond of a polarised B–H bond in which the boron atom is more positive. This is supported by the stereochemistry of the reaction (p. 218) and by the directive effect of polar substituents. Thus, in allyl derivatives and in nuclear substituted styrenes the proportion of product formed by addition of boron to the α-carbon atom increases with the electronegativity of the substituent (Brown and Sharp, 1966; Brown and Gallivan, 1968) (5.14).

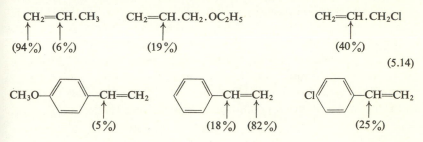

$$(5.14)$$

Hydroboration of substituted olefins. Hydroboration is readily effected with olefins containing many types of functional groups, thus greatly extending the usefulness of the reaction in synthesis. Where the other group is not reduced by diborane, hydroboration proceeds without difficulty, except in cases where the organoborane produced has a good leaving group in the β position, when an elimination may ensue. Thus, reaction of 5-chloropent-1-ene with diborane or, better, di(3-methylbut-2-yl)borane proceeds normally to give the trialkylborane, oxidation of which (see p. 224) gives the 1,5-chlorohydrin in almost quantitative yield (5.15). Similarly, a series of butenyl derivatives $CH_2{=}CH.CH_2.CH_2X$

$$Cl(CH_2)_3.CH{=}CH_2 \xrightarrow[\text{THF, 0°C}]{(C_5H_{11})_2BH} Cl(CH_2)_5.B(C_5H_{11})_2$$

$$\Big\downarrow NaOH, H_2O_2 \qquad (5.15)$$

$$ClCH_2.CH_2.CH_2.CH_2.CH_2OH$$

($X = OCH_3$, OC_6H_5, OH, $OCOCH_3$, Cl, NH_2 and SCH_3) was hydroborated without difficulty (Brown and Unni, 1968).

The organoborane from allyl chloride and di(3-methylbut-2-yl)borane was oxidised to trimethylenechlorohydrin and, interestingly, on treatment with alkali was converted into cyclopropane (5.16). Even better yields of cyclopropanes were obtained from γ-chloroalkyl-9-borabicyclo [3,3,1]nonanes. Thus, 3,4-dichlorobut-1-ene was converted into the

$$ClCH_2.CH{=}CH_2 \longrightarrow ClCH_2.CH_2.CH_2.B(C_5H_{11})_2$$

$$\downarrow \text{NaOH}$$

(5.16)

$$\underset{CH_2\text{------}CH_2}{\overset{CH_2}{\diagup \diagdown}} \quad (80\%)$$

highly reactive cyclopropylcarbinyl chloride in 81 per cent yield, and
gem-dimethylcyclopropane was obtained from α,α-dimethylallyl chloride
in 75 per cent yield (Brown and Rhodes, 1969).

In a slightly different procedure, cyclopropanol was obtained from
propargyl bromide by dihydroboration with 9-borabicyclo[3,3,1]nonane
followed by cyclisation with base and oxidation of the resulting cyclo-
propylboron intermediate (5.17). A similar sequence was used to obtain
cyclobutanol from but-3-ynol tosylate (Brown and Rhodes, 1969).

$$CH{\equiv}C.CH_2Br \xrightarrow[\text{THF}]{C_8H_{14}BH} \underset{\underset{B.C_8H_{14}}{|}}{\overset{\overset{B.C_8H_{14}}{|}}{H{-}C.CH_2.CH_2Br}}$$

$$\downarrow \text{NaOH or CH}_3\text{Li}$$ (5.17)

$$\underset{(65\%)}{\underset{CH_2}{\overset{CH_2}{\diagdown}}{>}CH.OH} \xleftarrow[\text{NaOAc}]{H_2O_2} \underset{CH_2}{\overset{CH_2}{\diagdown}}{>}CH{-}B.C_8H_{14}$$

Hydroboration of allyl chlorides and allyl tosylates with diborane
itself and subsequent transformation of the organoboranes, gives much
poorer yields of cyclopropanes and cyclopropanols than the procedures
described above. This is because the powerful directive influence of the
substituent leads to formation of considerable amounts of the secondary
alkylborane, which undergoes a rapid elimination of the neighbouring
boron and chlorine atoms, with formation of an olefin (5.18). This

$$ClCH_2.CH{=}CH_2 \xrightarrow{B_2H_6} \underset{(60\%)}{ClCH_2.CH_2.CH_2.B{<}} + \underset{\underset{/\backslash}{\overset{|}{B}}\ (40\%)}{ClCH_2.CH.CH_3}$$

$$\downarrow \text{fast}$$ (5.18)

$$CH_2{=}CH.CH_3 + ClB{<}$$

unwanted side reaction is largely circumvented by hydroboration with di(3-methylbut-2-yl)borane or 9-borabicyclo[3,3,1]nonane, which form only small amounts of the secondary alkylborane.

In crotyl derivatives the directive effect of the substituent becomes much more noticeable. Crotyl chloride, for example, on reaction with diborane and oxidation, gave a product consisting almost entirely of butan-1-ol, obviously formed by addition of boron to C-2 of the crotyl chloride followed by rapid elimination and addition of borane to the butene produced (5.19). Similar results were obtained with crotyl

$$CH_3.CH{=}CH.CH_2Cl \xrightarrow[\text{THF}]{B_2H_6} CH_3.CH_2.\underset{\underset{B<}{|}}{CH}.CH_2Cl \longrightarrow$$

$$(5.19)$$

$$CH_3.CH_2.CH{=}CH_2 \xrightarrow{B_2H_6} CH_3.CH_2.CH_2.CH_2.B< \longrightarrow$$

$$CH_3.CH_2.CH_2.CH_2OH$$

acetate, but with crotyl ethyl ether there was less tendency for elimination, ethoxy being a poor leaving group, and oxidation gave the ethoxy alcohols almost quantitatively. With allyl alcohols the hydroxy group directs the boron atom to the adjacent carbon atom, and high yields of 1,2-glycols can be obtained by subsequent oxidation. In these reactions elimination is prevented by protection of the hydroxyl group as a borinate ester or by some other means (Brown and Gallivan, 1968) (5.20).

$$C_6H_5.CH{=}CH.CH_2OH \xrightarrow[\text{THF, 0°C}]{(C_5H_{11})_2BH} C_6H_5.CH{=}CH.CH_2O.B(C_5H_{11})_2$$

$$\downarrow \begin{array}{l}\text{(1) } B_2H_6 \\ \text{(2) } H_2O_2, \text{NaOH}\end{array}$$

$$C_6H_5.CH_2.CHOH.CH_2OH$$

$$(92\%) \qquad (5.20)$$

$$(90\%)$$

Hydroboration of olefins containing functional groups which are normally reduced by diborane is more difficult, but can sometimes be achieved by use of the stoichiometric amount of diborane or, better, with disiamylborane, since addition to olefinic bonds is so much faster than to most other unsaturated groups. Thus, it has been possible to convert a series of terminally unsaturated olefinic esters into the ω-hydroxy esters by hydroboration with diborane and subsequent oxidation, although special conditions had to be used to prevent simultaneous reduction of the ester group (5.21). Better results were obtained with

$$CH_2{=}CH.CH_2.CO_2C_2H_5 \xrightarrow[\text{THF, 0°C}]{(C_5H_{11})_2BH} {\gt}B{-}CH_2.CH_2.CH_2.CO_2C_2H_5$$

$$\downarrow \text{H}_2\text{O}_2, \text{NaOH} \qquad (5.21)$$

$$HOCH_2.CH_2.CH_2.CO_2C_2H_5$$

(78%)

disiamylborane. This reagent does not reduce ester groups, and the ω-hydroxy esters were obtained in excellent yield virtually free from the isomeric secondary alcohol. Unsaturated carboxylic acids can similarly be hydroborated with disiamylborane without reduction of the acid group, but the carbonyl group of unsaturated aldehydes and ketones must be protected as the acetal or ketal.

Stereochemistry of hydroboration. Hydroboration of olefins and acetylenes is highly stereoselective and takes place by *cis* addition, in the case of olefins to the least hindered side of the multiple bond. Thus, reduction of acetylenes by hydroboration followed by hydrolysis affords *cis* olefins (see p. 223) and 1-alkylcycloalkenes, on hydroboration and oxidation, yield *trans*-2-alkylcycloalkanols almost exclusively (5.22)

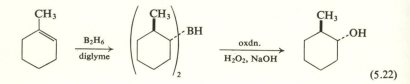

(5.22)

(compare p. 209). Both the oxidation of the boron–carbon bond to form an alcohol and hydrolysis to form a hydrocarbon have been found to occur with retention of configuration, thus establishing the *cis* stereochemistry of the original addition.

The hydroboration–oxidation sequence has been widely used to prepare alcohols of known stereochemistry. Thus, cholesterol gives cholestan-3β,6α-diol, and pinene is converted into isopinocampheol, in each case addition taking place from the least hindered side of the double bond (5.23).

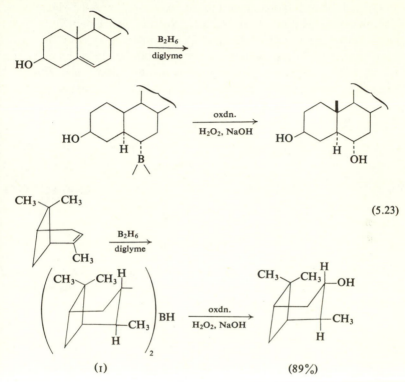

(5.23)

The diisopinocampheylboranes (I) obtained from (+)- and (−)-α-pinene are themselves optically active and have been used in the asymmetric synthesis of alcohols. Thus, hydroboration of *cis*-but-2-ene, *cis*-hex-3-ene and norbornene with the reagent and oxidation of the resulting organoboranes produced the corresponding alcohols with optical purities of 70–90 per cent. The absolute configuration of diisopinocampheylborane may be deduced from the known configuration of α-pinene, and on the basis of a simple model for hydroboration it is possible to predict the absolute configuration of the optically active alcohols obtained.

Hydroboration of dienes and acetylenes. Reaction of dienes with diborane can be controlled to give either mono- or di-hydroborated products, oxidation of which affords unsaturated alcohols or diols.

Monohydroboration of acyclic dienes is not practicable if the double bonds are conjugated, because the olefin formed in the initial reaction is more reactive than the original diene, but it can be achieved with non-conjugated dienes. Thus, reaction of excess of hexa-1,5-diene with diborane and oxidation of the product gave hex-5-en-1-ol in 26 per cent yield. Better results were obtained by hydroboration with the more selective disiamylborane, and where the double bonds in the diene are not identical high yields of unsaturated alcohols have been obtained using this reagent (5.24). Even with conjugated dienes good yields of

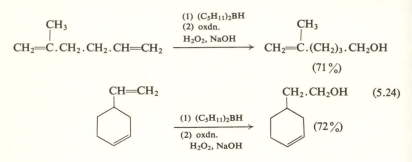

unsaturated alcohols have been obtained by taking advantage of the selective reactivity of disiamylborane towards different kinds of double bonds (5.25).

$$CH_3.CH\!=\!CH.CH\!=\!CH_2 \xrightarrow[\text{(2) oxdn.}]{\text{(1) } (C_5H_{11})_2BH} CH_3.CH\!=\!CH.CH_2.CH_2OH \quad (5.25)$$
$$(74\%)$$

Dihydroboration of dienes can be effected by reaction with excess of diborane. 1,3-Butadiene reacts to form a complex polymer which, on oxidation, affords a mixture of butan-1,3- (18 per cent) and 1,4- (57 per cent) diols, but 2,3-dimethylbutadiene gives the 1,4-diol exclusively. Penta-1,4-diene, on hydroboration with diborane and oxidation, affords a mixture of pentan-1,4- and 1,5-diols. The 1,4-isomer is thought to arise through the intermediacy of a five-membered cyclic organoborane (5.26).

Once again more selective reaction is attained with disiamylborane. With this reagent penta-1,4-diene affords pentan-1,5-diol in 85 per cent yield, and hexa-1,5-diene gives the 1,6-diol almost exclusively.

Cyclic and bicyclic organoboranes can be obtained readily by reaction of suitable dienes with 2,3-dimethylbut-2-ylborane, and have been used

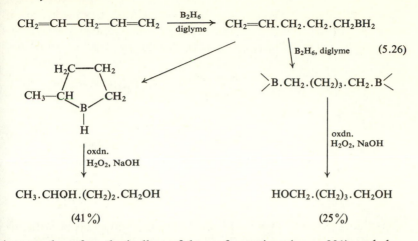

$$CH_2{=}CH{-}CH_2{-}CH{=}CH_2 \xrightarrow[\text{diglyme}]{B_2H_6} CH_2{=}CH.CH_2.CH_2.CH_2BH_2$$

$$\Big\downarrow B_2H_6, \text{diglyme} \qquad (5.26)$$

$${>}B.CH_2.(CH_2)_3.CH_2.B{<}$$

oxdn.
H_2O_2, NaOH

CH$_3$.CHOH.(CH$_2$)$_2$.CH$_2$OH   HOCH$_2$.(CH$_2$)$_3$.CH$_2$OH

(41 %)   (25 %)

in a number of synthetically useful transformations (see p. 231), and also to achieve stereochemical control of the direction of hydroboration, as in the conversion (5.27) of D-(+)-limonene into essentially pure D-(−)-carvomenthol (Brown and Pfaffenberger, 1967).

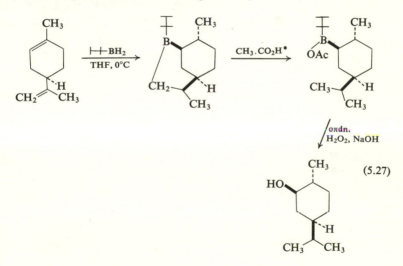

(5.27)

* Preferential protonolysis of the primary organoborane.

Hydroboration of disubstituted triple bonds proceeds readily with diborane and can be controlled to give mainly the vinylorganoborane. Even better results are obtained with disiamylborane or with 2,3-

dimethylbutylborane. Oxidation of the products with alkaline hydrogen peroxide affords ketones. The same result is, of course, achieved by acid-catalysed hydration of acetylenes (5.28). Reaction of terminal acetylenes

$$CH_3.CH_2.C{\equiv}C.CH_2.CH_3 \xrightarrow[\text{diglyme}]{(C_5H_{11})_2BH} CH_3.CH_2.C{=}CH.CH_2.CH_3$$

$$\begin{array}{c} | \\ B \\ / \ \backslash \end{array}$$

$$\downarrow \text{oxdn.} \qquad\qquad (5.28)$$

$$CH_3.CH_2.CO.CH_2.CH_2.CH_3$$

$$(68\%)$$

with diborane yields mainly dihydroboration products, but with di-siamylborane or trimethylamine-t-butylborane high yields of mono-hydroboration products are obtained. These compounds are synthetically useful because in them the boron atom is attached to the end of the chain and on oxidation they afford aldehydes. The series of reactions, hydro-boration followed by oxidation, thus complements the usual acid-catalysed hydration of terminal acetylenes which leads to methyl ketones (Zweifel and Brown, 1963) (5.29).

$$CH{\equiv}C.CH_2.CH_3 + \text{t-}C_4H_9\overset{\ominus}{B}H_2.\overset{\oplus}{N}(CH_3)_3$$

$$\downarrow$$

$$\overset{\ominus}{\underset{\overset{|}{\oplus N(CH_3)_3}}{{>}B}}{-}CH{=}CH.CH_2.CH_3 \xrightarrow{\text{oxdn.}} CHO.CH_2.CH_2.CH_3$$

$$(5.29)$$

$$CH{\equiv}C.(CH_2)_5.CH_3 \xrightarrow{(C_5H_{11})_2BH} (C_5H_{11})_2B{-}CH{=}CH.(CH_2)_5.CH_3$$

$$\downarrow \text{oxdn.}$$

$$CHO.(CH_2)_6.CH_3 \quad (70\%)$$

5.3. Reactions of organoboranes.   The usefulness of the hydroboration reaction in synthesis arises from the fact that the intermediate alkyl-boranes can be converted by further reaction into a variety of other products. On hydrolysis (protonolysis), for example, the boron atom is replaced by hydrogen, and under appropriate conditions the boranes are readily oxidised to alcohols or carbonyl compounds. A practical advantage of these reactions is that it is often unnecessary to isolate the intermediate organoborane.

Protonolysis. Protonolysis is best effected with an organic carboxylic acid (boiling propionic acid is often used), and provides a convenient non-catalytic method for reduction of carbon–carbon multiple bonds (5.30). Protonolysis is believed to take place with retention of con- figuration at the carbon atom concerned (Brown and Murray, 1961), and in accordance with the proposed mechanism of hydroboration, acetylenes are converted cleanly into *cis* olefins, and cyclic olefins undergo *cis* addition of hydrogen. An advantage of the hydroboration–protono- lysis procedure is that it can sometimes be used for the reduction of double or triple bonds in compounds which contain other easily reducible

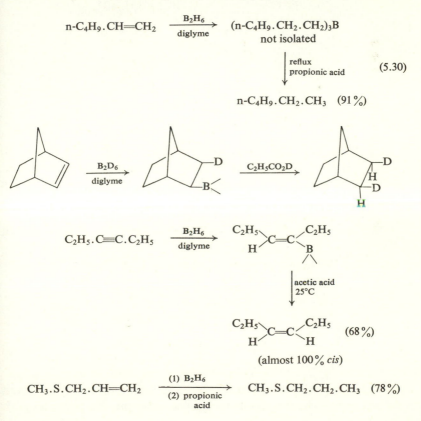

$$n\text{-}C_4H_9.CH{=}CH_2 \xrightarrow[\text{diglyme}]{B_2H_6} (n\text{-}C_4H_9.CH_2.CH_2)_3B$$
$$\text{not isolated}$$

reflux
propionic acid   (5.30)

$$n\text{-}C_4H_9.CH_2.CH_3 \quad (91\%)$$

$$C_2H_5.C{\equiv}C.C_2H_5 \xrightarrow[\text{diglyme}]{B_2H_6}$$

acetic acid
25°C

(68%)

(almost 100% *cis*)

$$CH_3.S.CH_2.CH{=}CH_2 \xrightarrow[\substack{(2)\ \text{propionic}\\ \text{acid}}]{(1)\ B_2H_6} CH_3.S.CH_2.CH_2.CH_3 \quad (78\%)$$

groups. Allyl methyl sulphide, for example, is converted into methyl propyl sulphide in 78 per cent yield.

Oxidation. Oxidation of organoboranes to alcohols is best effected with alkaline hydrogen peroxide, and several examples have already been given. The reaction is of wide applicability and many functional groups are unaffected by the reaction conditions, so that a variety of substituted olefins can be converted into alcohols by this procedure (Zweifel and Brown, 1963) (5.31). A useful feature of the reaction in synthesis is that it results in over-all anti-Markownikoff addition of water to the double or triple bond, and thus complements the more usual acid-catalysed hydration (see p. 222). This follows, of course, from the fact that in the hydroboration step the boron atom adds to the least substituted carbon of the multiple bond.

$$CH_2{=}CH.CH_2Cl \xrightarrow[\text{THF}]{(C_5H_{11})_2BH} (C_5H_{11})_2B.CH_2.CH_2.CH_2Cl$$

$$\downarrow \text{NaOH, H}_2\text{O}_2, \text{pH 7--8}$$

$$CH_2OH.CH_2.CH_2Cl \quad (77\%) \tag{5.31}$$

$$n\text{-}C_6H_{13}.C{\equiv}CH \xrightarrow[\text{(2) H}_2\text{O}_2, \text{NaOH}]{\substack{\text{(1) (C}_5\text{H}_{11})_2\text{BH} \\ \text{diglyme, 0}^\circ\text{C}}} C_6H_{13}.CH_2.CHO \quad (70\%)$$

Another noteworthy feature of the reaction, pointed out on p. 218, is that it leads to *cis* addition of the elements of water to the double bond. This is determined by the mechanism of the reaction and not by the stability of the product, for the thermodynamically more stable product is not invariably formed. Thus, hydroboration–oxidation of 1-methyl-cyclohexene affords the more stable *trans*-2-methylcyclohexanol, but 1,2-dimethylcyclohexene gives the less stable *cis*-1,2-dimethylcyclohexanol. β-Pinene similarly gives the less stable *cis*-myrtanol by addition of B–H to the less hindered side of the double bond (5.32).

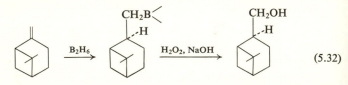

$$\tag{5.32}$$

Oxidation of the carbon–boron bond to form an alcohol is believed to take place with retention of configuration, and the reaction path

(5.33), involving intramolecular transfer of an alkyl group from boron to oxygen has been proposed.

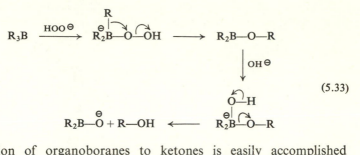

$$(5.33)$$

Oxidation of organoboranes to ketones is easily accomplished with aqueous chromic acid, and the sequence provides a very convenient method for conversion of an olefin into a ketone without isolation of the corresponding alcohol (Brown and Garg, 1961).

Other useful conversions which proceed in good yield are the reaction of trialkylboranes with hydroxylamine-*O*-sulphonic acid to give primary amines (Brown, Heydkamp, Breuer and Murphy, 1964), and with iodine in presence of sodium hydroxide to give iodides (Brown, Rathke and Rogic, 1968; see also Freeguard and Long, 1964).

Isomerisation and cyclisation of alkylboranes. An interesting and synthetically useful reaction of organoboranes is their ready isomerisation on gentle heating to compounds in which the boron atom is attached to the least hindered carbon of the alkyl group. Thus, tri(n-hex-3-yl)borane on heating at 150°C for 1 hour is converted into a mixture in which the major component has the boron atom attached to the terminal carbon atom (5.34). The reaction is strongly catalysed by

$$CH_3.CH_2.CH.CH_2.CH_2.CH_3 \xrightarrow[(B_2H_6)]{150°C} >B.CH_2.(CH_2)_4.CH_3 \quad (90\%)$$

$$+ CH_3.CH_2.CH.CH_2.CH_2.CH_3 \quad (4\%)$$

$$+ CH_3.CH.(CH_2)_3.CH_3 \quad (6\%)$$

$$(5.34)$$

diborane or other molecules containing boron–hydrogen bonds, and proceeds by a succession of eliminations and additions of borane, leading, eventually to the most stable alkylborane with the boron atom in the least hindered position. In line with this mechanism, it is found that the boron atom can migrate readily along a straight chain of carbon atoms and past a single alkyl branch, but it cannot pass a completely substituted carbon atom (5.35). This reaction can be used to

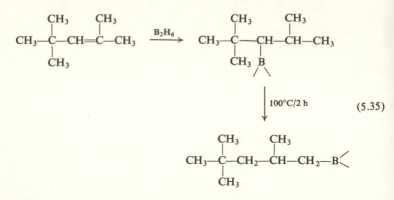

(5.35)

bring about 'contrathermodynamic isomerisation' of olefins, for after rearrangement the isomerised olefin can be displaced from the organoborane by heating it with a higher boiling olefin. Thus, 3-ethylpent-2-ene was converted into the thermodynamically less stable 3-ethylpent-1-ene in 82 per cent yield by thermal isomerisation of the borane and displacement of the olefin from the rearranged borane by heating it with dec-1-ene in diglyme solution (5.36). The reaction involves a slow dissociation of the organoborane into olefin and dialkylborane, which then reacts with the displacing olefin which is present in relatively large

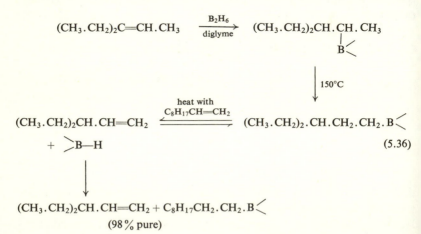

(5.36)

amount, with liberation of the more volatile olefin (Brown and Bhatt, 1966).

Oxidation of the rearranged organoboranes provides a means of converting internal olefins into primary alcohols, as illustrated in (5.37).

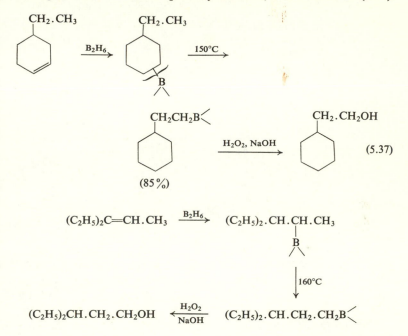

$$(C_2H_5)_2C{=}CH.CH_3 \xrightarrow{B_2H_6} (C_2H_5)_2.CH.CH.CH_3$$

Certain organoboranes, which cannot isomerise for structural reasons, cyclise at elevated temperatures to form cyclic organoboranes. This reaction appears to be especially prone to occur when a methyl group is in a position to form a five- or six-membered boron heterocycle (5.38). Oxidation of the cyclic products gives 1,4- and 1,5-diols. Related is the useful technique for converting olefins into 1,5-diols by oxidation of cyclic organoboranes obtained as illustrated in (5.39). Again reaction

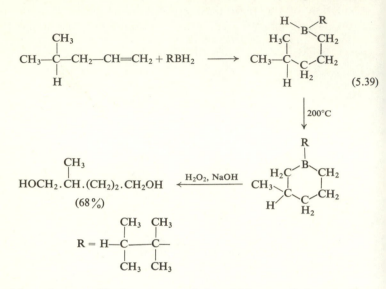

(5.39)

only takes place readily in compounds in which a methyl group is so disposed that cyclisation of the dialkylborane leads to a five- or six-membered cyclic organoborane (Brown, Murray, Müller and Zweifel, 1966).

Formation of cyclic compounds by pyrolysis of organoboranes has been extensively studied by Köster (1964).

Coupling reactions. In an interesting extension of the hydroboration reaction it has been found that treatment of the preformed trialkyl-boranes with alkaline silver nitrate leads to coupling of the alkyl groups. Reaction takes place under mild conditions and provides a useful new method for making carbon–carbon bonds (Brown, Verbrugge and Snyder, 1961). Hex-1-ene, for example, is converted into n-dodecane in 70 per cent yield by way of tri-n-hexylborane. 'Mixed' coupling can also be achieved in reasonable yield by using an excess of one of the olefins. For example, n-hexylcyclopentane was obtained in 45 per cent yield from cyclopentene and excess of n-hexene by hydroboration of the mixture of olefins followed by addition of silver nitrate and sodium hydroxide. The reaction is believed to take place by way of alkyl free radicals formed by breakdown of intermediate silver alkyls.

Carbonylation of organoboranes. One of the most useful reactions of organoboranes in synthesis is their reaction with carbon monoxide at 100–125°C which, under appropriate conditions, can be directed to give primary, secondary and tertiary alcohols, aldehydes, and open-chain, cyclic and polycyclic ketones (Brown, 1969).

It had earlier been observed by Hillman and others that trialkylboranes reacted with carbon monoxide in presence of water at high pressures and a temperature of 50–75°C to form compounds which could be oxidised to dialkylcarbinols. Brown and Rathke (1967) found, however, that at a slightly higher temperature a wide variety of organoboranes reacted with carbon monoxide at atmospheric pressure in diglyme solution to form intermediates which were oxidised by alkaline hydrogen peroxide to trialkylcarbinols in excellent yield (5.40). The

$$R_3B + CO \xrightarrow{\text{diglyme}} (R_3C.BO)_x \xrightarrow{\text{H}_2\text{O}_2,\ \text{NaOH}} R_3C.OH \quad (5.40)$$

reaction appears to be of wide applicability, and for trialkylcarbinols containing bulky groups gives much higher yields than any other method. Tricyclohexylcarbinol, for example, is obtained from cyclohexene in 85 per cent yield, whereas the Grignard method gives only 7 per cent.

The reaction obviously involves migration of alkyl groups from boron to the carbon atom of carbon monoxide, and this was shown to occur by an intramolecular process by the fact that carbonylation of an equimolecular mixture of triethylborane and tri-n-butylborane gave, after oxidation, only triethylcarbinol and tri-n-butylcarbinol; no 'mixed' carbinols were formed. Similarly, dicyclohexyl-n-octylborane gave only dicyclohexyl-n-octylcarbinol. The stepwise mechanism (5.41), involving three successive intramolecular transfers has been proposed. If the

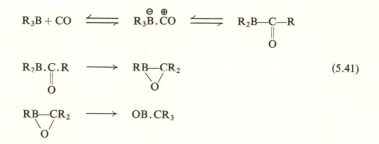

carbonylation is conducted in presence of a small amount of water, migration of the third alkyl group is inhibited. Oxidation of the intermediate thus produced then gives the dialkyl ketone instead of the trialkyl carbinol. Alkaline hydrolysis is said (Brown, 1969) to lead to the secondary alcohol, but no experimental details or examples of this transformation appear to have been published. The water apparently converts the bora-epoxide into the corresponding hydrate which less easily undergoes transfer of the third alkyl group (5.42). Yields obtained

$$RB\!\!-\!\!CR_2 \xrightarrow[\text{fast}]{H_2O} \quad R.B\!\!-\!\!CR_2 \xrightarrow{\text{slow}} \quad (HO)_2B.CR_3 \qquad (5.42)$$

with $RB\!\!-\!\!CR_2$ having a bridging O, and $R.B\!\!-\!\!CR_2$ having OH OH groups.

are generally high, and the sequence provides a very convenient synthetic route to ketones. Oct-1-ene, for example, was smoothly converted into di-n-octyl ketone in 80 per cent yield, and cyclopentene gave dicyclopentyl ketone (90 per cent).

The method can be extended to the synthesis of unsymmetrical ketones by using 'mixed' organoboranes, but the scope of this modification is limited by the availability of the mixed organoborane. A better procedure makes use of the highly hindered 2,3-dimethylbut-2-ylborane (p. 213). This compound reacts with olefins in two distinct stages to form, first, the dialkylborane and then the trialkylborane (5.43). Successive addition

$$\begin{array}{ccc}
CH_3 \;\; CH_3 & CH_3 \;\; CH & CH_3 \;\; CH_3 \\
\mid \quad\;\; \mid & \mid \quad\;\; \mid & \mid \quad\;\; \mid \\
CH\!\!-\!\!C\!\!-\!\!BH_2 \xrightarrow{\text{olefin A}} & CH\!\!-\!\!C\!\!-\!\!B\!\!<\!\!\begin{smallmatrix}R_A\\H\end{smallmatrix} \xrightarrow{\text{olefin B}} & CH\!\!-\!\!C\!\!-\!\!B\!\!<\!\!\begin{smallmatrix}R_A\\R_B\end{smallmatrix} \\
\mid \quad\;\; \mid & \mid \quad\;\; \mid & \mid \quad\;\; \mid \\
CH_3 \;\; CH_3 & CH_3 \;\; CH_3 & CH_3 \;\; CH_3
\end{array}$$

$$(5.43)$$

of two different olefins results in the formation of an organoborane with three different alkyl groups attached to the boron atom. In these reagents the 2,3-dimethylbut-2-yl group shows an exceptionally low aptitude for migration, and carbonylation in presence of water, followed by oxidation, leads to high yields of the ketone $R_A.CO.R_B$. Because of the bulky nature of the dimethylbutyl group, carbonylation of these compounds requires rather more vigorous conditions than usual and generally has to be effected under pressure (5.44). The presence of functional groups in the olefin does not interfere with the reaction, and the procedure can be used to synthesise a ketone from almost any two olefins (Brown and Negishi, 1967b).

Dienes similarly yield cyclic ketones, and in a notable extension of the reaction bicyclic ketones have been prepared, as illustrated below for the stereospecific conversion of cyclohexanone into the thermo-

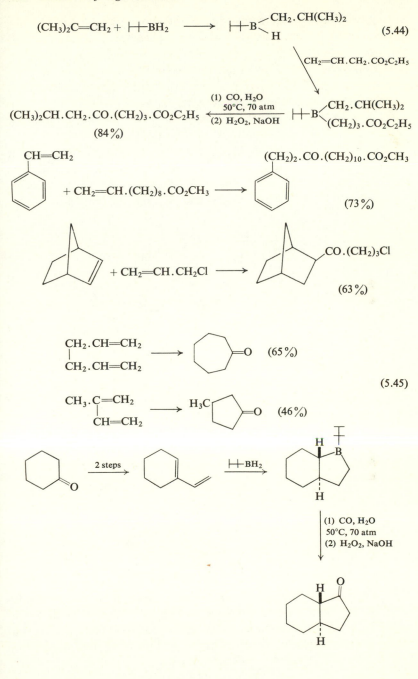

$$(CH_3)_2C{=}CH_2 + {\vdash}{-}BH_2 \longrightarrow {\vdash}{-}B\overset{CH_2.CH(CH_3)_2}{\underset{H}{\diagdown}} \qquad (5.44)$$

$$\downarrow {\scriptstyle CH_2=CH.CH_2.CO_2C_2H_5}$$

$$(CH_3)_2CH.CH_2.CO.(CH_2)_3.CO_2C_2H_5 \xleftarrow[\text{(2) } H_2O_2, \text{ NaOH}]{\substack{\text{(1) CO, } H_2O \\ 50°C, 70 \text{ atm}}} {\vdash}{-}B\overset{CH_2.CH(CH_3)_2}{\underset{(CH_2)_3.CO_2C_2H_5}{\diagdown}}$$

$$(84\%)$$

$$(73\%)$$

$$(63\%)$$

$$(65\%)$$

$$(5.45)$$

$$(46\%)$$

dynamically disfavoured *trans*-perhydroindanone (Brown and Negishi, 1967*b*; 1968) (5.45). *Trans*-1-decalone is obtained similarly from 1-allylcyclohexene, and the reaction appears to have considerable generality for the synthesis of *trans* fused polycyclic ketones with a carbonyl group adjacent to the *trans* ring junction. The high stereospecificity of these reactions, leading exclusively to the *trans* fused compounds, is a result of the mechanism of hydroboration which requires *cis* addition of the boron–hydrogen bond to the olefinic double bond.

A series of remarkable syntheses of polycyclic systems, involving the formation of two or three new carbon–carbon bonds, has also been achieved, and further developments in this field are to be expected (Knights and Brown, 1968*c*; Brown and Negishi, 1967*b*) (5.46).

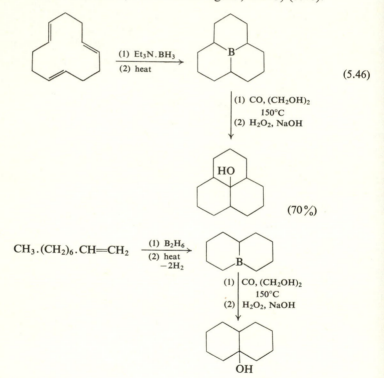

(5.46)

The carbonylation reaction can also be modified to produce aldehydes and primary alcohols. In presence of lithium hydridotrimethoxyaluminate the rate of reaction of carbon monoxide with organoboranes at atmospheric pressure is greatly increased, and the products, on oxidation with buffered hydrogen peroxide, afford aldehydes. Alkaline hydrolysis, on the other hand, gives the corresponding primary alcohol. The reaction is believed to proceed by reduction of one of the intermediates $R_3B.CO$ or $R_2B.CO.R$ (see p. 229) by the complex hydride thus precluding migration of further alkyl groups (5.47). Cyclohexene, for

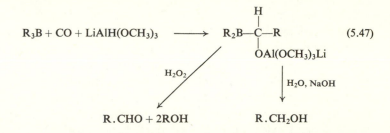

$$R_3B + CO + LiAlH(OCH_3)_3 \longrightarrow R_2B—\overset{\overset{\displaystyle H}{|}}{\underset{\underset{\displaystyle OAl(OCH_3)_3Li}{|}}{C}}—R \qquad (5.47)$$

example, was readily converted into cyclohexanecarboxaldehyde and cyclohexylcarbinol.

A disadvantage of this direct procedure is that only one of the three alkyl groups of the trialkylborane is converted into the required derivative, so that even under ideal conditions a maximum yield of only 33 per cent is possible. This difficulty has been overcome by hydroboration with 9-borabicyclo[3,3,1]nonane (see p. 213). Reaction of the resulting trialkylborane with carbon monoxide and lithium hydridotrimethoxyaluminate results in preferential migration of the boron-alkyl group, and by this method high yields of aldehydes containing a variety of functional groups have been obtained from variously substituted olefins. In some cases, where the olefin contains a reducible substituent, better yields of the required aldehyde are obtained with the weaker reducing agent lithium hydridotri-t-butoxyaluminate (Brown, Knights and Coleman, 1969; Brown and Coleman, 1969). Some representative conversions are shown in (5.48). Dienes can be converted into the corresponding diols

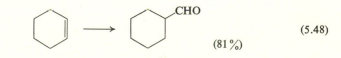

(5.48)

$$CH_2{=}CH.CH_2.CO_2C_2H_5 \longrightarrow OCH.(CH_2)_3.CO_2C_2H_5 \quad (83\%)$$

or dialdehydes, and by taking advantage of selective hydroboration with 9-borabicyclo[3,3,1]nonane it is sometimes possible to effect the conversion of a diene into an unsaturated aldehyde (5.49).

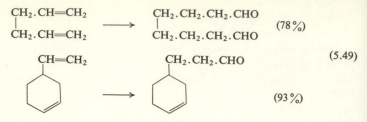

(5.49)

The reactions are highly stereoselective, the original boron–carbon bond being replaced by a carbon–carbon bond with retention of the original stereochemistry. 1-Methylcyclopentene, for example, was converted entirely into *trans*-2-methylcyclopentylcarbinol.

Another method for the homologation of trialkylboranes by one carbon atom involves reaction of the borane with dimethylsulphonium methylide and oxidation of the product, but it also suffers from the disadvantage that only one of the three alkyl groups of the trialkylborane is converted into the product (Tufariello, Wojtkowski and Lee, 1967) (5.50).

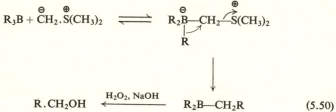

(5.50)

1,4-Addition to αβ-unsaturated carbonyl compounds. Trialkylboranes undergo a remarkably fast addition to many αβ-unsaturated carbonyl compounds, such as acrolein, methyl vinyl ketone and 2-methylenecyclo-alkanones, to form aldehydes and ketones in which the chain has been extended by three or more carbon atoms (Brown, Rogic, Rathke and

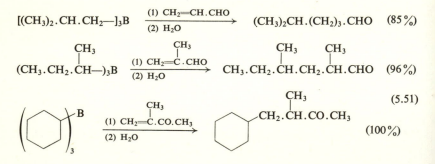

Kabalka, 1967) (5.51). These reactions proceed with retention of configuration at the carbon of the organoborane. Tri(*trans*-2-methylcyclopentyl)borane, for example, with acrolein gives *trans*-2-methylcyclopentylpropionaldehyde (Brown, Rogic, Rathke and Kabalka, 1969). They are believed to take place by a free-radical pathway to form first the enol borinate, hydrolysis of which yields the aldehyde or ketone (Kabalka, Brown, Suzuki, Honma, Arase and Itoh, 1970) (5.52).

$$R\cdot + CH_2{=}CH.CHO \longrightarrow R.CH_2.\overset{\cdot}{C}H.CHO \longleftrightarrow R.CH_2CH{=}CHO\cdot$$

$$\downarrow R_3B$$

$$R.CH_2.CH_2.CHO \overset{H_2O}{\longleftarrow} R.CH_2.CH{=}CHOBR_2 + R\cdot$$

$$(5.52)$$

It is found in practice that only aldehydes and ketones containing the group $CH_2{=}\overset{|}{C}{-}\overset{|}{C}{=}O$ undergo the spontaneous reaction easily. If there is one or more alkyl substituent on the terminal carbon, as in *trans*-crotonaldehyde or cyclohexenone, spontaneous reaction does not take place, probably as a result of a shorter radical chain length and inefficient initiation. In presence of catalytic amounts of diacetyl peroxide or of

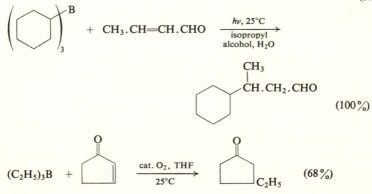

$$(C_2H_5)_3B + CH_3.CH{=}CH.CO.CH_3 \xrightarrow[\substack{25°C,\ 24\ h}]{\substack{\text{diacetyl peroxide,}\\ \text{diglyme, }H_2O}}$$

$$\overset{CH_3}{\overset{|}{C_2H_5.CH.CH_2.CO.CH_3}} \quad (88\%)$$

$$(5.53)$$

oxygen, however, or by irradiation of the reaction mixture with ultra-violet light, all of which favour the formation of radicals from the tri-alkylborane, these 'inert' compounds too undergo rapid addition of trialkylboranes to form enol borinates which are readily hydrolised to the carbonyl compounds (5.53). The sequence provides a most useful and versatile method for the synthesis of aldehydes and ketones from practically any combination of olefin and conjugated aldehyde or ketone (Brown and Kabalka, 1970).

The reaction has been extended to cyclic organoboranes, obtainable by hydroboration of dienes (p. 220), to give products which on oxidation afford ω-hydroxyaldehydes or ω-hydroxyalkyl methyl ketones (5.54).

$$
\begin{array}{c}
\text{CH}_2\text{—CH}_2 \\
| \quad\quad\quad \diagdown \\
\quad\quad\quad\quad\quad \text{B—} \\
| \quad\quad\quad \diagup \\
\text{CH}_2\text{—CH}_2
\end{array}
\xrightarrow[\text{ROH}]{\text{CH}_2\text{=CH.CO.CH}_3}
\quad \text{ROB.(CH}_2)_6.\text{CO.CH}_3
$$

$$\Big\downarrow \text{H}_2\text{O}_2,\ \text{NaOH} \qquad (5.54)$$

$$\text{HOCH}_2.(\text{CH}_2)_5.\text{CO.CH}_3 \quad (95\%)$$

Reaction with α-bromoketones and α-bromoesters. Ketones can also be obtained from organoboranes by reaction with α-bromoketones in presence of potassium t-butoxide, or, better, the hindered base 2,6-di-t-butylphenoxide ion (Brown, Nambu and Rogic, 1969). 2-Ethylcyclo-hexanone, for example, is readily obtained from triethylborane and 2-bromocyclohexanone. The reaction path (5.55) has been suggested.

$$
\text{R.CO.CH}_2\text{Br} + \text{base} \longrightarrow \underset{\underset{\text{Br}}{|}}{\text{R.CO.}\overset{\ominus}{\text{CH}}} \xrightarrow{\text{BR}^1_3} \underset{\underset{\text{Br}}{|}}{\text{R.CO.CH.}\overset{\ominus}{\text{BR}^1_3}}
$$

$$\Big\downarrow \qquad\qquad (5.55)$$

$$
\text{R.CO.CH}_2\text{R}^1 + \text{R}^1_2\text{BOC}_2\text{H}_5 \xleftarrow{\text{C}_2\text{H}_5\text{OH}} \underset{\underset{\text{R}^1}{|}}{\text{R.CO.CH.BR}^1_2} + \text{Br}^\ominus
$$

The usefulness of the direct reaction using the trialkylboranes is limited, however, by the fact that organoboranes containing highly branched groups do not react and, in addition, only one of the three alkyl groups in the trialkylborane takes part in the reaction. Both of these difficulties can be circumvented by using a *B*-alkyl-9-borabicyclo-[3,3,1]nonane reagent, readily available from 9-borabicyclo[3,3,1]nonane and the appropriate olefin, instead of the trialkylborane. By this technique good yields of α-substituted ketones are obtained from α-bromo-ketones (Brown, Nambu and Rogic, 1969) (5.56).

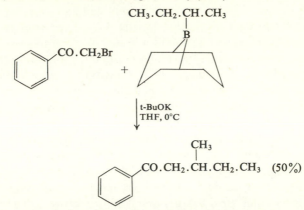

$$(5.56)$$

Mono- and di-halogenoacetates undergo similar reactions, providing a useful alternative to the malonic ester route to substituted acetic acids. Again best yields are obtained using 9-aryl- or 9-alkyl-borabicyclo[3,3,1]-nonanes as starting materials (Brown and Rogic, 1969). Thus, isobutylene was converted into ethyl 4-methylpentanoate in 53 per cent yield by reaction of the borane with ethyl bromoacetate in presence of potassium t-butoxide, and cyclopentene similarly gave ethyl cyclopentylacetate (5.57). 9-Arylborabicyclononanes (p. 213) were smoothly converted into ethyl arylacetates.

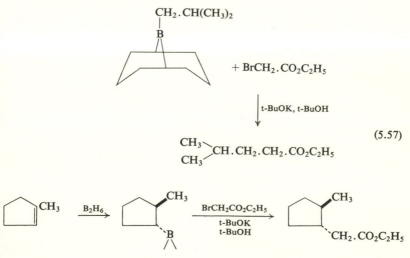

$$(5.57)$$

The reaction can be extended to dibromoacetates, and can be controlled to yield either α-bromoacetates or disubstituted acetates. With one equivalent of base only one bromine is replaced and an α-bromo-ester results; with two equivalents of base the dialkylated ester is formed. Moreover, the dialkylation can be effected in two successive steps, allowing the introduction of two different organic groups into the acetic acid moiety (Brown and Rogic, 1969). Reaction again takes place with retention of configuration at the carbon atom of the organoborane.

α-Chloronitriles also react rapidly with organoboranes in presence of potassium 2,6-di-t-butylphenoxide, to give the alkylated nitrile.

Reaction with diazo compounds. Another route from organoboranes to ketones and esters, which depends on the nucleophilic addition of a diazo compound to the organoborane, has been reported by Hooz and Linke (1968). Diazoketones, for example, react with organoboranes to form products which yield ketones on alkaline hydrolysis. The mechanism (5.58) has been suggested, in which migration of R from boron to

$$R_3B + N_2CH.CO.CH_3 \longrightarrow R_3\overset{\ominus}{B}.\underset{\underset{N_2^{\oplus}}{|}}{CH}.CO.CH_3$$

$$\downarrow -N_2 \qquad\qquad (5.58)$$

$$R.CH_2.CO.CH_3 \overset{\text{hydrolysis}}{\longleftarrow} R_2B.\underset{\underset{R}{|}}{CH}.CO.CH_3$$

carbon is facilitated by expulsion of nitrogen. The reaction appears to be fairly general and yields of 36–89 per cent are claimed.

Diazoesters and nitriles react similarly to give the corresponding homologated esters and nitriles (5.59).

$$(\text{n-}C_6H_{13})_3B + N_2CH.CO_2C_2H_5 \longrightarrow$$

$$(\text{n-}C_6H_{13})_2B.\underset{\underset{\text{n-}C_6H_{13}}{|}}{CH}.CO_2C_2H_5 \overset{\text{hydrolysis}}{\longrightarrow} \text{n-}C_6H_{13}.CH_2.CO_2C_2H_5$$

$$(5.59)$$

6 Oxidation

For practical purposes most organic chemists mean by 'oxidation' either addition of oxygen to the substrate, as in the conversion of ethylene into ethylene oxide, removal of hydrogen as in the oxidation of ethanol to acetaldehyde or removal of one electron as in the conversion of phenoxide anion to the phenoxy radical.

Of the wide variety of agents available for the oxidation of organic compounds probably the most widely used are potassium permanganate and derivatives of hexavalent chromium. Permanganate, a derivative of heptavalent manganese, is a very powerful oxidant. Its reactivity depends to a great extent on whether it is used under acid, neutral or basic conditions. In acid solution it is reduced to the divalent manganous ion Mn^{2+} with net transfer of five electrons ($Mn^{VII} \rightarrow Mn^{II}$), while in neutral or basic media manganese dioxide is usually formed, corresponding to a three electron change ($Mn^{VII} \rightarrow Mn^{IV}$). Permanganate is generally used in aqueous solution and this restricts its usefulness since not many organic compounds are sufficiently soluble in water and only a few organic solvents are resistant to the oxidising action of the reagent. Solutions in acetic acid, t-butanol or dry acetone or pyridine can sometimes be employed.

Chromic acid, a derivative of hexavalent chromium, is one of the most versatile of the available oxidising agents, and reacts with almost all types of oxidisable groups. The reactions can often be controlled to yield largely one product, and for this reason chromic acid oxidation is a useful process in synthesis. In oxidations chromium is reduced from the hexavalent to the trivalent state ($Cr^{VI} \rightarrow Cr^{III}$) with production of a chromic salt. The most common reagents are chromium trioxide and sodium or potassium dichromate. Chromium trioxide is a polymer which dissolves in water with depolymerisation to form chromic acid.

$$(CrO_3)_n + H_2O \longrightarrow HO-\overset{\overset{\displaystyle O}{\|}}{\underset{\underset{\displaystyle O}{\|}}{Cr}}-OH$$

It is commonly employed in solution in dilute sulphuric acid, sometimes containing acetic acid to aid dissolution of the substrate; a comparable solution is obtained by adding sodium or potassium dichromate to aqueous sulphuric acid. These solutions contain an equilibrating mixture of the acid chromate and the dichromate ions.

$$2HCrO_4^- \; \rightleftharpoons \; H_2O + Cr_2O_7^{2-}$$

Chromium trioxide may also be used in solution in acetic anhydride, t-butanol or in pyridine. In these solutions the reactive species present are chromyl acetate, t-butyl chromate and the pyridine–chromium trioxide complex, as shown (6.1).

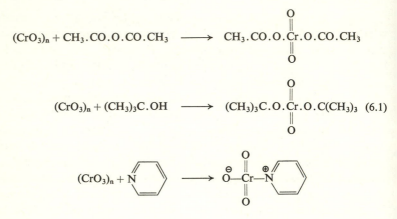

In oxidations of acid-sensitive compounds, and in cases where the initial product is susceptible to further oxidation, it is often advantageous to employ an immiscible solvent, such as benzene or methylene chloride, which serves to reduce contact between the organic compounds and the acid oxidising solution. Thus, cyclopentendiol can be converted into cyclopentendione in 70–80 per cent yield by oxidation with chromium trioxide and sulphuric acid in presence of methylene chloride (6.2).

The most common lower oxidation state of chromium is Cr^{III} and most oxidations with chromic acid lead to this state by a net transfer of three electrons. However, each stage in the oxidation of most organic compounds involves transfer of only two electrons, and it is evident that most reactions must give rise to Cr^V or Cr^{IV} species as intermediates, as exemplified in the following scheme for the oxidation of a secondary alcohol.

$$Cr^{VI} + R_2CHOH \longrightarrow R_2C{=}O + 2H^+ + Cr^{IV}$$

$$Cr^{IV} + Cr^{VI} \longrightarrow 2Cr^V$$

$$Cr^V + R_2CHOH \longrightarrow R_2C{=}O + 2H^+ + Cr^{III}$$

The Cr^{IV} and Cr^V ions are themselves powerful oxidants and may give rise to unwanted side reactions and to products different from those formed in the original oxidation by Cr^{VI}.

6.1. Oxidation of hydrocarbons with chromic acid

Paraffin hydrocarbons. Under vigorous conditions both chromic acid and permanganate attack alkanes, but the reaction is of little synthetic use for usually mixtures of products are obtained in low yield. The reaction is of importance in the Kuhn–Roth estimation of methyl groups. This depends on the fact that a methyl group is rarely attacked (the relative rates of oxidation of primary, secondary and tertiary C–H bonds are 1:110:7000) and is eventually converted into acetic acid. The usual method is to boil the substance with chromic acid in aqueous sulphuric acid and determine the amount of steam volatile acid formed (almost entirely acetic acid) by titration with alkali. The sesquiterpene hydrocarbon cadinene, for example, under these conditions gives three molecules of acetic acid, one from each of the ring methyl substituents and one other from the isopropyl group (6.3). Under less vigorous

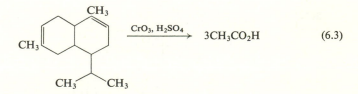

$$(6.3)$$

conditions intermediate oxidation products can sometimes be isolated, and this may be useful in degradative work. Thus the structure of the C-35 hydrocarbon from the macrocyclic antibiotic fungichromin was determined by oxidation with chromium trioxide in acetic acid and identification of the neutral fragments formed by gas–liquid chromatography and mass spectrometry (6.4).

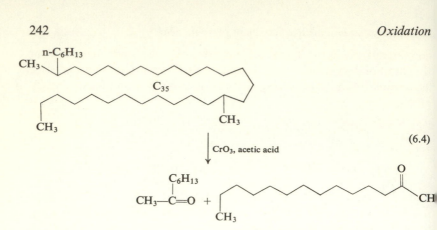

(6.4)

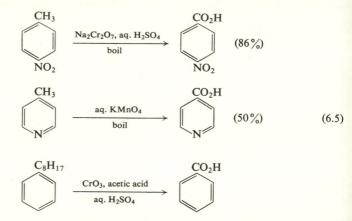

Aromatic hydrocarbons. In the absence of activating hydroxyl or amino substituents benzene rings are only slowly attacked by chromic acid or permanganate, but alkyl side chains are degraded with formation of benzene carboxylic acids (6.5). This is a useful method for the preparation of benzene carboxylic acids, and also for determining the orientation pattern of unknown polyalkylbenzenes which are degraded to benzene polycarboxylic acids of known orientation. Nuclear hydroxyl or amino substituents, if present, must be converted into their methyl ethers or acetyl derivatives, for otherwise they activate the ring to attack, and quinones or, with excess of reagent, carbon dioxide and water are formed. With longer side chains than methyl, initial attack is thought

to take place at the benzylic carbon atom. This is suggested by the fact that t-butylbenzene is very resistant to oxidation and ethylbenzene gives some acetophenone as well as benzoic acid. The rate-determining step

in these chromic acid oxidations is known to be cleavage of the benzylic
C–H bond, but it is uncertain at present whether attack on the benzylic
hydrogen involves initial removal of a hydrogen atom to form a radical,
or a concerted abstraction of hydride ion to give a chromate (6.6).

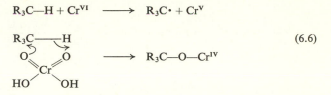

With polycyclic aromatic hydrocarbons oxidation with chromic acid
leads to the formation of quinones, but with *aqueous* solutions of sodium
dichromate at 200–300°C high yields of carboxylic acids are obtained
from alkyl derivatives (6.7). Interestingly, alkyl groups other than

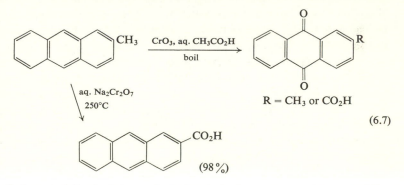

methyl are oxidised at the end of the chain by aqueous sodium dichromate. Ethylbenzene, for example, is converted into phenylacetic acid
(96 per cent).

The conversion of methyl groups attached to benzene rings into the
formyl group can be achieved by an oxidation with chromium trioxide
in acetic anhydride in the presence of a strong acid, or with a solution
of chromyl chloride in carbon disulphide or carbon tetrachloride (the
Étard reaction) (6.8). The success of the first reaction is due to the
initial formation of the di-acetate which protects the aldehyde group
against further oxidation. In the Étard reaction a complex of composition
1 hydrocarbon : $2CrO_2Cl_2$ is first formed and is converted into the
aldehyde by treatment with water. Another useful reagent for this purpose is ceric ion which readily oxidises aromatic methyl groups to

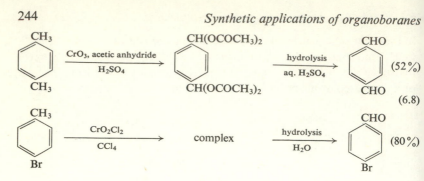

aldehyde in acidic media. The aldehyde group is not oxidised further and, in a polymethyl compound only one methyl group is oxidised under normal conditions (Syper, 1966). Mesitylene, for example, gave 3,5-dimethylbenzaldehyde quantitatively, and acetyl-*p*-toluidine was converted into *p*-acetamidobenzaldehyde (94 per cent).

Unsaturated hydrocarbons. Oxidation of unsaturated hydrocarbons may take place both at the double bond and at the adjacent allylic positions and important synthetic reactions of both types, involving oxidations with permanganate, per-acids, ozone and other reagents are discussed below. With chromic acid, mixtures of products are often produced, and for this reason oxidation of olefinic compounds with chromic acid is of limited value in synthesis. For example, with chromic acid in acetic acid cyclohexene gives a mixture of cyclohexenone and adipic acid, although with straight chain alkenes allylic oxidation appears to be a relatively minor reaction. With acetic acid as solvent epoxides are also frequently produced, and with aqueous acid more complex mixtures including rearranged products may result (Wiberg, 1965).

6.2. Oxidation of alcohols

Chromic acid. One of the most important uses of chromic acid in synthesis is in the oxidation of alcohols, and particularly in the oxidation of secondary alcohols to ketones. This reaction is commonly effected with a solution of the alcohol and aqueous acidic chromic acid in acetic acid, or with aqueous acidic chromic acid in a heterogeneous mixture. If no complicating structural features are present high yields of ketone are usually obtained. In some cases, particularly with phenyl alkyl carbinols, small amounts of cleavage products may be formed as well (6.9).

Oxidation of primary alcohols to aldehydes with acidic solutions of

$$C_6H_5.CHOH.CH\diagdown\begin{matrix}CH_3\\CH_3\end{matrix} \xrightarrow[\text{CH}_3\text{CO}_2\text{H, 30°C}]{\text{CrO}_3,\ \text{H}_2\text{O}} C_6H_5.CO.CH\diagdown\begin{matrix}CH_3\\CH_3\end{matrix} \quad (6.9)$$

$$+ C_6H_5.CHO \quad (6\%)$$

chromic acid is usually less satisfactory because the aldehyde is easily oxidised further to the carboxylic acid and, more importantly, because under the acidic conditions the aldehyde reacts with unchanged alcohol to form a hemiacetal which is rapidly oxidised to an ester (6.10). Satis-

$$n\text{-}C_3H_7.CH_2OH \xrightarrow[\substack{\text{H}_2\text{O, H}_2\text{SO}_4\\<20°\text{C}}]{\text{Na}_2\text{Cr}_2\text{O}_7} n\text{-}C_3H_7.CHO$$

$$\Big\updownarrow n\text{-}C_3H_7.CH_2OH \qquad (6.10)$$

$$n\text{-}C_3H_7.CO.OC_4H_9\text{-}n \longleftarrow n\text{-}C_3H_7.CHOH.OC_4H_9\text{-}n$$

factory yields of aldehyde can be obtained in favourable cases by removing the aldehyde from the reaction medium by distillation as it is formed or, better, by use of the pyridine–chromium trioxide complex or of t-butyl chromate as oxidising agent. The latter reagent is prepared by careful addition of chromium trioxide to t-butanol and is reported to give excellent yields of aldehydes by oxidation of primary alcohols in petroleum ether solution (Wiberg, 1965) (6.11).

$$CH_3.CH_2.CH_2OH \xrightarrow[\substack{\text{H}_2\text{O, H}_2\text{SO}_4\\\text{boil}}]{\text{K}_2\text{Cr}_2\text{O}_7} CH_3.CH_2.CHO \quad (49\%)$$

removed by continuous
distillation

$$C_{15}H_{31}.CH_2OH \xrightarrow[\substack{\text{petroleum ether}\\\text{t-butanol, }<60°\text{C}}]{\text{t-butyl chromate}} C_{15}H_{31}.CHO \quad (94\%)$$

$$(6.11)$$

$$\xrightarrow[\substack{\text{petroleum ether}\\\text{t-butanol}}]{\text{t-butyl chromate}} \qquad (85\%)$$

In general tertiary alcohols are unaffected by chromic acid, but tertiary 1,2-diols are rapidly cleaved, provided they are sterically capable of forming cyclic chromate esters. *cis*-1,2-Dimethylcyclopentandiol, for example, is oxidised 17×10^3 times faster than the *trans* isomer (6.12).

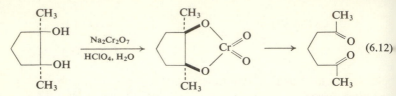

(6.12)

Oxidation of alcohols by chromic acid is believed to take place by initial formation of a chromate ester, followed by breakdown of the ester, as shown for oxidation of isopropanol. Whether proton abstraction from the ester takes place by an intermolecular or intramolecular process is uncertain (6.13).

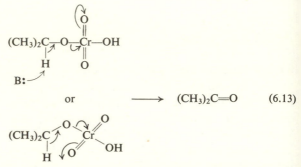

(6.13)

With unhindered alcohols the initial reaction to form the chromate ester is fast, and the subsequent decomposition of the ester is the rate-controlling step. But in cases where formation of the ester results in steric overcrowding, ester decomposition is accelerated because steric strain is relieved in going from reactant to product. In extreme cases the initial esterification may become rate-determining. In the cyclohexane series it is found that axial hydroxyl groups are generally oxidised more rapidly than equatorial by a factor of about 3, presumably because of 1,3-diaxial interactions in the axial ester. This has been used in the determination of configurations of steroidal alcohols. Relative oxidation rates for a number of epimeric alcohols are given in (6.14). In the borneol–isoborneol pair, isoborneol is oxidised more rapidly because the hydroxyl group, and even more so the chromate ester, is sterically crowded by the *gem*-dimethyl group. In norborneol, however, the *gem*-dimethyl group is absent and the *endo*-compound now has the more hindered hydroxyl group and reacts more rapidly.

Oxidation with acid solutions of chromic acid is unsuitable for alcohols which contain acid-sensitive groups or other easily oxidisable groups

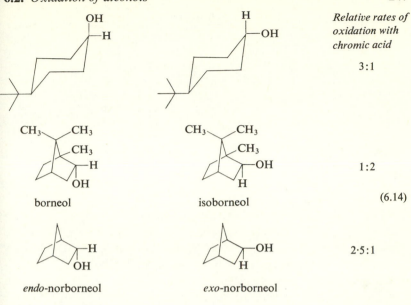

Relative rates of oxidation with chromic acid

3:1

borneol · isoborneol · 1:2

(6.14)

endo-norborneol · *exo*-norborneol · 2·5:1

such as olefinic bonds or allylic or benzylic C–H bonds elsewhere in the molecule. A milder method which often allows selective oxidation of a hydroxyl group in such molecules is by dropwise addition of the stoichiometric amount of a solution of chromium trioxide in aqueous sulphuric acid (the Jones reagent) to a cooled (0–20°C) solution of the alcohol in acetone (Bowers, Halsall, Jones and Lemin, 1953). Over-oxidation is thus lessened or prevented, and selective oxidation of unsaturated secondary alcohols to unsaturated ketones without appreciable oxidation or rearrangement of double bonds can often be achieved in good yield. Primary alcohols may give either aldehydes or carboxylic acids. In many cases the 'end point' is easily observed by the persistence of the red colour of the chromic acid after addition of the theoretical amount of oxidant (6.15).

Chromium trioxide–pyridine complex. A useful reagent for oxidation of alcohols that contain acid-sensitive functional groups is the chromium trioxide–pyridine complex $CrO_3.2C_5H_5N$ (Poos, Arth, Beyler and Sarett, 1953). It is prepared by addition of chromium trioxide to pyridine (caution: never the other way round) to form a pyridine solution of the complex, or, more simply, by adding a concentrated aqueous solution of chromium trioxide to pyridine. Reaction is effected at room temperature in pyridine solution by addition of the alcohol to the

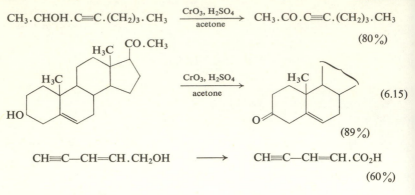

$$CH_3.CHOH.C\equiv C.(CH_2)_3.CH_3 \xrightarrow[\text{acetone}]{CrO_3, H_2SO_4} CH_3.CO.C\equiv C.(CH_2)_3.CH_3$$

(80%)

(6.15)

(89%)

$$CH\equiv C-CH=CH.CH_2OH \longrightarrow CH\equiv C-CH=CH.CO_2H$$

(60%)

previously prepared reagent. The complex is apparently still reactive enough to esterify a primary or secondary alcoholic group, but not to attack double bonds or other weakly electron-donating groups. Primary and, particularly, secondary alcohols are converted into the corresponding carbonyl compounds in good yield. Acid-sensitive protecting groups are unaffected. The tricyclic diol (I), for example, is converted into the dione without cleavage of the ketal, migration of the double bond or epimerisation at C-4a, (6.16). Polyhydroxy compounds can sometimes

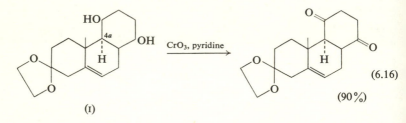

(6.16)

(90%)

(I)

be selectively oxidised at one position by protection of the other hydroxyl groups by acetal formation, as in the example (6.17), (Cornforth, Cornforth and Popják, 1962).

Another very useful application of the reagent is in the oxidation of allylic or benzylic primary alcohols to the corresponding aldehydes. Thus, cinnamyl alcohol is smoothly converted into cinnamyl aldehyde at room temperature, and nerol gives neral. Yields of aldehydes from non-allylic primary alcohols by the original procedure were capricious, but a recent modification by Collins, Hess and Frank (1968) gives high yields in a rapid reaction at room temperature. He made and isolated separately the complex $CrO_3.2C_5H_5N$ and effected oxidation with a

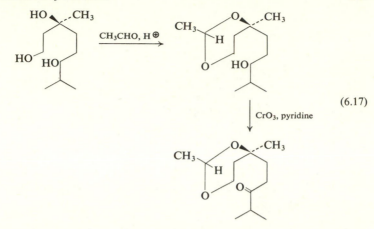

(6.17)

solution of this material in methylene chloride under anhydrous conditions. n-Heptanol gave n-heptanal under these conditions in 93 per cent yield, and 5,9-dimethyldeca-5,9-dienal was obtained in 92 per cent yield from the corresponding alcohol.

Similar oxidations have been effected with solutions of chromium trioxide in dimethylformamide, and this reagent is said to be superior to chromium trioxide in pyridine for some reactions (Snatzke, 1961).

Manganese dioxide and argentic oxide. Another useful mild reagent for the oxidation of primary and secondary alcohols to carbonyl compounds is manganese dioxide (Evans, 1959). The advantage of this reagent is that it is specific for allylic and benzylic hydroxyl groups, and reaction takes place under mild conditions (room temperature) in a neutral solvent (water, benzene, petroleum ether, chloroform). The general technique is simply to stir a solution of the alcohol in the solvent with the manganese dioxide for some hours. The manganese dioxide has to be specially prepared to obtain maximum activity. The best method appears to be by action of manganous sulphate with potassium permanganate in alkaline solution; the hydrated manganese dioxide obtained is highly active, but whether the actual oxidising agent is manganese dioxide itself or some other manganese compound adsorbed on the surface of the dioxide is not clear at present.

Olefinic and acetylenic bonds are unaffected by the reagent as illustrated in the examples (6.18), and the reaction has been widely used for oxidation of polyunsaturated alcohols in the carotenoid and vitamin A series. Hydroxyl groups adjacent to acetylenic bonds or cyclopropane rings are also easily oxidised, but under ordinary conditions saturated

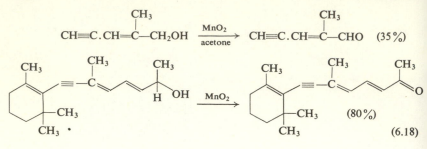

alcohols are not attacked (although they may be under more vigorous conditions), allowing selective oxidation of activated hydroxyl groups in appropriate cases (6.19).

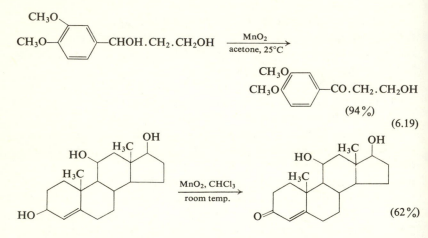

Although in general oxidation of allylic primary alcohols with manganese dioxide takes place without significant further oxidation to carboxylic acids, it has recently been found that, in presence of cyanide ions and an alcohol, very high yields of carboxylic esters can be obtained from the corresponding aldehydes (Corey, Gilman and Ganem, 1968). Thus, in methanol solution, cinnamaldehyde is converted into methyl cinnamate in >95 per cent yield, and geranial gives methyl geranate in 85–95 per cent yield. Reaction is thought to proceed through the cyanhydrin (6.20). An important feature of the reaction is that oxidation takes place without any *cis-trans* isomerisation of the $\alpha\beta$-olefinic bond. The traditional method of oxidising an aldehyde to a carboxylic acid using alkaline silver(I) oxide (Ag_2O) is relatively unsatisfactory for $\alpha\beta$-

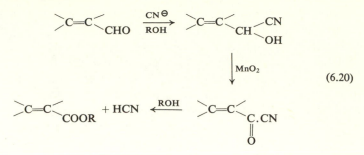

(6.20)

unsaturated aldehydes, since appreciable *cis-trans* αβ-isomerisation and other base-catalysed side reactions can occur.

Non-conjugated aldehydes are not converted into esters by the action of cyanide ion and manganese dioxide in methanol, even though cyanhydrin formation occurs readily. But with the more powerful reagent argentic oxide, AgO, prepared by action of alkaline permanganate on silver nitrate, oxidation to the corresponding *carboxylic acid* takes place readily. Thus, cinnamaldehyde with argentic oxide and sodium cyanide in methanol led to cinnamic acid in >90 per cent yield, and the non-conjugated aldehyde cyclohexenyl-3-carboxaldehyde was smoothly transformed into cyclohexenyl-3-carboxylic acid. Cinnamaldehyde cyanhydrin was likewise converted into cinnamic acid, and the catalytic effect of cyanide ions was indicated by the fact that little oxidation of cinnamaldehyde to the acid occurred in the absence of cyanide. The difference between the two reagents, manganese dioxide and argentic oxide, may be due to the presence of nucleophilic hydroxide or oxide ions on the surface of the argentic oxide, which strongly catalyse heterogeneous hydrolysis of the acyl cyanide in this system.

An even simpler method for the conversion of aldehydes into carboxylic acids, which is especially applicable to non-conjugated aldehydes, is oxidation with argentic oxide in tetrahydrofuran–water at 25°C under neutral conditions. Dodecenal and cyclohexenyl-3-carboxaldehyde, for example, were oxidised to the corresponding carboxylic acids in >90 per cent yield by this method. The oxidation of conjugated aldehydes is slower under these conditions, and cyanide-catalysed oxidation is preferable in such cases.

Oxidation of αβ-unsaturated aldehydes to αβ-unsaturated carboxylic acids can also be effected in high yield with hydrogen peroxide and selenium dioxide in t-butanol solution (Smith and Holm, 1957).

Silver carbonate. An excellent reagent for oxidising primary and secondary alcohols to aldehydes and ketones in high yield under mild and essentially neutral conditions is silver carbonate precipitated on celite (Fétizon and Golfier, 1968). The reaction is easily effected in boiling benzene and the product is recovered, usually in a high state of purity, by simply filtering off the spent reagent and evaporating off the solvent. Other functional groups are unaffected. Under these conditions nerol, for example, is converted into neral in 95 per cent yield. Highly hindered hydroxyl groups are not attacked, allowing selective oxidation in appropriate cases as in the first example below, where attack on the C-6 hydroxyl group is prevented by the steric effect of the *gem*-dimethyl group at C-4. Primary alcohols are oxidised more slowly than secondary which are themselves much less reactive than benzylic and allylic alcohols, and in acetone or methanol solution selective oxidation of benzylic or allylic hydroxyl groups can be effected (6.21).

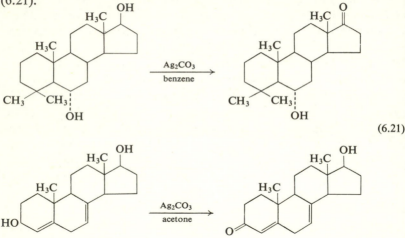

(6.21)

The behaviour of diols is different and depends on their structure. Butan-1,4-diols, pentan-1,5-diols and hexan-1,6-diols are converted, in an interesting reaction, into the corresponding γ-, δ- and ϵ-lactones (6.22). Other diols give hydroxy-ketones or -aldehydes. Vicinal diols are oxidised to α-hydroxy-ketones and β-glycols give β-hydroxy-ketones, usually in high yield (6.23). Butan-1,3-diol forms 1-hydroxybutan-3-one exclusively; no significant amount of hydroxy-aldehyde is found, in line with the observation that secondary alcohols are oxidised more rapidly than primary with this reagent. Hydroxy-aldehydes are obtained from hexan-1,7-diol and octan-1,8-diol (6.24).

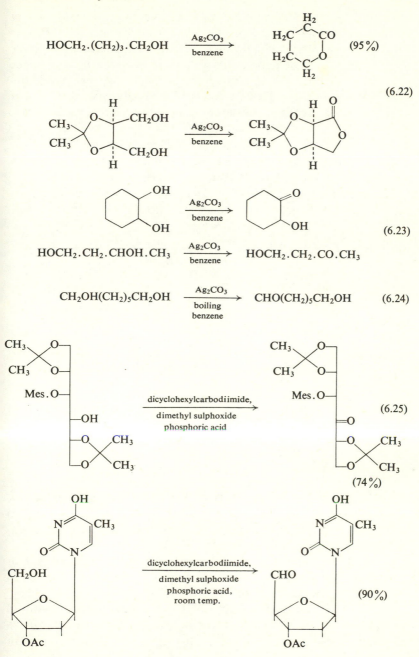

Dimethyl sulphoxide. Another convenient method for oxidation of primary and secondary alcohols to aldehydes and ketones is by reaction with dicyclohexylcarbodiimide and dimethyl sulphoxide in presence of phosphoric acid or pyridinium trifluoroacetate as a proton source (Epstein and Sweat, 1967). Conditions are mild and high yields are obtained in many cases. This method has been used to oxidise a number of sensitive compounds in the steroid, alkaloid and carbohydrate series. Thus, the 'isolated' hydroxyl group of the mannitol derivative in (6.25) was smoothly oxidised to the ketone in 74 per cent yield, and 3'-*O*-acetyl-thymidine was converted into the 5'-aldehyde in 90 per cent yield, with no trace of the carboxylic acid.

The mechanism of the oxidation has been elucidated by tracer experiments and is thought to involve initial formation of a sulphoxide–

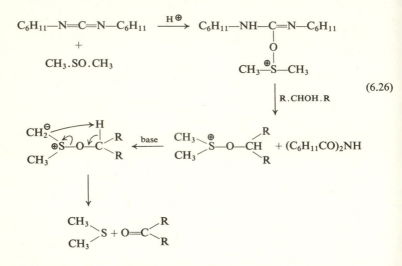

(6.26)

carbodiimide adduct which reacts with the alcohol to give an alkoxy-sulphonium ion. This then undergoes proton abstraction to form an ylid which collapses to the ketone and dimethyl sulphide by an intra-molecular concerted process (6.26).

A related method uses dimethyl sulphoxide and acetic anhydride or phosphorus pentoxide to oxidise primary and secondary alcohols to carbonyl compounds. Here the acetic anhydride acts as the 'activating' group. The mechanism established by tracer studies again involves an ylid (6.27). Methylthiomethyl ethers formed by alternative breakdown of the acyloxysulphonium ion are often obtained as by-products in

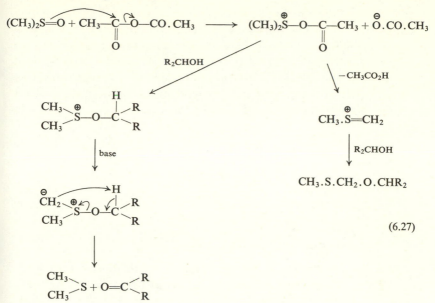

(6.27)

oxidations with this reagent. Thus testosterone gave 70 per cent of androsten-3,17-dione and 30 per cent of the methythio compound (6.28). The method is, however, widely used for oxidation of hydroxyl groups in the carbohydrate series (Brimacombe, 1969).

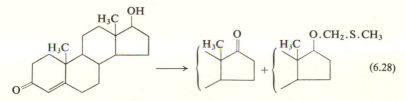

(6.28)

Related to these reactions is the oxidation of alkyl halides and toluene-*p*-sulphonates to carbonyl compounds with dimethyl sulphoxide. This useful reaction is simply effected by dissolving the halide or sulphonate in dimethyl sulphoxide, generally in presence of a proton acceptor such as sodium bicarbonate or collidine. In most cases reaction proceeds readily at room temperature. Oxidation never proceeds beyond the carbonyl stage and other functional groups are generally unaffected (Epstein and Sweat, 1967). The reaction has been applied to phenacyl halides, benzyl halides, primary sulphonates and iodides and a limited number of secondary sulphonates. Primary bromides and chlorides give

only poor yields, but may be converted *in situ* into the corresponding
sulphonates and oxidised without purification. With secondary sulphonates and halides elimination becomes an important side reaction and the
reaction is less useful with such compounds. Where elimination is not
possible good yields of ketones are obtained. It is thought that this
reaction also probably proceeds through an ylid, but direct proton
abstraction from the alkoxysulphonium ion may operate in some cases
(6.29).

$$(CH_3)_2S=O + R_2CHX \longrightarrow (CH_3)_2\overset{\oplus}{S}-O-CR_2$$
$$\underset{H}{|}$$

(6.29)

$$\overset{\ominus}{CH_2}\overset{\oplus}{\underset{CH_3}{S}}-O-CR_2 \longrightarrow \overset{CH_3}{\underset{CH_3}{>}}S + O=C\overset{R}{\underset{R}{<}}$$

Other methods. Among a number of other useful methods for selective
oxidation of primary and secondary alcohols to aldehydes and ketones
under mild conditions are the Oppenauer oxidation, oxidation with lead
tetra-acetate and catalytic oxidation with oxygen and platinum.

The Oppenauer oxidation (Djerassi, 1951) with aluminium alkoxides
in acetone is the reverse of the Meerwein–Pondorff–Verley reduction
discussed on p. 352. It has been widely used in the steroid series, particularly for the oxidation of allylic secondary hydroxyl groups to $\alpha\beta$-
unsaturated ketones (6.30). $\beta\gamma$-Double bonds generally migrate into
conjugation with the carbonyl group under the conditions of the reaction. The aluminium alkoxide serves only to form the aluminium salt

$$\overset{CH_3}{\underset{|}{}}$$
$$CH_3.CHOH.CH=CH.CH=C.CH=CH_2$$

Al t-butoxide, acetone,
boiling benzene

$$\overset{CH_3}{\underset{|}{}}$$
$$CH_3.CO.CH=CH.CH=C.CH=CH_2 \quad (80\%) \qquad (6.30)$$

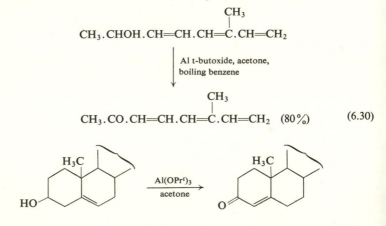

of the alcohol which is then oxidised through a cyclic transition state at the expense of the acetone. By use of excess of acetone the equilibrium is forced to the right (6.31).

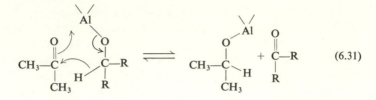

(6.31)

Lead tetra-acetate in refluxing benzene, hexane or chloroform is a good reagent for oxidation of primary and secondary alcohols to the corresponding aldehyde or ketone, provided there is no δ C–H group in the molecule. In this circumstance high yields of tetrahydrofuran derivatives are obtained (see p. 197). It has recently been found, however, that in pyridine solution lead tetra-acetate oxidises a variety of primary and secondary alcohols to the carbonyl compounds in good yield at room temperature whether they contain δ C–H groups or not. Both allylic and saturated alcohols are oxidised and cyclisation products appear not to be formed (Partch, 1964) (6.32).

$$CH_3.(CH_2)_3CH_2OH \xrightarrow[\text{pyridine}]{\text{Pb(OAc)}_4} CH_3(CH_2)_3CHO \quad (70\%)$$

$$CH_3CHOH.CH_2.CH_2.CHOH.CH_3 \longrightarrow CH_3CO.CH_2CH_2COCH_3 \quad (89\%)$$

$$C_6H_5.CH{=}CH.CH_2OH \longrightarrow C_6H_5.CH{=}CH.CHO \quad (91\%)$$

(6.32)

Catalytic oxygenation with a platinum catalyst and molecular oxygen is another valuable method for oxidation of primary and secondary hydroxyl groups under mild conditions. With primary alcohols the reaction can be regulated to give aldehydes or acids. Double bonds, in general, are not affected and unsaturated alcohols such as tiglic alcohol (2-methylbut-2-en-1-ol) can be oxidised catalytically to the unsaturated aldehyde (Heyns and Paulsen, 1963) (6.33).

$$CH_3(CH_2)_{10}CH_2OH \xrightarrow[\text{heptane} \atop \frac{1}{2}\text{h}]{O_2/Pt} CH_3(CH_2)_{10}CHO \quad (77\%)$$

(6.33)

$$\xrightarrow{2\text{ h}} CH_3(CH_2)_{10}COOH \quad (96\%)$$

$$CH_3.CH=C(CH_3)-CH_2OH \xrightarrow[\text{heptane}]{O_2/Pt} CH_3.CH=C(CH_3).CHO$$

In general, primary hydroxyl groups are attacked before secondary, and in cyclic secondary alcohols axial groups appear to react before equatorial. The method has been widely used in the carbohydrate series to effect selective oxidation of specific hydroxyl groups (Brimacombe, 1969). Thus, L-sorbose is oxidised at 30°C to 2-keto-L-gulonic acid, an intermediate in a synthesis of ascorbic acid (6.34).

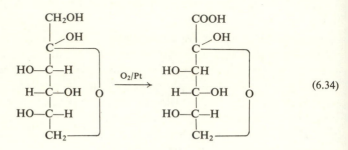

$$(6.34)$$

Other useful reagents are the *N*-halo-imides, *N*-bromosuccinimide and *N*-bromo- and *N*-chloro-acetamide. In aqueous acetone or aqueous dioxan they readily oxidise many primary and secondary alcohols to aldehydes and ketones (Filler, 1963). In general, secondary alcohols are oxidised more readily than primary, and among cyclic secondary alcohols axial hydroxyl groups are oxidised more easily than equatorial ones. This has led to the extensive use of these reagents for selective oxidation of nuclear hydroxyl groups in the steroid series. Thus, aetiocholan-$3\alpha,11\alpha,17\beta$-triol with *N*-bromo-acetamide in aqueous acetone is converted into aetiocholan-11α-ol-3,17-dione by oxidation of the two axial hydroxyl groups (6.35). By contrast, chromic acid gave the 3,11,17-trione. Benzylic or allylic alcohols are particularly easily oxidised.

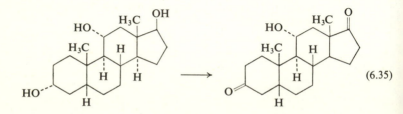

$$(6.35)$$

Steroidal allylic alcohols have also been selectively oxidised to the corresponding αβ-unsaturated ketones in excellent yield with the high-potential quinone 2,3-dichloro-5,6-dicyanobenzoquinone (6.36). Saturated alcohols are unaffected.

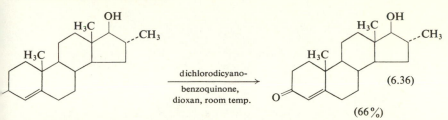

(6.36)

(66%)

Recently Rees and Storr (1969) have shown that 1-chlorobenzotriazole, which is readily obtained from benzotriazole with sodium hypochlorite, is an excellent reagent for the oxidation of alcohols to carbonyl compounds under mild conditions. Oxidation of alcohols to aldehydes and ketones in neutral solution at room temperature has also been effected in good yield with 4-phenyl-1,2,4-triazoline-3,5-dione (Cookson, Stevens and Watts, 1966) (6.37).

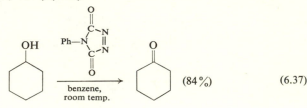

(84%)   (6.37)

6.3. Oxidation of olefinic bonds

Hydroxylation. Hydroxylation of olefinic double bonds

is useful in degradation and synthesis, and can be accomplished stereospecifically with a number of different reagents. For *cis*-hydroxylation the best methods are reaction with potassium permanganate, with osmium tetroxide or with iodine and moist silver acetate. The most important method of *trans*-hydroxylation is reaction with per-acids, but the Prévost reaction, that is the action of iodine and silver acetate under anhydrous conditions, is also useful.

The olefin may, of course, have either the *cis* or *trans* configuration

TABLE 6.1   *Relation between olefins of different configuration and diols produced by* cis- *and* trans-*hydroxylation*

Olefin	Hydroxylation	
	cis	*trans*
cis	*cis, meso, erythro*	*trans, ±, threo*
trans	*trans, ±, threo*	*cis, meso, erythro*

and by *cis*- or *trans*-hydroxylation can give rise to isomeric diols which may be described by the terms *cis* or *trans*, *erythro* or *threo*, or *meso* or (±), depending on the nature of the other groups attached to the glycol system. The relation between the olefins and the diols produced is shown in Table 6.1.

Potassium permanganate. Oxidation with permanganate is a widely used method for *cis*-hydroxylation of olefins, but needs careful control

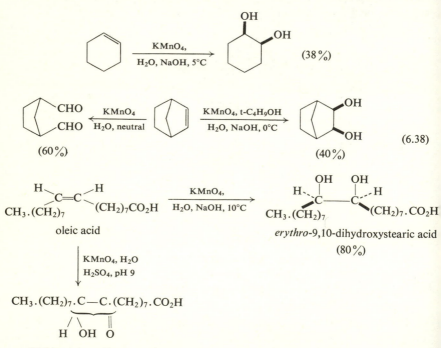

erythro-9,10-dihydroxystearic acid
(80%)

(75%, mixture of two products)

to avoid over-oxidation. Best results are obtained in alkaline solution, using water or aqueous organic solvents (acetone, ethanol or t-butanol); in acid or neutral solution α-ketols or even cleavage products are formed. The method is particularly suitable for hydroxylation of unsaturated acids, which dissolve in the alkaline solution (6.38).

These reactions are believed to proceed through the formation of cyclic manganese esters. The mode of addition of the hydroxyl groups is shown to be *cis* by the conversion of maleic acid into *meso*-tartaric acid and of fumaric acid into (±)-tartaric acid, and the cyclic ester mechanism is supported by studies with ^{18}O which show that transfer of oxygen from permanganate to the substrate occurs. Competition between ring-opening of the cyclic ester by hydroxyl ion and further oxidation by permanganate accounts for the effect of pH on the distribution of products (Stewart, 1965) (6.39).

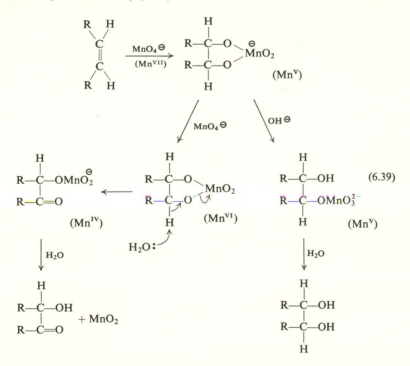

Osmium tetroxide. Reaction with osmium tetroxide is another good method for *cis*-hydroxylation of double bonds, but is suitable only for small scale work with valuable compounds because of the expense and toxicity of the reagent; for example it has been much used in the steroid hormone series (Gunstone, 1960; Fieser and Fieser, 1967) (6.40). Some

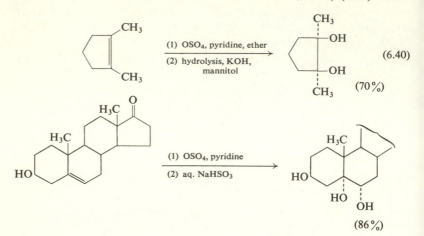

polycyclic aromatic hydrocarbons are also attacked (6.41). Reaction is accelerated by tertiary bases, especially pyridine, and pyridine is often added to the reaction medium. Brightly coloured complexes in which

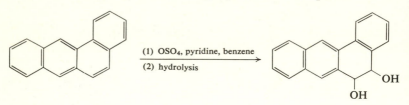

osmium is co-ordinated with two molecules of base separate in almost quantitative yield (6.42). Hydrolysis of the osmate esters to the diols

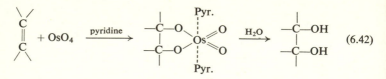

is effected by treatment with sodium hydroxide and mannitol, sodium sulphite or hydrogen sulphide.

Osmium tetroxide has also been used as a catalyst in conjunction with other oxidising agents, notably with chlorates and with hydrogen peroxide. Thus, the furan–maleic anhydride adduct on treatment with hydrogen peroxide in t-butanol in presence of a catalytic amount of osmium tetroxide (the Milas reagent) gave the *exo*-diol in 50 per cent yield. Similarly pent-2-enoic acid, with aqueous barium chlorate and a

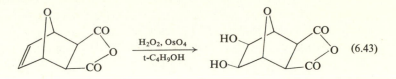

(6.43)

catalytic amount of osmium tetroxide was converted into the corresponding diol. In these reactions the initial osmate ester is oxidatively hydrolysed by the oxidising agent with regeneration of osmium tetroxide which continues the reaction, so that a small amount suffices.

Oxidation with iodine and silver carboxylates. Many of the difficulties attending the oxidation of olefins to 1,2-glycols with other reagents can be avoided by using Prévost's reagent – a solution of iodine in carbon tetrachloride together with an equivalent of silver acetate or silver benzoate. Under anhydrous conditions this oxidant directly yields the diacyl derivative of the *trans* glycol (Prévost conditions), while in presence of water the mono-ester of the *cis* glycol is obtained (Woodward conditions). Thus, cyclohexene on treatment with iodine and silver benzoate in boiling carbon tetrachloride under anhydrous conditions gives the dibenzoate of *trans*-1,2-dihydroxycyclohexane. With iodine and silver acetate in moist acetic acid, however, the mono-acetate of *cis*-1,2-dihydroxycyclohexane is formed. Similarly, oleic acid on oxidation under Prévost conditions gives *threo*-9,10-dihydroxystearic acid, while by the Woodward procedure the *erythro*-isomer results (Gunstone, 1960) (6.44).

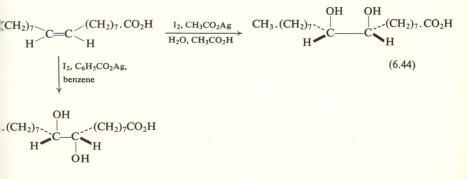

(6.44)

The value of these reagents is due to their specificity and to the mildness of the reaction conditions; free iodine, under the conditions used, hardly affects other sensitive groups in the molecule. Reaction proceeds through formation of an iodonium cation which, in presence of acyloxy and silver ions, forms the resonance stabilised cation (II) (6.45). Attack on the cation by acetate ion in a bimolecular process gives the *trans*-diacyl compound. In presence of water, however, a hydroxy acetal is formed which on hydrolysis affords the *cis*-hydroxy-acyloxy compound. With conformationally rigid molecules the *cis* diol obtained

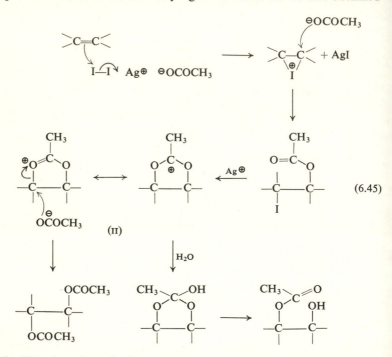

(6.45)

by the Woodward method may have the opposite configuration from that obtained with osmium tetroxide. Thus, with the tricyclic compound (III) oxidation with osmium tetroxide gives the α-*cis* glycol, whereas with silver acetate and iodine in aqueous acetic acid the β-*cis* glycol is formed, through initial attack of I⊕ from the least hindered side of the double bond (6.46).

Oxidation with per-acids. Oxidation of olefins with per-acids gives rise to epoxides or to *trans* 1,2-glycols, depending on the experimental conditions (Swern, 1953).

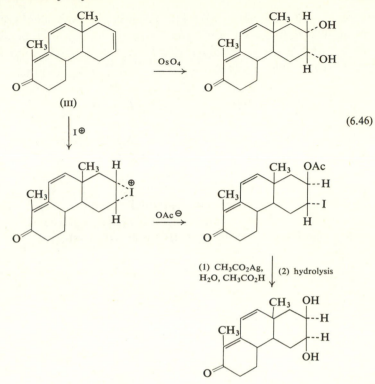

(6.46)

The per-acids most commonly used in the laboratory are perbenzoic, perphthalic, performic, peracetic and trifluoroperacetic acid. Recently, *m*-chloroperbenzoic acid has come into use; it is commercially available and is an excellent reagent for epoxidation of olefinic double bonds. The other reagents are rather unstable and are generally prepared freshly before use. Perbenzoic acid is obtained from benzoyl peroxide and sodium methoxide or by reaction of benzoic acid with hydrogen peroxide in methanesulphonic acid; the sulphonic acid serves both as the solvent and an acid catalyst. Performic and peracetic acids are often prepared *in situ*, and not isolated, by action of hydrogen peroxide on the acids.

$$R.COOH + H_2O_2 \rightleftharpoons H_2O + R.CO_3H$$

With acetic acid the equilibrium is attained only slowly, and sulphuric acid is usually added as a catalyst to hasten formation of the per-acid. Solutions prepared in this way from acetic acid and formic acid are widely used for the preparation of *trans*-glycols from olefins. Trifluoro-peracetic acid, the most powerful of the common reagents, is obtained mixed with trifluoroacetic acid by action of hydrogen peroxide on trifluoroacetic anhydride (6.47).

$$(CF_3.CO)_2O + H_2O_2 \xrightarrow{CH_2Cl_2} CF_3.CO_2H + CF_3.CO_3H \qquad (6.47)$$

It is probable that reaction of all per-acids with olefins gives rise to the epoxide in the first place, but unless proper conditions are chosen the epoxide may be converted directly into an acyl derivative of the α-glycol. The best reagents for the preparation of the epoxide appear to be the three aromatic per-acids, although solutions of peracetic acid buffered with sodium acetate to neutralise the sulphuric acid used in their preparation, are also frequently employed (6.48).

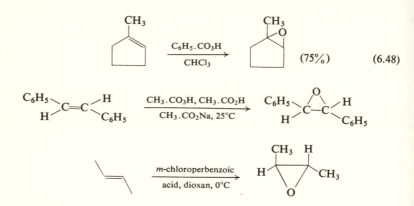

Reaction is believed to take place by electrophilic attack of the per-acid on the double bond, as illustrated in the equation (6.49), (Lee

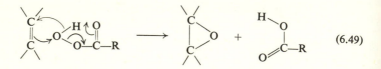

and Uff, 1967). In accordance with this mechanism the rate of epoxi-dation is increased by electron-withdrawing groups in the per-acid (for example, trifluoroperacetic acid is more reactive than peracetic acid) or electron-donating substituents on the double bond. Terminal mono-olefins react only slowly with most per-acids, but the rate of reaction increases with the degree of alkyl substitution. 1,2-Dimethylcyclohexa-1,4-diene for example, reacts entirely at the tetrasubstituted double bond (6.50). On the other hand, conjugation of the olefin with other

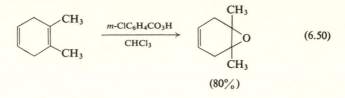

(6.50)

(80%)

unsaturated groups reduces the rate of epoxidation because of delocalisation of the π electrons. $\alpha\beta$-Unsaturated acids or esters, for example, require the strong reagent trifluoroperacetic acid for successful reaction. These reactions are carried out in presence of a buffer to prevent hydrolysis of the oxide by trifluoroacetic acid (6.51). With $\alpha\beta$-unsaturated

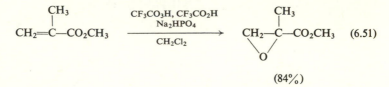

(6.51)

(84%)

ketones (6.52), reaction is complicated by simultaneous, or even preferential, Baeyer–Villiger type oxidation at the carbonyl group (see p. 287).

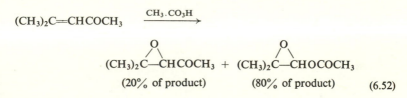

(20% of product)   (80% of product)   (6.52)

Epoxides of $\alpha\beta$-unsaturated aldehydes and ketones are best made by action of nucleophilic reagents such as hydrogen peroxide or t-butyl hydroperoxide in alkaline solution. The reaction is believed to take the course shown in (6.53). Thus, acrolein is readily converted into glycidaldehyde by this method, and cyclohexenone gives 2,3-epoxycyclohexanone.

Epoxidations with per-acids are highly stereoselective and take place by *cis* addition to the olefinic bond. This follows from the results of numerous experiments and has been shown unequivocally by X-ray

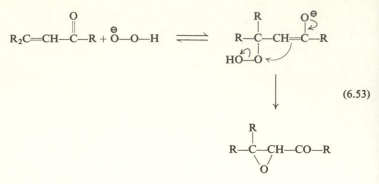

(6.53)

diffraction and infrared studies of the products obtained by epoxidation of oleic and elaidic acid with perbenzoic and peracetic acid. Thus, oleic acid gave *cis*-9,10-epoxystearic acid as shown (6.54), whereas elaidic acid gave the isomeric *trans* compound.

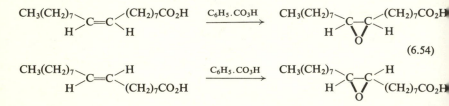

(6.54)

With cyclic olefins the reagent usually approaches from the least hindered side of the double bond, to give the least hindered epoxide, as illustrated in the epoxidation of norbornene shown in (6.55). But where

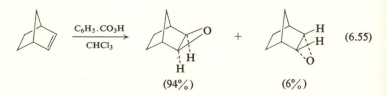

(6.55)

there is a polar substituent in the allylic position this may influence the direction of attack by the per-acid (Henbest, 1965). Thus, whereas cyclohex-2-enyl acetate gives mainly the *trans* epoxide as expected, the free hydroxy compound gives the *cis*-epoxide in 80 per cent yield under the same conditions (6.56). Similarly Δ^1-cholestene forms the α-epoxide while with Δ^1-cholestenol the β-epoxide is obtained. The rate of reaction is faster with the hydroxy compounds and it has been suggested

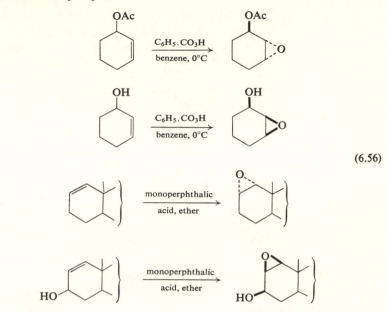

(6.56)

that hydrogen bonding causes association of the reactants in an orientation favourable for *cis* epoxidation.

Epoxides are synthetically useful because they react with a variety of reagents with opening of the epoxide ring (Parker and Isaacs, 1959; Dittius, 1965). Hydrolysis with dilute mineral acids affords *trans* 1,2-diols and with carboxylic acids mono-esters of *trans* 1,2-diols are obtained. Consequently, epoxidations with per-acids carried out in presence of an excess of the corresponding carboxylic acid or of a mineral acid, frequently yield hydroxy-esters or diols derived from the initially formed epoxide (6.57). This is especially the case for reactions run in formic acid and for reaction mixtures which contain a strong mineral acid.

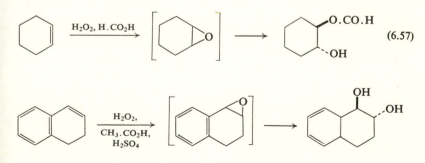

(6.57)

Another useful reaction of epoxides is reduction with lithium aluminium hydride to form alcohols (6.58). In general reaction takes place at the least substituted carbon atom to give the more substituted carbinol (see p. 360).

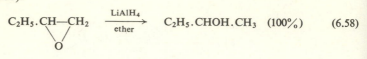

$$C_2H_5.CH\!-\!CH_2 \xrightarrow[\text{ether}]{\text{LiAlH}_4} C_2H_5.CHOH.CH_3 \quad (100\%) \qquad (6.58)$$

These ring-opening reactions are nucleophilic substitutions and take place in most cases with inversion of configuration at the carbon atom attacked, resulting in overall *trans* addition to the double bond. Thus hydrolysis with dilute mineral acid affords the *trans* glycol (6.59), and

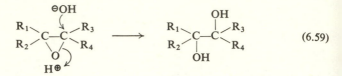

$$\tag{6.59}$$

the sequence of epoxidation followed by hydrolysis of the epoxide represents a widely used method for the preparation of *trans* diols from olefins, and is a convenient alternative to the Woodward–Prévost route discussed on p. 263. *trans*-1-Methylcyclopentan-1,2-diol, for example, is readily obtained by hydrolysis of 1-methylcyclopentene epoxide (6.60).

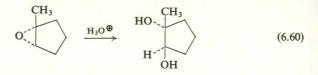

$$\tag{6.60}$$

Similarly reduction of 1,2-dimethylcyclopentene epoxide affords *trans*-1,2-dimethylcyclopentanol.

In conformationally rigid cyclohexane derivatives it is found that reactions leading to opening of the epoxide ring take place with formation of the *trans*-diaxial product, rather than the *trans*-diequatorial one, in preponderant amount. Thus, acid-catalysed hydrolysis of 2,3-epoxy-*trans*-decalin gives a 90 per cent yield of the *trans*-diaxial-diol as the only isolated product (6.61). Numerous other examples are known in the steroid series. Reaction of $2\alpha,3\alpha$-epoxycholestane with toluene-*p*-sulphonic acid, for example, gives the 3α-hydroxy-2β-tosyloxy derivative (diaxial) and reaction with hydrobromic acid leads to the diaxial bromohydrin (6.62).

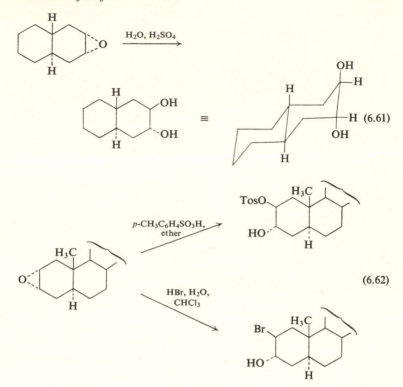

A few cases are known in which opening of the epoxide ring takes place with retention of configuration. It is thought that these reactions may involve a double inversion, as in the reaction of dypnone oxide with hydrochloric acid (6.63). In other cases either retention or inversion of configuration may be observed depending on the conditions.

Epoxidation of enol acetates gives rise to epoxyacetates which easily

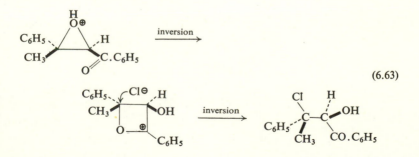

rearrange with heat or by chromatography on silica gel to α-acetoxy-ketones. The series of reactions provides a useful method for conversion of a ketone into its α-acetoxy-, and hence α-hydroxy-, derivative. Thus, α-tetralone is converted into 2-acetoxytetralone by the steps shown (6.64) and cholestanone similarly gives 2-acetoxycholestanone. Thermal

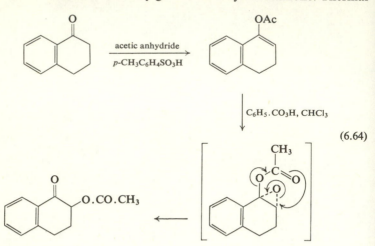

(6.64)

rearrangement of a mixture of 1-acetoxy-1,2-epoxy-4-methylcyclohexane and 1,2-epoxy-1-propionoxycyclohexane showed that they rearranged intramolecularly; no 'mixed product' was obtained.

Enol ethers can also be converted into epoxides, but in many cases these react further with the carboxylic acid in the reaction mixture to form α-acyloxy derivatives of the original ketones (Stevens and Dykstra, 1953) (6.65).

Another useful synthetic reaction of epoxides is their acid-catalysed

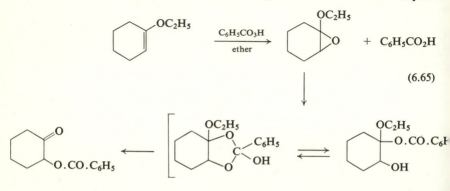

(6.65)

rearrangement to carbonyl compounds (Parker and Isaacs, 1959), and the reaction series provides a route for the conversion of olefins into carbonyl compounds (6.66). Mineral acids or Lewis acids such as boron

$$>C=C< \longrightarrow >\overset{\overset{\displaystyle O}{\diagup\!\!\!\!\diagdown}}{C-C}< \longrightarrow -\overset{|}{\underset{|}{C}}-C=O \qquad (6.66)$$

trifluoride etherate or magnesium bromide are frequently used as catalysts. In some cases products formed by rearrangement of the carbon skeleton are obtained. Thus, 1-methylcyclohexene epoxide gives 2-methylcyclohexanone or 1-formyl-1-methylcyclopentane depending on the experimental conditions (6.67). With magnesium bromide as catalyst *trans*-bromohydrins may also be formed. The method has been used

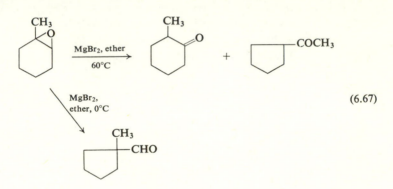

(6.67)

in synthesis for the conversion of cyclohexene derivatives into the corresponding cyclohexanones, as with the steroid epoxide shown in (6.68).

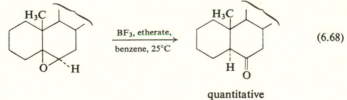

(6.68)

quantitative

Acetylenic bonds react much more slowly than olefinic bonds with per-acids and a conjugated enyne is attacked at the double bond only. This was exploited by Raphael (1949) in his synthesis of (±)-arabitol and (±)-ribitol, in which the *trans*-pentenynol (IV) was hydroxylated at the olefinic bond to give the *erythro*-diol (compare p. 260), and the acetylenic group was converted into a diol in another way, after semi-hydrogenation (6.69).

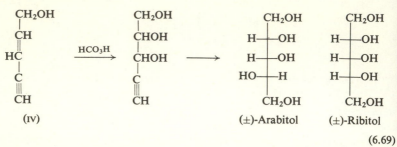

(6.69)

Ozonolysis. Ozonolysis, that is reaction of an olefin with ozone followed by splitting of the resulting ozonide, is a very convenient method for oxidative cleavage of an olefinic double bond.

Recent physical measurements have shown that the ozone molecule is a resonance hybrid of the structures shown in (6.70). It is an electro-

(6.70)

philic reagent and reacts with olefinic double bonds to form ozonides which can be cleaved oxidatively or reductively to carboxylic acids, ketones or aldehydes; the nature of the products formed depends on the method used and on the structure of the olefin (Bailey, 1958). The reaction is usually carried out by passing a stream of oxygen containing 2–10 per cent ozone into a solution or suspension of the compound in a suitable solvent, such as methylene chloride or methanol, at or below room temperature. Oxidation of the crude ozonisation product, without isolation, by hydrogen peroxide or other reagent, leads normally to carboxylic acids or ketones or both, depending on the degree of sub-stitution of the olefin (6.71). Thus oleic acid gives azelaic and pelargonic acids (6.72).

$$R\underset{H}{\overset{}{>}}C{=}C\underset{R^2}{\overset{R^1}{<}} \xrightarrow{O_3} \text{ozonide} \xrightarrow{H_2O_2} R.CO_2H + O{=}C\underset{R^2}{\overset{R^1}{<}}$$

(6.71)

$$R\underset{H}{\overset{}{>}}C{=}C\underset{H}{\overset{R^1}{<}} \xrightarrow{O_3} \text{ozonide} \xrightarrow{H_2O_2} R.CO_2H + HO_2C.R^1$$

$$CH_3.(CH_2)_7.CH{=}CH.(CH_2)_7.CO_2H$$

$$\downarrow \begin{array}{l} \text{(1) } O_3, \\ \text{(2) oxidation} \end{array} \qquad\qquad (6.72)$$

$$CH_3(CH_2)_7.CO_2H + HO_2C(CH_2)_7CO_2H$$

Reductive decomposition of the crude ozonide leads to aldehydes and ketones (6.73). Various methods of reduction have been used including

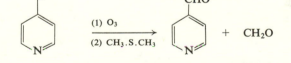

$$(6.73)$$

catalytic hydrogenation and reduction with zinc and acids or with triethyl phosphite, but, in general, yields of aldehydes have not been high. Recently, reaction with dimethyl sulphide in methanol has been found to give excellent results (Pappas and Keaveney, 1966) and this reagent appears to be superior to all others previously used. Reaction takes place under neutral conditions and the reagent is highly selective; nitro and carbonyl groups, for example, elsewhere in the molecule are

$$CH_2{=}CH(CH_2)_5CH_3 \xrightarrow{\ O_3\ } \text{ozonide}$$

$$\downarrow \begin{array}{l} CH_3.S.CH_3 \\ \text{methanol} \end{array}$$

$$CHO.(CH_2)_5.CH_3$$
$$(75\%)$$

$$(6.74)$$

not affected (6.74). The procedure hinges on the fact that hydroperoxides are rapidly and cleanly reduced to alcohols by sulphides (6.75). Ozonis-

$$(CH_3)_3C{-}O{-}OH \xrightarrow{(CH_3)_2S} (CH_3)_3C.OH + (CH_3)_2SO \qquad (6.75)$$

ation of an olefin in methanol solution gives rise to a hydroperoxide (see p. 277) which is reduced in the same way by dimethyl sulphide to the hemi-acetal as shown (6.76).

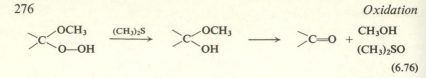

$$(6.76)$$

Criegee and Gunther (1963) have found recently that carbonyl compounds can be obtained directly from olefins by reaction of ozone with a 1:1 mixture of the olefin and tetracyanoethylene in acetic acid (6.77).

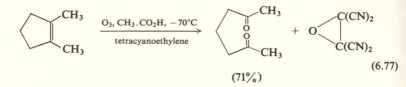

$$(71\%)$$

$$(6.77)$$

Similar results were obtained by reaction in methylene chloride solution in presence of a small amount of pyridine (Conia and Leriverend, 1960). Under these conditions cyclobutanone, for example, was obtained directly from methylenecyclobutane in high yield. In the steroid derivative shown in (6.78) selective attack on the side chain double bond was achieved by ozonisation in methylene chloride solution in presence of a small amount of pyridine.

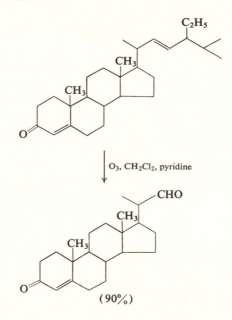

$$(6.78)$$

$$(90\%)$$

Largely through the work of Criegee (1957) it now seems clear that most normal ozonolyses proceed by formation of a primary ozonide of uncertain structure which decomposes to give a zwitterion and a carbonyl compound. The fate of the zwitterion depends on its structure, on the structure of the carbonyl compound and on the solvent. In an inert (non participating) solvent, if the carbonyl compound is reactive, it may react with the zwitterion to form an ozonide (v); otherwise the zwitterion may dimerise to the peroxide (vi) or give ill-defined polymers. In nucleophilic solvents such as methanol or acetic acid, however, hydroperoxides of the type of (vii) are formed (6.79). Strong evidence

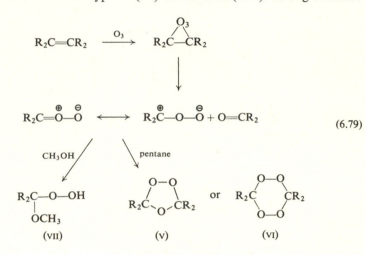

(6.79)

for the intermediacy of a zwitterion was found by Criegee in the ozonolysis of tetramethylethylene. In an inert solvent the cyclic peroxide and acetone were obtained. But when formaldehyde was added to the re-action mixture the known ozonide of isobutene was isolated. In the first case the intermediate zwitterion has dimerised; but in the second it has reacted preferentially with the highly reactive carbonyl compound (6.80). Criegee's general scheme is supported by much recent experi-mental work, although it appears that in some cases the exact mechanism is sensitive to a number of experimental factors (Murray, 1968; Fliszar and Carles, 1969).

Ozonolysis is widely used both in degradative work, for locating the position of a double bond in a molecule, and in synthesis, for the prep-aration of aldehydes, ketones and carboxylic acids from olefins. Thus, ω-aldehydic acids can be obtained from cycloalkanone enol ethers (6.81).

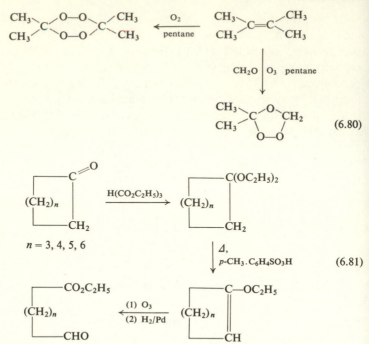

(6.80)

(6.81)

Another useful application is in the cleavage of cyclic ketones without loss of carbon by ozonolysis of the benzylidene or furfurylidene derivatives (6.82).

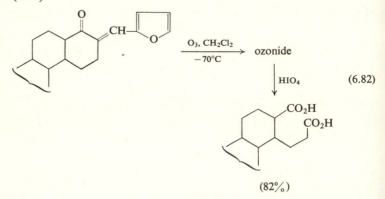

(6.82)

(82%)

With some substrates ozonolysis follows an 'abnormal' course. Olefins which are sterically hindered on one side of the double bond often give rise to epoxides, or rearrangement products thereof, containing the same number of carbon atoms as the starting material. In the example (6.83) the epoxide is formed, supposedly, by way of a π complex which, because of steric hindrance, can stabilise itself more easily by losing a molecule of oxygen than by collapsing to the 'normal' zwitterion.

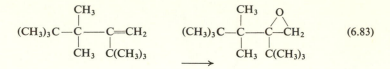

(6.83)

$\alpha\beta$-Unsaturated ketones or acids generally give products containing fewer than the expected number of carbon atoms. Thus, the tricyclic $\alpha\beta$-unsaturated ketone shown in (6.84) is converted into the keto-acid

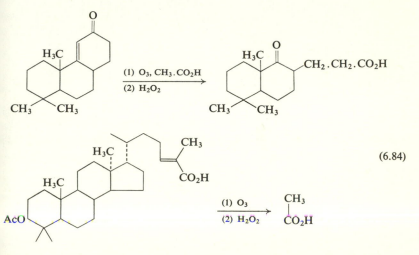

(6.84)

with loss of a carbon atom, and the triterpenoid carboxylic acid is degraded with formation of one molecule of acetic acid. The following general mechanism involving a release of electrons by O—H heterolysis has been suggested to account for these results (6.85). Abnormal results have also been observed in the ozonolysis of allylic alcohols, and amines (Bailey, 1958).

A useful alternative to ozonolysis is oxidation with periodate in presence of a catalytic amount of potassium permanganate (6.86). With this reagent double bonds are cleaved to give ketones and aldehydes or, more usually, carboxylic acids formed by further oxidation of the aldehydes (Lemieux and von Rudloff, 1955). Thus, citronellal is

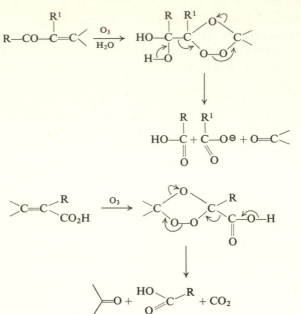

$$(6.85)$$

oxidised to acetone and 3-methyladipic acid in high yield, and the unsaturated alcohol shown (6.86), is cleaved without attack on the acetoxy

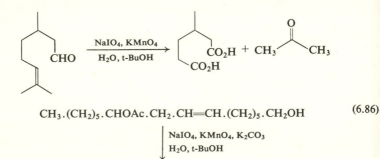

$$CH_3 . (CH_2)_5 . CHOAc . CH_2 . CH{=}CH . (CH_2)_5 . CH_2OH \qquad (6.86)$$

$$\downarrow \;\; \text{NaIO}_4, \text{KMnO}_4, \text{K}_2\text{CO}_3$$
$$\text{H}_2\text{O, t-BuOH}$$

$$CH_3(CH_2)_5 . CHOAc . CH_2 . CO_2H + HO_2C(CH_2)_5 . CH_2OH$$

and primary alcohol groups. Reaction is conducted at pH 7–8, and under these conditions the α-hydroxyketone or the 1,2-diol formed from the olefin with permanganate is cleaved by the periodate to carbonyl compounds. The permanganate is reduced only to the manganate stage at this pH and is re-oxidised by the periodate, which itself does not attack the double bond. Only catalytic amounts of permanganate are thus needed.

Similar results are obtained with a combination of periodate and osmium tetroxide, and this technique has the advantage that it does not proceed beyond the aldehyde stage. It produces the same result as ozonolysis followed by reductive cleavage of the ozonide. Cyclohexene, for example, is converted into adipaldehyde in 77 per cent yield. The osmium tetroxide oxidises the olefin to the diol, which is cleaved by the periodate. Catalytic amounts of osmium tetroxide are sufficient because the periodate oxidises the reduced osmium back to the tetroxide.

Another excellent method for cleaving olefinic double bonds is by action of ruthenium tetroxide in combination with sodium periodate (see below). Carboxylic acids are usually produced from disubstituted olefins with this reagent.

6.4. Oxidations with ruthenium tetroxide and nickel peroxide. Ruthenium tetroxide and nickel peroxide, two very powerful oxidising agents, have been introduced recently for oxidation of a variety of functional groups at room temperature. Reactions with ruthenium tetroxide are generally effected in solution in carbon tetrachloride (it attacks most other organic solvents) either with the pure tetroxide, which is expensive, or with a catalytic amount of tetroxide in presence of sodium periodate which serves to oxidise the reduced dioxide back to the active tetroxide (Piatak *et al.*, 1969). By either method secondary alcohols are oxidised smoothly to ketones in excellent yield at room temperature (6.87). Ketones are stable to further oxidation, but aldehydes are rapidly

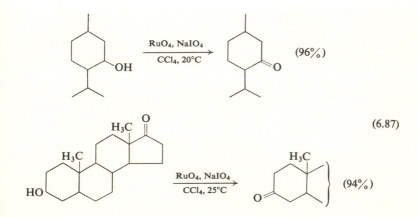

attacked, and oxidation of primary alcohols gives the corresponding carboxylic acids.

In the carbohydrate series the reagent has been used effectively for oxidation of secondary hydroxyl groups in suitably protected derivatives. Thus, 1,2:5,6-di-*O*-isopropylidene-α-D-*ribo*-hexofuranos-3-ulose was readily obtained from the alcohol (VIII) in 75 per cent yield (6.88).

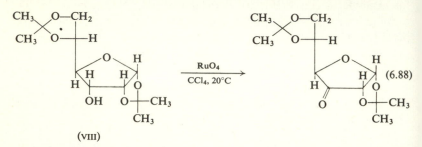

(VIII)

A remarkable reaction is the smooth conversion of ethers into esters or lactones. Thus, dibutyl ether gave n-butyl butyrate in quantitative yield and tetrahydrofuran similarly yielded butyrolactone. This reaction has been exploited in the steroid series to convert the ether (IX) into the corresponding lactone (6.89).

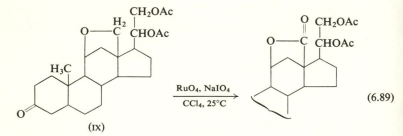

(IX)

Unlike osmium tetroxide, ruthenium tetroxide does not react with olefinic double bonds to form diols. Instead, cleavage takes place to give carboxylic acids or ketones, and oxidation with ruthenium tetroxide and sodium periodate provides a convenient alternative to ozonolysis. Ruthenium tetroxide–sodium periodate is found, for example, to be an excellent reagent for cleavage of the diphenylethylenes obtained in the Barbier–Wieland degradation sequence (Stork, Meisels and Davies, 1963) (6.90).

αβ-Unsaturated ketones are degraded in the same way as with ozone. Thus, testosterone acetate gives the ring-A cleaved product (6.91), and

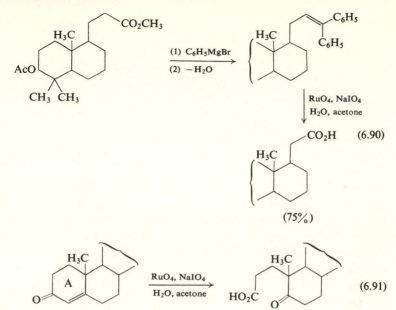

another excellent method for cleavage of cyclo-alkanones is illustrated in (6.92). Even benzene rings are attacked. Oestrone, for example, is

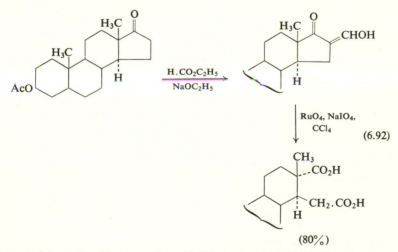

converted into the dicarboxylic acid (x) in good yield (6.93). Benzene itself is reported to react with ruthenium tetroxide with explosive violence.

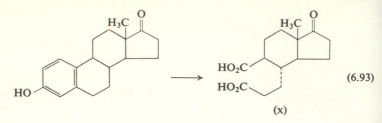

(6.93)

(x)

Nickel peroxide has also proved useful for the oxidation of a variety of functional groups. It is obtained as a black powder, insoluble in water and organic solvents, by action of alkaline sodium hypochlorite on nickel sulphate (Nakagawa, Konaka and Nakata, 1962). The composition of the reagent has been given as $NiO_{2.77}H_{2.85}$. It is generally used in a little excess of the stoichiometric amount and the 'available oxygen' is determined for each batch of reagent by iodometry. In many of its reactions it resembles manganese dioxide, but is a more powerful oxidant.

The oxidising power for alcohols of nickel peroxide is affected by the alkalinity of the solvent and the temperature. In aqueous alkaline solution saturated aliphatic primary alcohols are oxidised to the corresponding carboxylic acids in good yield (6.94). In some cases isolated olefinic double bonds are partially cleaved, but triple bonds appear to be unaffected. Under neutral conditions in organic solvents (benzene, ether)

$$CH_3.CH_2.CH_2.CH_2OH \xrightarrow[\text{30°C, H}_2\text{O}]{\text{Ni peroxide}} CH_3.CH_2.CH_2.CO_2H \quad (94\%)$$

$$C_6H_5.CH{=}CH.CH_2OH \xrightarrow[\text{H}_2\text{O, 50°C}]{\text{Ni peroxide}} C_6H_5.CH{=}CH.CO_2H \quad (70\%)$$

(6.94)

$$CH{\equiv}C.CH_2OH \xrightarrow[\text{H}_2\text{O, 5°C}]{\text{Ni peroxide}} CH{\equiv}C.CO_2H \quad (50\%)$$

allylic and benzylic alcohols are readily oxidised to aldehydes and ketones, and for this purpose the reagent is more active than manganese dioxide and has the advantage that only a slight excess is required to complete the reaction (6.95). Propynyl alcohols are also readily oxidised to aldehydes. But saturated aliphatic alcohols are largely unaffected under these conditions.

In aprotic solvents α-glycols and α-hydroxy-acids are cleaved to aldehydes and ketones, and α-keto-alcohols and acids similarly give carboxylic acids in aqueous alkaline solution.

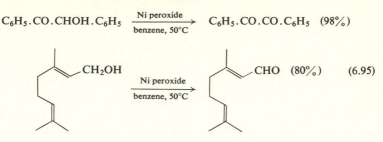

$$C_6H_5.CO.CHOH.C_6H_5 \xrightarrow[\text{benzene, 50°C}]{\text{Ni peroxide}} C_6H_5.CO.CO.C_6H_5 \quad (98\%)$$

By oxidation at −20°C in an ethereal solution of ammonia, aromatic and allylic aldehydes can be converted into amides in good yield. At higher temperatures yields of amide decrease and nitriles are the main products. The reaction is thought to take the pathway shown in (6.96).

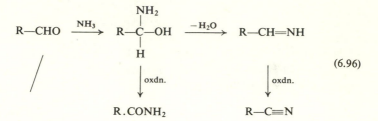

Thus, *p*-chlorobenzaldehyde is converted into *p*-chlorobenzamide or *p*-chlorobenzonitrile (6.97). Benzylic and allylic alcohols are similarly

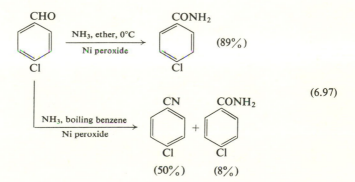

converted directly into amides or nitriles through preliminary oxidation to the aldehydes.

Other useful reactions which can be effected in good yield with nickel peroxide are the oxidative coupling of phenols and the oxidation of hydrazones to diazo compounds. Thus, 2,6-di-t-butylphenol is converted into tetra-t-butyldiphenoquinone, and benzophenone hydrazone gives diphenyldiazomethane. Studies using electron spin resonance spectroscopy have shown that oxidation of 2,6-di-t-butylphenol takes place by way of the phenoxy radical, and it is believed that all oxidations with nickel peroxide proceed by a free radical mechanism probably induced by free hydroxyl radicals on the catalyst surface (Konaka, Terabe and Kuruma, 1969). Isotopic studies suggest that the conversion of benzhydrol into benzophenone takes the course shown in (6.98). The

$$C_6H_5.CHOH.C_6H_5 \longrightarrow C_6H_5.\overset{\displaystyle\cdot}{C}.C_6H_5 \longrightarrow C_6H_5.\overset{\displaystyle\cdot}{C}.C_6H_5$$
$$\qquad\qquad\qquad\qquad\qquad\qquad |\qquad\qquad\qquad\qquad |$$
$$\qquad\qquad\qquad\qquad\qquad\qquad OH\qquad\qquad\qquad O\cdot$$

(6.98)

$$C_6H_5.CO.C_6H_5$$

formation of free radicals serves to explain the ready dimerisation of diphenylacetonitrile with nickel peroxide, and the conversion of chloroform into hexachloroethane (6.99).

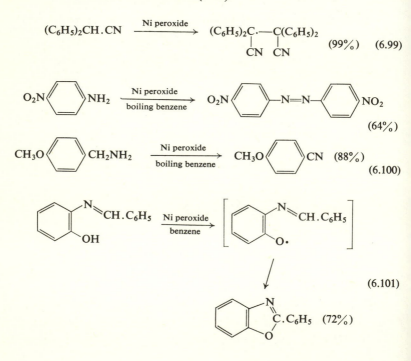

$$(C_6H_5)_2CH.CN \xrightarrow{\text{Ni peroxide}} (C_6H_5)_2C\!\cdot\!\!-\!\!C(C_6H_5)_2 \quad (99\%) \quad (6.99)$$
$$\qquad\qquad\qquad\qquad\qquad\qquad\qquad CN \quad CN$$

(64%)

CH₃O⟨⟩CN (88%)

(6.100)

(6.101)

(72%)

Other useful reactions effected by nickel peroxide include the conversion of aniline derivatives into azobenzenes, and of benzylamines and alkylamines into nitriles (6.100). In an interesting and useful cyclisation reaction *o*-hydroxy Schiff's bases are converted into 2-phenylbenzoxazoles (6.101).

6.5. Baeyer–Villiger oxidation of ketones. On oxidation with per-acids, ketones are converted into esters or lactones. This reaction was discovered in 1899 by Baeyer and Villiger who found that reaction of a number of alicyclic ketones, including menthone, with Caro's acid (permonosulphuric acid) led to the formation of lactones (6.102).

$$\text{menthone} \xrightarrow{\text{HO}_2\text{SO}_3\text{H}} \text{lactone} \qquad (6.102)$$

Better yields are obtained with organic per-acids such as peracetic acid in acetic acid containing sulphuric acid, or one of the stronger per-acids such as permaleic or trifluoroperacetic acid. With the latter acid a buffer such as disodium hydrogen phosphate is often employed to prevent transesterification by reaction of the newly formed ester with the trifluoroacetic acid always present in the reaction medium. The reaction occurs under mild conditions and has been widely used both in degradative work and in synthesis. It is applicable to open chain and cyclic ketones as well as to aromatic ketones, and has been used to prepare a variety of steroidal and terpenoid lactones, as well as medium and

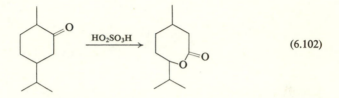

$$CH_3.CO.CH_2.CH_3 \xrightarrow[\text{Na}_2\text{HPO}_4, \text{CH}_2\text{Cl}_2]{\text{CF}_3.CO_3\text{H}, \text{CF}_3.CO_2\text{H}} CH_3.CO.OC_2H_5 \quad (72\%)$$

(6.103)

large ring lactones which are otherwise virtually unobtainable. It also provides a route to alcohols from ketones, by hydrolysis of the esters formed (Hassall, 1957; Lee and Uff, 1967), (6.103). With unsaturated ketones Baeyer–Villiger reaction often takes place in preference to oxidation of the double bond, with formation of an unsaturated lactone (6.104).

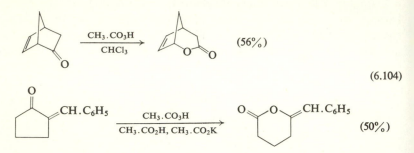

(6.104)

An unsymmetrical ketone could obviously give rise to two different products in this reaction. Cyclohexyl phenyl ketone, for example, on reaction with peracetic acid gives both cyclohexyl benzoate and phenyl cyclohexanecarboxylate by migration of both the cyclohexyl and the phenyl group. It is found that the relative ease of migration of different groups in the reaction is in the order

t-alkyl > cyclohexyl ~ s-alkyl ~ benzyl ~
~ phenyl > primary alkyl > methyl

That is, in the alkyl series migratory aptitudes are in the series tertiary > secondary > primary; among benzene derivatives migration is facilitated by electron-releasing *para* substituents, and hindered by electron-withdrawing ones. The methyl group shows the least tendency to migrate, so that methyl ketones always give acetates in the Baeyer–Villiger reaction. Thus phenyl *p*-nitrophenyl ketone gives only phenyl *p*-nitrobenzoate and methyl s-butyl ketone is converted into s-butyl acetate. It has been suggested that the ease of migration is related to the ability of the migrating group to accommodate a partial positive charge in the transition state (see below), but it seems that in some cases steric effects may be involved, and the experimental conditions may influence the result as well. Thus, while 1-methylnorcamphor affords the expected lactone on oxidation with peracetic acid, the product obtained from camphor itself depends on the conditions. With buffered peracetic acid a mixture containing a considerable proportion of the 'abnormal' product, formed by migration of the primary carbon, is obtained; but in presence of sulphuric acid only the 'abnormal' product is formed (6.105). Epicamphor gives only the 'abnormal' product in 94 per cent yield (Sauers and Ahearn, 1961).

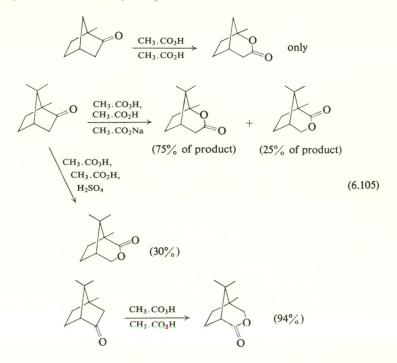

(75% of product)   (25% of product)

(6.105)

(30%)

(94%)

The reaction is thought to take place by a concerted intramolecular process involving migration of a group from carbon to oxygen, possibly by way of a cyclic transition state (Smith, 1963; Lee and Uff, 1967) (6.106). In presence of acid there may be addition of per-acid to the protonated ketone, but acid is not needed and in its absence addition may take place to the ketone itself. The general mechanism is supported by the fact that the reaction is catalysed by acid and is accelerated by electron-releasing groups in the ketone and by electron-withdrawing groups in the acid. In an elegant experiment using ^{18}O benzophenone Doering and Dorfman (1953) showed that the phenyl benzoate obtained had the same ^{18}O content as the ketone and that the ^{18}O was contained

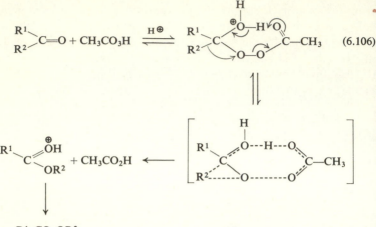

$$(6.106)$$

$$R^1 . CO . OR^2$$

entirely in the carbonyl oxygen. The intramolecular concerted nature of the reaction is supported also by several demonstrations of complete retention of configuration in the migrating carbon atom. Thus optically active methyl α-phenethyl ketone was converted into α-phenethyl acetate with no loss of chirality.

Oxidation of aldehydes with per-acids is not so synthetically useful as oxidation of ketones, and generally gives either carboxylic acids or formate esters. But reaction of *o*- and *p*-hydroxy-benzaldehydes or -acetophenones with alkaline hydrogen peroxide is a useful method for making catechols and quinols (the Dakin reaction). With benzaldehyde itself only benzoic acid is formed, but salicylaldehyde gives catechol

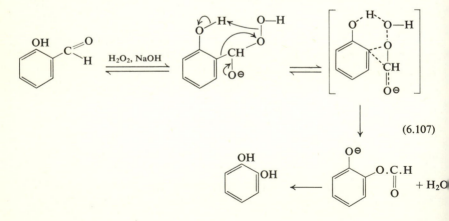

$$(6.107)$$

almost quantitatively, and 3,4-dimethyl catechol was obtained by oxidation of 2-hydroxy-3,4-dimethylacetophenone, possibly by way of a cyclic transition state similar to that postulated for the Baeyer–Villiger reaction (Lee and Uff, 1967) (6.107).

6.6. Photosensitised oxidation of olefins. Irradiation of dilute solutions of olefins and conjugated dienes in presence of oxygen and a sensitiser gives rise to hydroperoxides and cyclic peroxides (6.108). The addition

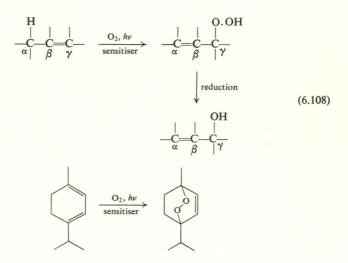

(6.108)

to dienes is analogous to the Diels–Alder reaction and is discussed further on p. 125.

Photosensitised oxygenation of mono-olefins has been widely studied and provides a convenient method for introducing oxygen in a highly specific fashion into such compounds (Schenck, 1957; Gollnick, 1968). The first products of the reaction are allylic hydroperoxides which may be isolated if desired or reduced directly to the corresponding allylic alcohols. For example, α-pinene is converted into *trans*-pinocarveyl hydroperoxide which on reduction gives *trans*-pinocarveol, and 2-methylpent-2-ene affords, after reduction of the hydroperoxides, a mixture of two methylpentenols corresponding to attack at each end of the double bond (6.109).

The oxidations are conducted in solution in benzene, pyridine or a lower aliphatic alcohol. Common sensitisers may be organic dyes such

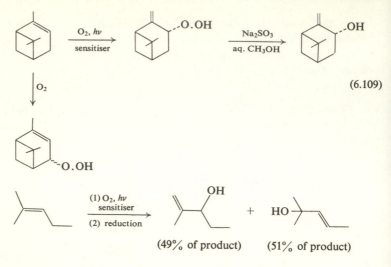

(6.109)

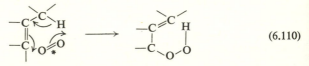

(49% of product)   (51% of product)

as fluorescein derivatives, methylene blue and certain porphyrin derivatives. Essentially no reaction takes place if the olefin lacks an allylic hydrogen atom or if sensitiser, light or oxygen is excluded. In every case the oxygen molecule adds to one carbon of the double bond and a hydrogen atom from the allylic position migrates to the oxygen with concomitant shift of the double bond. A concerted cyclic mechanism, analogous to that postulated for the 'ene' reaction (p. 168) has been suggested (6.110).

$$\text{(6.110)}$$

The reaction should be distinguished from the familiar *radical* oxidation of olefins which can also be initiated by sensitisers (such as benzophenone) which abstract hydrogen and which may also give rise to allylic hydroperoxides. It is only in exceptional cases, however, that the products of these autoxidations are the same as those of photo-oxygenation. Autoxidation of α-pinene, for example, leads to verbenyl hydroperoxide and not to pinocarveyl hydroperoxide. The photo-sensitised reactions do not involve free radical intermediates, as autoxidations do. This is shown, for example, by the fact that the alcohol (XII) obtained as one product from photosensitised oxygenation of (+)-

limonene is optically active (6.111). A (symmetrical) allylic free-radical intermediate (XIII) would have given a racemic product.

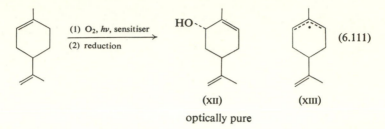

(6.111)

(XII)

(XIII)

optically pure

Photosensitised oxygenation is useful in synthesis for the specific introduction of an oxygenated functional group at the site of a double bond, and has been employed in the synthesis of a number of natural products. Thus, the compound (XIV), an analogue of the diterpene alkaloid garryfoline, was readily obtained by photo-oxygenation of an olefinic precursor (6.112). Again, a key step in the synthesis of the stereo-

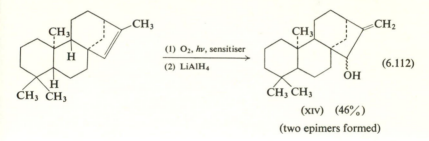

(6.112)

(XIV) (46%)

(two epimers formed)

isomeric 'rose-oxides' (XV), the active principles of rose scent, was photo-oxygenation of citronellol (6.113). An interesting reaction is the conversion of the triene (XVI) into a hydroxy-allene as well as the more conventional Diels–Alder type addition product (6.114). A similarly disposed allene unit occurs in certain carotenoids and it is possible that it may be formed in a similar manner in nature.

Photo-oxygenation of allylic alcohols gives rise to $\alpha\beta$-epoxyketones. Thus, Δ^4-cholesten-3β-ol led stereospecifically to the α-epoxyketone, while the 3α-ol gave the epimeric β-epoxide (6.115). From the known *cis*-stereochemistry of the reaction (see below) the hydroperoxides (XVII) and (XVIII) are expected intermediates; collapse of these to the epoxy-ketones accounts for the observed stereospecificity. The $\alpha\beta$-unsaturated ketone Δ^4-cholesten-3-one was also formed in both reactions in an

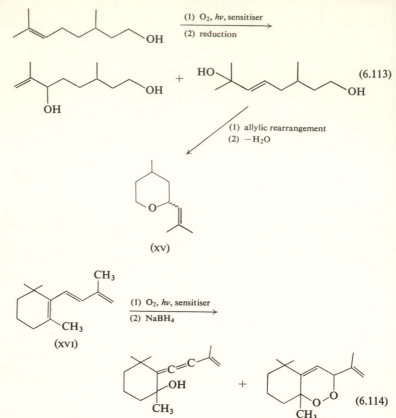

$$(6.113)$$

(xv)

(xvi)

$$(6.114)$$

amount dependent on the sensitiser used. This has been shown to be due to the differing ratios of the two forms of singlet oxygen produced (Kearns, Hollins, Khan, Chambers and Radlick, 1967).

As already indicated photo-oxygenations in many cases show a high degree of stereoselectivity. Experiments using steroidal olefins have revealed that, in agreement with the proposed cyclic reaction path, it is a *cis* reaction, that is the new C–O bond is formed on the same side of the molecule as the C–H bond which is broken. This is shown, for example, by photo-oxygenation of the deuterated cholesterols (6.116). The α-deutero compound gave, after reduction of the hydroperoxide, a diol which retained only 8 per cent of the original deuterium, whereas the diol from the β-deutero compound retained 95 per cent of the deuterium. In these reactions approach of the activated oxygen takes

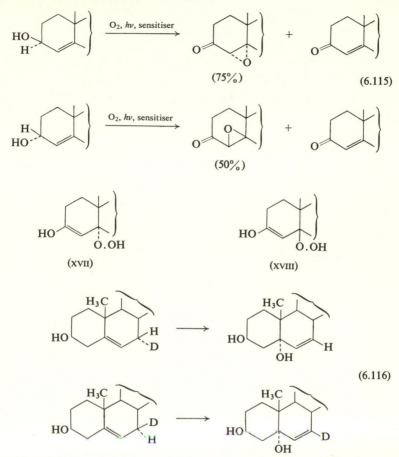

place from the less hindered α-side of the steroid molecule, with migration of the *cis* H or D from C-7 to oxygen.

In all these sensitised photo-oxygenations attack of the activated oxygen on the π orbital of the double bond takes place in a direction perpendicular to the olefinic plane, and the allylic hydrogen atom must be suitably oriented to allow transfer to oxygen. This requirement is best met when the C–H bond is perpendicular to the plane of the double bond, for this favours overlap of the C–H orbital with the *p* orbital of the double bond (6.117). A consequence of this is that in cyclohexenoid systems, a *quasi* axial hydrogen atom should be better disposed for reaction than a *quasi* equatorial one. In agreement, 5α-cholest-3-ene,

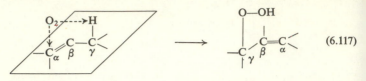

(6.117)

with a *quasi* axial C-5–H bond, is much more readily oxygenated than is 5β-cholesten-3-ene, in which the C-5–H is *quasi* equatorial (6.118).

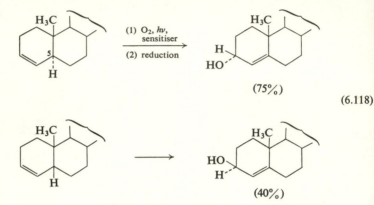

(6.118)

Again, in the photo-oxygenation of α-pinene the small amount of *trans*-2-hydroperoxy-Δ^3-pinene formed is explained by assuming that the rigidity of the α-pinene system does not allow the α-hydrogen atom at C-4 to assume the necessary *pseudo* axial position for efficient transfer to oxygen (6.119). The exclusive formation of products by oxygenation

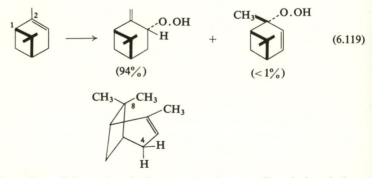

(6.119)

from the α-face of the molecule (i.e. *trans* to the *gem*-dimethylmethylene bridge) is due to the steric effect of the C-8 methyl group which prevents attack on the β-face of the double bond.

Consideration of the stereoelectronic course of the reaction also serves to explain some apparently anomalous results obtained with open chain olefins. 1,1-Dimethyl-2-isopropylethylene, for example, affords the secondary hydroperoxide in more than 95 per cent yield; almost none of the tertiary hydroperoxide is formed. The most stable conformation of the olefin is (xix) in which the hydrogen atom at C-3 almost eclipses the double bond. The alternative conformation (xx) necessary for photo-oxygenation at C-1, in which the C-3–H bond is perpendicular to the double bond, is disfavoured by steric interaction between a C-3-methyl group and the dimethyl vinyl group. Consequently very little tertiary hydroperoxide is formed (6.120). In contrast, 1,1-dimethylbut-1-

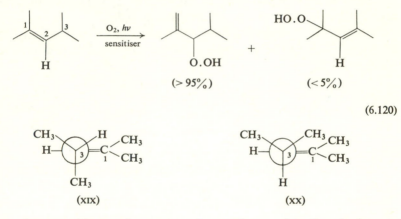

(6.120)

(XIX)   (XX)

ene gives roughly equal amounts of secondary and tertiary hydroperoxides because the conformation (xxi) appropriate to tertiary peroxide formation is no longer destabilised by steric factors (6.121).

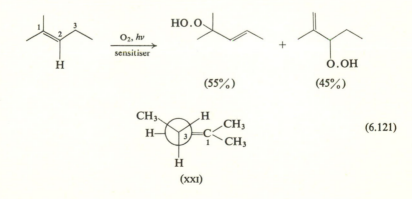

(6.121)

(XXI)

The reactive species in these reactions is believed to be singlet oxygen, formed by energy transfer from the light-energised triplet sensitiser to oxygen or an oxygen-sensitiser complex (cf. Wagner and Hammond, 1968). In line with the suggestion that singlet oxygen is involved it has been shown (Foote, 1968) that chemically generated singlet oxygen, obtained by reaction of hypochlorites with hydrogen peroxide, gives product distributions on oxidation of olefins which are identical with those obtained in sensitised photo-oxygenations. The sensitisers used in the photo-oxidations are very bulky molecules and would be expected to exert a strong influence on the stereochemistry of the reaction intermediate if they were present in the transition state. No such effect is observed, and this would seem to tell against the suggestion that the active species is an oxygen–triplet sensitiser complex.

Photosensitised oxygenation of hydrazones to hydroperoxides of azo compounds can be effected under the same conditions as that of olefins (6.122).

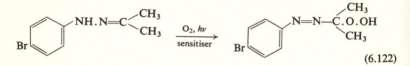

(6.122)

Another useful reaction is the cleavage of enamines to a carbonyl compound and an amide, possibly by way of a four-membered cyclic intermediate (6.123). Yields are good and the reaction provides a convenient method for cleaving carbonyl compounds between the carbonyl group and the α-carbon atom.

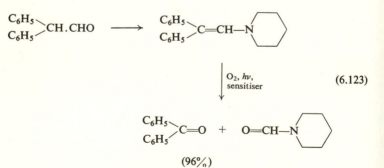

(6.123)

(96%)

7 Reduction

There must be few organic syntheses of any complexity which do not involve a reduction step at some stage. Reduction is here used in the sense of addition of hydrogen to an unsaturated group such as an olefinic bond, a carbonyl group or an aromatic nucleus, or addition of hydrogen with concomitant fission of a bond between two atoms, as in the reduction of a disulphide to a thiol or of an alkyl halide to a hydrocarbon.

Reductions are generally effected either chemically or by catalytic hydrogenation, that is by the addition of molecular hydrogen to the compound under the influence of a catalyst. Each method has its advantages. In many reductions either method may be used equally well. Complete reduction of an unsaturated compound can generally be achieved without undue difficulty, but the aim is often selective reduction of one group in a molecule in the presence of other unsaturated groups. Both catalytic and chemical methods of reduction offer considerable scope in this direction, and the method of choice in a particular case will often depend on the selectivity required and on the stereochemistry of the desired product.

7.1. Catalytic hydrogenation. Of the many methods available for reduction of organic compounds catalytic hydrogenation is one of the most convenient. Reaction is easily effected simply by stirring or shaking the substrate with the catalyst in a suitable solvent, or without a solvent if the substance being reduced is a liquid, in an atmosphere of hydrogen in an apparatus which is arranged so that the uptake of hydrogen can be measured. At the end of the reaction the catalyst is filtered off and the product recovered from the filtrate, often in a high state of purity. The method is easily adapted for work on a micro scale, or on a large, even industrial, scale. In many cases reaction proceeds smoothly at or near room temperature and at atmospheric or slightly elevated pressure. In other cases high temperatures (100–200°C) and pressures (100–300 atmospheres) are necessary, requiring special high pressure equipment.

Detailed descriptions of equipment suitable for hydrogenations under different conditions are given by Schiller (1955) and by Augustine (1965).

Catalytic hydrogenation may result simply in the addition of hydrogen to one or more unsaturated groups in the molecule, or it may be accompanied by fission of a bond between atoms. The latter process is known as hydrogenolysis.

Most of the common unsaturated groups in organic chemistry, such as $>C=C<$, $-C\equiv C-$, $>C=O$, $-CO.OR$, $-C\equiv N$, $-NO_2$ and aromatic and heterocyclic nuclei can be reduced catalytically under appropriate conditions, although they are not all reduced with equal ease. Certain groups, notably allylic and benzylic hydroxyl and amino groups and carbon–halogen and carbon–sulphur single bonds, readily undergo hydrogenolysis, resulting in cleavage of the bond between carbon and the hetero-atom. This may be advantageous in certain circumstances. Thus, much of the usefulness of the benzyloxycarbonyl protecting group in peptide chemistry is due to the ease with which it can be removed by hydrogenolysis over a palladium catalyst, as in the conversion of (I) into (II) (7.1). Similarly, hydrogenolysis of chloro-

$$p\text{-HO}.C_6H_4.CH_2.\overset{\overset{\displaystyle NH.CO_2CH_2C_6H_5}{|}}{CH}.CO_2CH_3 \xrightarrow[\text{HCl}]{\text{H}_2, \text{ Pd, } C_2H_5OH}$$

(I)

$$p\text{-HO}.C_6H_4.CH_2.\overset{\overset{\displaystyle NH_2.HCl}{|}}{CH}.CO_2CH_3$$

(II)

$$+ C_6H_5CH_3 + CO_2$$

(7.1)

derivatives is a useful method for the synthesis of some nitrogen heterocyclic compounds (7.2).

$$\text{(7.2)}$$

The catalyst. Many different catalysts have been used for catalytic hydrogenations; they are mainly finely divided metals, metallic oxides or sulphides. The most commonly used in the laboratory are the platinum metals (platinum, palladium and, to a lesser extent rhodium and ruthenium), nickel and copper chromite. The catalysts are not specific and with the exception of copper chromite may be used for a variety of different reductions. The most widely used are the platinum metal catalysts and much of what can be accomplished by hydrogenation is best done with these catalysts. They are exceptionally active catalysts and promote the reduction of most functional groups under mild conditions, with the notable exception of the carboxyl, carboxylic ester and amide groups. They are used either as the finely divided metal or, more commonly, supported on a suitable carrier such as asbestos, activated carbon, alumina or barium sulphate. In general, supported metal catalysts, since they have a larger surface area, are more active than the unsupported metal, but the activity is influenced strongly by the support and by the method of preparation, and this provides a means of preparing catalysts of varying activity. Platinum is very often used in the form of its oxide PtO_2, 'Adams' catalyst', which is reduced to metallic platinum by hydrogen in the reaction medium.

The platinum metal catalysts are usually prepared by reduction of a metallic salt, in presence of the support if a supported catalyst is wanted. Detailed instructions are given by Kimmer (1955). It has recently been found that very effective catalysts can be obtained by reduction of various metal salts with either sodium borohydride (Brown and Brown, 1962) or a trialkylsilane (Eaborn, Pant, Peeling and Taylor, 1969). Salts of platinum, palladium, rhodium and ruthenium are reduced to the finely divided metals which may be used directly as catalysts or adsorbed on carbon. Reaction of excess of the sodium borohydride with acid provides a source of hydrogen, allowing direct hydrogenation of easily reducible functional groups such as nitro and unhindered olefin (7.3). The selectivity of hydrogenation is increased and hydrogenolysis is suppressed by conducting the reactions at low temperature (Brown, 1969).

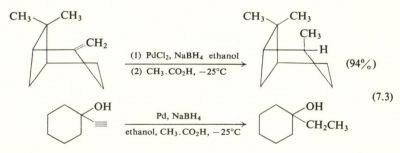

Most platinum metal catalysts, with the exception of Adams' catalyst, are stable and can be kept for many years without appreciable loss of activity, but they are deactivated by many substances, particularly compounds of divalent sulphur. For best results in a hydrogenation, therefore, particularly with platinum catalysts, it is necessary to use pure materials and pure solvents. On the other hand, catalytic activity is often increased by addition of small amounts of promoters, usually platinum or palladium salts or mineral acid. The activity of platinum catalysts derived from platinum oxide is often markedly increased in presence of mineral acid; this may simply be due to neutralisation of alkaline impurities in the catalyst.

For hydrogenations at high pressure the most common catalysts are Raney nickel and copper chromite. Raney nickel is a porous, finely divided nickel obtained by treating a powdered nickel–aluminium alloy with sodium hydroxide. It is generally used at high temperatures and pressures, but with the more active catalysts many reductions can be effected at atmospheric pressure and normal temperature. Nearly all unsaturated groups can be reduced with Raney nickel but it is most frequently used for reduction of aromatic rings and hydrogenolysis of sulphur compounds (see p. 368). When freshly prepared it contains 25–100 ml adsorbed hydrogen per gram of nickel. The more adsorbed hydrogen the more active is the catalyst, and Raney nickel catalysts of graded activities can be obtained by variation of the preparative procedure (Augustine, 1965; Kimmer, 1955). Raney nickel catalysts are alkaline and may only be used for hydrogenations which are not adversely affected by basic conditions. They are deactivated by acids. Another useful catalyst which is said to resemble Raney nickel in activity is nickel boride, which is easily prepared by reduction of nickel salts in aqueous solution with sodium borohydride (Brown and Brown, 1963).

Copper chromite, $CuCr_2O_4$, is prepared by thermal decomposition of copper ammonium chromate; a more active catalyst is obtained by adding barium nitrate to the reaction mixture. It is a relatively inactive

$$C_6H_5.CH_2.CO_2Et \xrightarrow[250°C,\ 200\ atm]{H_2,\ CuCr_2O_4} C_6H_5.CH_2.CH_2OH$$

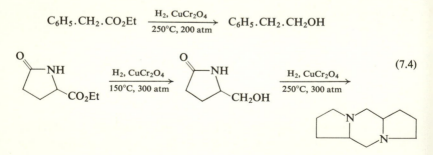

(7.4)

catalyst and is only effective at high temperatures (100–200°C) and pressures (200–300 atmospheres). It does not reduce aromatic rings under these conditions as Raney nickel does, and is principally used for reduction of esters to alcohols and of amides to amines (7.4).

Hydrogenation using copper chromite is not so frequently used now because of the introduction of mixed hydride reducing agents (p. 357) which can accomplish the same reactions under much milder conditions, but it is still useful for reactions on a large scale.

Selectivity of reduction. Many hydrogenations proceed satisfactorily under a wide range of conditions, but where a selective reduction is wanted, conditions may be more critical.

The choice of catalyst for a hydrogenation is governed by the activity and selectivity required. Selectivity is a property of the metal, but it also depends to some extent on the activity of the catalyst and on the reaction conditions. In general, the more active the catalyst the less discriminating it is in its action, and for greatest selectivity reactions should be run with the least active catalyst and under the mildest possible conditions consistent with a reasonable rate of reaction. The rate of a given hydrogenation may be increased by raising the temperature, by increasing the pressure or by an increase in the amount of catalyst used, but all these factors may result in a decrease in selectivity. For example, hydrogenation of ethyl benzoate with copper chromite catalyst under the appropriate conditions leads to benzyl alcohol by reduction of the ester group, while Raney nickel gives ethyl hexahydrobenzoate by selective attack on the benzene ring (7.5). At higher temperatures, however,

$$C_6H_5.CH_2OH \xleftarrow[160°C, 250\ atm]{H_2, CuCr_2O_4} C_6H_5.CO_2Et \xrightarrow[50°C, 100\ atm]{H_2, Raney\ Ni} C_6H_{11}.CO_2Et$$

$$(7.5)$$

the selective activity of the catalysts is lost and mixtures of the two products and toluene are obtained from both reactions.

Both the rate and, sometimes, the course of a hydrogenation may be influenced by the solvent used. The most common solvents are methanol, ethanol and acetic acid. Many hydrogenations over platinum metal catalysts are favoured by strong acids. For example, reduction of β-nitrostyrene in acetic acid–sulphuric acid is rapid and affords 2-phenylethylamine in 90 per cent yield; but in absence of sulphuric acid reduction is slow and the yield of amine poor.

Not all functional groups are reduced with equal ease. Table 7.1, due to House (1965), shows the approximate order of decreasing ease

TABLE 7.1   *Approximate order of reactivity of functional groups in catalytic hydrogenation*

Functional group	Reduction product
R—CO.Cl	R—CHO, R—CH$_2$OH
R—NO$_2$	R—NH$_2$
R—C≡C—R	$\overset{H}{\underset{R}{}}C=C\overset{H}{\underset{R}{}}$, RCH$_2CH_2$R
R—CHO	R—CH$_2$OH
R—CH=CH—R	R—CH$_2$.CH$_2$—R
R—CO—R	R—CHOH—R, R—CH$_2$—R
C$_6$H$_5$.CH$_2$OR	C$_6$H$_5$CH$_3$ + ROH
R—C≡N	R—CH$_2$NH$_2$
Polycyclic aromatic hydrocarbons	Partially reduced products
R—COOR′	R—CH$_2$OH + R′OH
R—CONHR′	R—CH$_2$NHR
⬡ R—CO$_2$$^\ominus$Na$^\oplus$	⬡ inert

H.O. House, *Modern synthetic reactions*, copyright 1965, W. A. Benjamin, Inc., Menlo Park, California.

of catalytic hydrogenation of a number of common groups. This order is not invariable and is influenced to some extent by the structure of the compound being reduced and by the catalyst employed. In general, groups near the top of the list can be selectively reduced in the presence of groups near the bottom, but reduction of groups at the bottom in presence of the more reactive groups at the top is more difficult. For example, reduction of an unsaturated ester or ketone to a saturated ester or ketone is, in most cases, readily accomplished by hydrogenation over palladium or platinum, but selective reduction of the carbonyl group to form an unsaturated alcohol is difficult to achieve by catalytic hydrogenation and is generally effected by chemical reduction. Similarly, nitrobenzene is easily converted into aniline, but selective reduction to nitrocyclohexane is not possible.

Reduction of functional groups: Olefins. Hydrogenation of olefinic double bonds takes place easily, and in most cases can be effected under mild conditions. Only a few highly hindered olefins are resistant to hydrogenation, and even these can generally be reduced under more vigorous conditions. Platinum and palladium are the most frequently used catalysts. Both are very active and the preference is determined by the nature of other functional groups in the molecule and by the degree of selectivity required; platinum usually brings about a more exhaustive reduction. Raney nickel may also be used in certain cases. Thus cinnamyl alcohol is reduced to the dihydro compound with Raney nickel in ethanol at 20°C, and 1,2-dimethylcyclohexene with hydrogen and platinum oxide in acetic acid is converted mainly into *cis*-1,2-dimethyl-cyclohexane (cf. p. 317) (7.6).

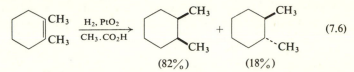

$$(82\%) \qquad (18\%) \qquad (7.6)$$

Rhodium and ruthenium catalysts have not so far been much used in hydrogenation of olefins, but they sometimes show useful selective properties. Rhodium is particularly useful for hydrogenation of olefins when concomitant hydrogenolysis of an oxygen function is to be avoided. Thus, the plant toxin, toxol (III), on hydrogenation over rhodium–alumina in ethanol was smoothly converted into the dihydro compound; with platinum and palladium catalysts, on the other hand, extensive hydrogenolysis took place and a mixture of products was formed (7.7).

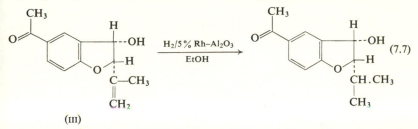

(III)

The ease of reduction of an olefin decreases with the degree of sub-stitution of the double bond, and this sometimes allows selective re-duction of one double bond in a polyolefin. Thus, limonene can be converted into Δ^1-*p*-menthene in almost quantitative yield by hydro-genation over platinum oxide if the reaction is stopped after absorption of one molecule of hydrogen (7.8). In contrast, the isomeric $\Delta^{1(7),8}$-*p*-

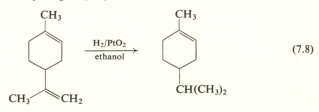

menthadiene, in which both double bonds are disubstituted, gives only the completely reduced product.

Catalytic hydrogenation of olefins over platinum metal catalysts is often accompanied by migration of the double bond, but unless tracers are used or special products result, or the new bond is resistant to hydrogenation, no evidence of migration remains on completion of the reduction. A not uncommon result in suitable structures is formation of a tetrasubstituted double bond which resists further reduction. Thus, in the reduction of 3,4-dimethylcycloheptatrienone the main product was 3,4-dimethylcycloheptenone (7.9). Similarly, the tetracyclic triterpene

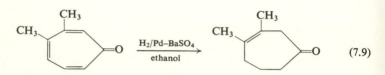

$$\text{(7.9)}$$

derivative methyl dihydromasticadienol acetate, with platinum oxide and deuterium in deuteroacetic acid gave the isomer (IV); the location of the original double bond was deduced from the position of the deuterium atom in this product (7.10).

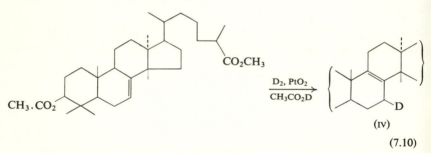

(IV)

$$\text{(7.10)}$$

A further indication that catalytic hydrogenation of an olefin need not occur by straightforward addition of two hydrogen atoms at the site of the original double bond is found in the fact that catalytic deuteration generally results in the formation of products containing more and fewer than two atoms of deuterium per molecule. Deuteration of hex-1-ene, for example, over platinum oxide was found by Smith and Burwell (1962) to afford a mixture of products containing from one to six deuterium atoms per molecule.

Selective reduction of olefinic double bonds in presence of other unsaturated groups can usually be accomplished, except in presence of triple bonds, aromatic nitro groups and acyl halides. Palladium is usually the best catalyst. Thus, 2-benzylidenecyclopentanone is readily converted into 2-benzylcyclopentanone with hydrogen and palladium in methanol; with a platinum catalyst, however, benzylcyclopentanol is formed (7.11). Unsaturated nitriles and aliphatic nitro compounds

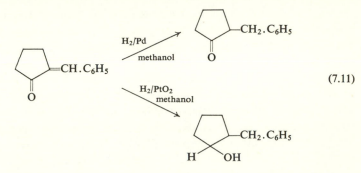

(7.11)

are also reduced selectively at the double bond with a palladium catalyst.

With palladium and platinum catalysts hydrogenation of allylic or vinylic alcohols, ethers and esters, is often accompanied by hydrogenolysis of the oxygen function, and this reaction is facilitated by acids. Hydrogenolysis can often be avoided with rhodium or ruthenium catalysts. Thus, 3,7-diacetoxycycloheptene undergoes hydrogenolysis over platinum oxide to give a mixture of products, but with ruthenium the dihydro compound is readily obtained (7.12).

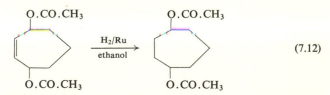

(7.12)

Hydrogenation of acetylenes. Catalytic hydrogenation of acetylenes takes place in a stepwise manner, and both the olefin and the paraffin can be isolated. Complete reduction of acetylenes to the saturated compound is easily accomplished over platinum, palladium or Raney nickel. A complication which sometimes arises, particularly with platinum catalysts, is the hydrogenolysis of propargylic hydroxyl groups (7.13).

More useful from a synthetic point of view is the partial hydrogenation of acetylenes to *cis* olefins. This reaction can be effected in high yield

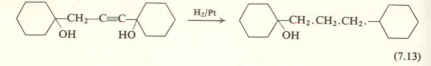

$$(7.13)$$

with a palladium–calcium carbonate catalyst which has been partially deactivated by addition of lead acetate (Lindlar's catalyst) or quinoline. It is aided by the fact that the more electrophilic acetylenic compounds are adsorbed on the electron-rich catalyst surface more strongly than the corresponding olefins. An important feature of these reductions is their high stereoselectivity. In most cases the product consists very largely of the thermodynamically less stable *cis* olefin and partial catalytic hydrogenation of acetylenes provides one of the most convenient routes to *cis* olefins. Thus stearolic acid on reduction over Lindlar's catalyst in ethyl acetate solution affords a product containing 95 per cent of the *cis* olefin oleic acid (7.14). Partial reduction of acetylenes

$$CH_3.(CH_2)_7.C\equiv C.(CH_2)_7.CO_2H$$

$$\downarrow \quad H_2/\text{Lindlar catalyst}$$

$$(7.14)$$

$$\underset{H}{\overset{CH_3(CH_2)_7}{\diagdown}}C=C\underset{H}{\overset{(CH_2)_7CO_2H}{\diagup}}$$

with Lindlar's catalyst has been invaluable in the synthesis of carotenoids and many other natural products with *cis* disubstituted double bonds.

Hydrogenation of aromatic compounds. Reduction of aromatic compounds by catalytic hydrogenation is more difficult than with most other functional groups, and selective reduction is not easy. Generally, at least one ring is completely reduced, for the olefinic bonds in a partially reduced product would be more easily reduced than the aromatic compound itself. The commonest catalysts are platinum and rhodium, which can be used at ordinary temperatures, and Raney nickel or ruthenium which require high temperatures and pressures.

Benzene itself can be reduced to cyclohexane with platinum oxide in acetic acid solution. Derivatives of benzene such as benzoic acid, phenol or aniline, are reduced more easily. For large-scale work the most convenient method is hydrogenation over Raney nickel at 150–200°C and 100–200 atmospheres. A typical case is the reduction of β-naphthol in alcohol to the tetrahydro compounds (v) and (vi) (7.15). Under alkaline conditions the alcohol (VI) becomes the major product of the reaction. Hydrogenation of phenols, followed by oxidation of the resulting cyclohexanols is a convenient method for the preparation of substituted cyclohexanones. Good yields of cyclohexanones may also be

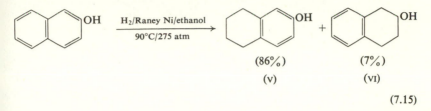

(86%)
(V)

(7%)
(VI)

(7.15)

obtained directly by selective hydrogenation of phenols. Thus 2,4-dimethylcyclohexanone is obtained by hydrogenation of 2,4-dimethylphenol over a palladium–carbon catalyst (7.16).

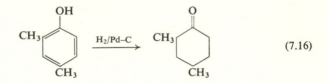

(7.16)

Reduction of benzene derivatives carrying oxygen or nitrogen functions in benzylic positions is complicated by the easy hydrogenolysis of such groups, particularly over palladium catalysts (7.17). Preferential

$$C_6H_5.CH_2O.CO.CH_3 \xrightarrow{H_2/Pd-C} C_6H_5.CH_3 \qquad (7.17)$$

reduction of the benzene ring in such compounds is best achieved with ruthenium, or preferably with rhodium, catalysts which can be used under mild conditions. Thus mandelic acid is readily converted into hexahydromandelic acid over rhodium–alumina in methanol solution, whereas with palladium, hydrogenolysis to phenylacetic acid is the main reaction (7.18).

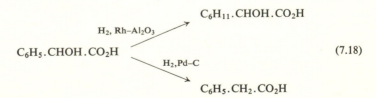

(7.18)

With polycyclic aromatic compounds it is often possible, by varying the conditions, to obtain either partially or completely reduced products. Thus, naphthalene can be converted into the tetrahydro- or decahydro-compound over Raney nickel (7.19). With anthracene and phenanthrene

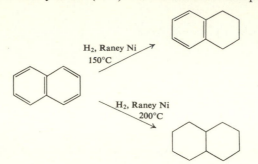

$$\text{(7.19)}$$

the 9,10-dihydro-compounds are obtained by hydrogenation over copper chromite, although, in general, aromatic rings are not reduced with this catalyst. To obtain more fully hydrogenated compounds more powerful catalysts must be used.

Hydrogenation of aldehydes and ketones. Hydrogenation of the carbonyl group of aldehydes and ketones is easier than that of aromatic rings but not so easy as that of most olefinic double bonds.

For aliphatic aldehydes and ketones reduction to the alcohol is usually effected under mild conditions over platinum or the more active forms of Raney nickel. Ruthenium is also an excellent catalyst for reduction of aliphatic aldehydes and can be used to advantage with aqueous solutions (7.20). Palladium is not very active for hydrogenation of aliphatic carbonyl compounds.

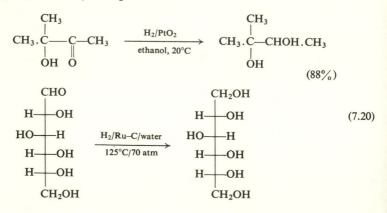

Selective hydrogenation of a carbonyl group in presence of olefinic double bonds is difficult and in most cases is best effected with mixed hydride reducing agents or by the Meerwein–Pondorff method (see p. 352). If the double bonds are highly substituted selective reduction of the carbonyl group can sometimes be achieved.

It has recently been found that reduced osmium supported on alumina or activated charcoal is an excellent catalyst for reduction of $\alpha\beta$-unsaturated aldehydes to unsaturated alcohols. With this catalyst cinnamyl aldehyde gave cinnamyl alcohol in 95 per cent yield (Rylander and Steele, 1969) (7.21). The catalyst is apparently not effective for the

$$C_6H_5.CH{=}CH.CHO \xrightarrow[100°C/30\ atm]{H_2/Os{-}C} C_6H_5.CH{=}CH.CH_2OH \quad (7.21)$$

selective reduction of unsaturated ketones, however, for mesityl oxide was reduced to methyl isobutyl ketone. Selective reduction of a carbonyl group in presence of an aromatic nucleus is often possible with a copper chromite catalyst. Palladium is the most effective catalyst for the reduction of aromatic aldehydes and ketones, and excellent yields of the carbinols can be obtained if the reaction is interrupted after absorption of one mole of hydrogen. Prolonged reaction, particularly at elevated temperatures or in presence of acid, leads to hydrogenolysis with formation of the methyl or methylene compound and hydrogenation over a palladium catalyst in presence of acid is a convenient method for the reduction of aromatic ketones to methylene compounds. Thus the tetralone (VII) on interrupted reduction over palladised charcoal affords the carbinol (VIII), but if reaction is allowed to proceed to completion at 80°C the tetralin (IX) is obtained (7.22). Similarly the

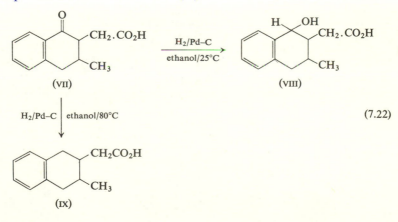

(7.22)

indanone (x) is rapidly converted into the indane derivative (xi) by hydrogenation over palladised charcoal in presence of sulphuric acid (7.23). Both the Clemmensen and Huang-Minlon methods failed with this compound.

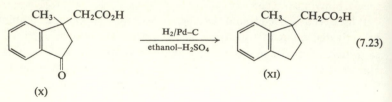

(7.23)

(x)

When no other reducible groups are present, or when complete reduction of the carbonyl compound is required, hydrogenation over Raney nickel at high temperature and pressure is the most effective procedure for large-scale work.

Nitriles, oximes and nitro compounds. Functional groups with multiple bonds to nitrogen are also readily reduced by catalytic hydrogenation. Nitriles, oximes, azides and nitro compounds, for example, are all smoothly converted into primary amines. Reduction of nitro compounds takes place very easily and is generally faster than reduction of olefinic bonds or carbonyl groups. Raney nickel or any of the platinum metals can be used as catalyst, and the choice will be governed by the nature of other functional groups in the molecule. Thus β-phenylethylamines, useful for the synthesis of isoquinolines, are conveniently obtained by catalytic reduction of $\alpha\beta$-unsaturated nitro compounds (7.24).

$$C_6H_5.CH{=}CH.NO_2 \xrightarrow[\substack{ethanol/25°C \\ H_2SO_4}]{H_2/Pd{-}C} C_6H_5.CH_2.CH_2.NH_2 \qquad (7.24)$$

Nitriles are conveniently reduced with platinum or palladium at room temperature, or with Raney nickel under pressure. Unless precautions are taken, however, large amounts of secondary amines may be formed in a side reaction of the amine with the intermediate imine (7.25). With

$$R.C{\equiv}N \xrightarrow{H_2} R.CH{=}NH \xrightarrow{H_2} R.CH_2.NH_2$$

$$R.CH_2NH_2 + RCH{=}NH \rightleftharpoons$$

$$R.CH_2.NH.CHR \underset{NH_2}{\overset{-NH_3}{\rightleftharpoons}} R.CH_2N{=}CHR$$

$$R.CH_2.N{=}CHR \xrightarrow{H_2} R.CH_2.NH.CH_2.R \qquad (7.25)$$

the platinum metal catalysts this reaction can be suppressed by conducting the hydrogenation in acid solution or in acetic anhydride, which removes the amine from the equilibrium as its salt or as its acetate. For reactions with Raney nickel, where acid cannot be used, secondary amine formation is prevented by addition of ammonia (7.26). Hydro-

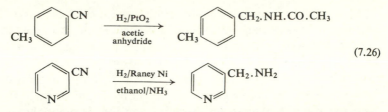

(7.26)

genation of nitriles containing other functional groups may lead to cyclic compounds. Thus, hydrogenation of γ-cyanoesters over copper chromite in an alcohol as solvent has been used as a route to N-alkyl-piperidines; the N-alkyl group is derived from the alcohol used as solvent. Suitably oriented aminonitriles may also afford cyclic products on hydrogenation. Thus indolizidine and quinolizidine derivatives have been obtained by reductive cyclisation of ω-(2-pyridyl)-propionitriles and -butyronitriles over platinum. The reaction follows the path shown in (7.27).

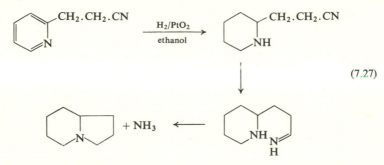

(7.27)

Reduction of oximes to primary amines takes place under conditions similar to those used for nitriles, with palladium or platinum in acid solution, or with Raney nickel under pressure. Similar conditions are used in the reductive alkylation of aldehydes and ketones to amines. In this reaction a mixture of the carbonyl compound, ammonia or a primary or secondary amine, is hydrogenated over Raney nickel or palladium (7.28). The product is formed, presumably, by reduction of the intermediate imine (Emerson, 1948).

n-C$_6$H$_{13}$.CHO $\xrightarrow[\text{Raney Ni, H}_2]{\text{C}_6\text{H}_5\text{NH}_2\text{, ethanol}}$ [n-C$_6$H$_{13}$.CH=N.C$_6$H$_5$]

(7.28)

n-C$_6$H$_{13}$.CH$_2$.NH.C$_6$H$_5$ (65%)

Stereochemistry and mechanism. Hydrogenation of an unsaturated compound takes place by adsorption of the compound on to the catalyst surface, followed by transfer of hydrogen from the catalyst to the side of the molecule which is adsorbed on it. Adsorption on to the catalyst is largely controlled by steric factors, and it is found in general that hydrogenation takes place by *cis* addition of hydrogen atoms to the less hindered side of the unsaturated centre. However, it is not always easy to decide which is the least hindered side and in such cases it may be difficult to predict what the steric course of a hydrogenation will be.

Thus, hydrogenation of the *trans*-stilbene derivative (XII) forms the (±)-dihydro compound (XIII) by *cis* addition of hydrogen, while the *cis*-stilbene (XIV) gives the *meso* isomer (XV) (7.29). Hydrogenation of the

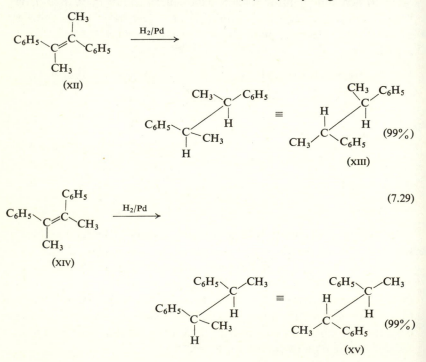

(7.29)

pinene derivative (XVI) and of the trimethylcyclohexanone (XVII) gave products formed by *cis* addition of hydrogen to the more accessible side of the double bonds (7.30). Similarly in the hydrogenation of the diter-

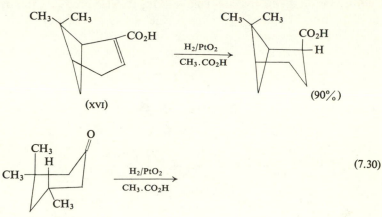

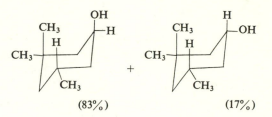

(7.30)

pene derivative (XVIII) addition of hydrogen takes place on the α face of the molecule which is not hindered by the two axial methyl groups at C-4 and C-10 (7.31). In all these examples the molecule possesses

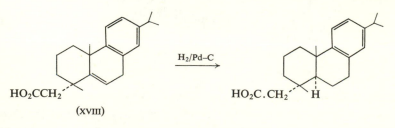

(7.31)

a certain degree of rigidity. Where this rigidity is absent it becomes more difficult to decide which side of the molecule will be more easily adsorbed on the catalyst and to predict the steric course of a hydrogenation which, in such cases, may be influenced by the experimental conditions or by functional groups far removed from the centre of unsaturation. Thus, hydrogenation of the hydroxy-enone (xix, R = H) produces the *trans*-decalone by addition of hydrogen to the α-face of the molecule; but under the same conditions the acetate (xix, R = Ac) gives the *cis*-decalone (7.32). In other cases the steric course of a hydrogenation may

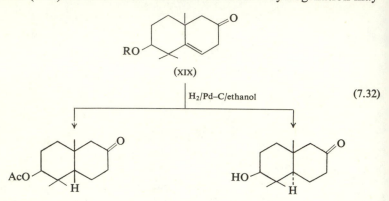

(xix)

H_2/Pd–C/ethanol   (7.32)

be influenced by 'anchor effects', that is the affinity of a particular functional group in the molecule for the catalyst or the catalyst support. This may sometimes be used to advantage in synthesis. Thus hydrogenation of the exocyclic double bond in the amino acid derivative (xx) over platinum or palladium on the usual supports led to mixtures containing only a small proportion of the desired *trans* isomer (xxi) (7.33). The high yield of *cis* product showed that the molecule was

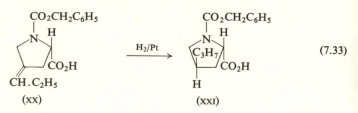

(xx)   (xxi)

being adsorbed on to the catalyst surface on the side opposite to the carboxyl group. This effect was reversed and a high yield of the *trans* isomer was obtained by employing as catalyst support a basic ion-exchange resin on which the carboxyl group was 'anchored' by salt formation. Transfer of hydrogen from the catalyst surface to the double bond then took place mainly *cis* to the carboxyl group.

The hydrogenation of substituted monocyclic olefins is anomalous in many cases in that substantial amounts of *trans* addition-products are formed, particularly with palladium catalysts. Thus, $\Delta^{9(10)}$-octalin on hydrogenation over palladium in acetic acid affords mainly *trans*-decalin, and other examples are shown in the equations (7.34). Similarly,

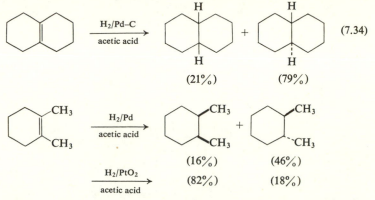

in the hydrogenation of the isomeric xylenes over platinum oxide, the *cis*-dimethylcyclohexanes are the main products, but some *trans* isomer is always produced. The reason for the formation of the *trans* products is not completely clear, but it has been suggested that it may be due to isomerisation of the double bond on the catalyst surface (which would be favoured by palladium) followed by desorption and random readsorption (Siegel and Smith, 1960).

The rate and steric course of many hydrogenations are altered by the presence of acid or base in the reaction mixtures. Hydrogenation of Δ^4-cholestene in neutral solution affords coprostane while in acid solution cholestane is obtained (7.35). Similarly for 3-oxo-Δ^4-steroids the course of hydrogenation is appreciably influenced by the solvent (McQuillin,

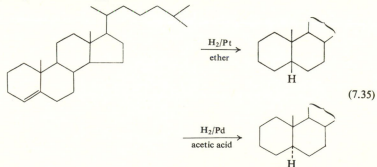

Ord and Simpson, 1963). Alkaline conditions and a hydroxylic solvent strongly favour β- addition of hydrogen leading to *cis* A/B ring fusion, whereas with acid in an aprotic less polar solvent more product of α-addition (*trans* A/B ring fusion) is formed (7.36). For the simpler

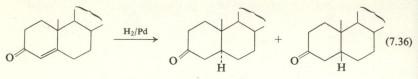

(variable proportions depending on conditions)

unsaturated ketone (xxɪɪ) the proportion of *cis* and *trans* products obtained on reduction varied with the solvent as shown in (7.37).

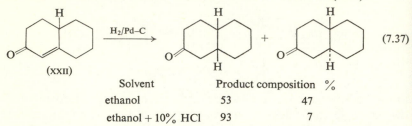

Solvent	Product composition %	
ethanol	53	47
ethanol + 10% HCl	93	7

The stereochemical course of the hydrogenation of the carbonyl group of cyclohexanones is also influenced by the nature of the solvent. The von Auwers–Skita rule, as modified by Barton (1953) is often used as a guide, and predicts that hydrogenation in strongly acid solution, which is rapid, will lead to formation of the axial alcohol, whereas in alkaline or neutral medium the equatorial alcohol will predominate if the ketone is not hindered; strongly hindered ketones will again give the axial

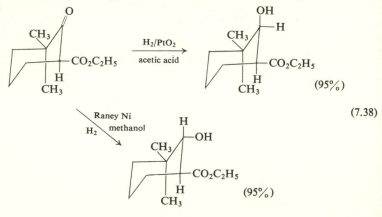

alcohol. Thus, hydrogenations with Raney nickel, which is alkaline, or with Adams' catalyst in a neutral solvent, often leads to the equatorial alcohol, while with the Adams' catalyst in acid solution the axial alcohol is formed (7.38). However, the rule must be used with caution, and the effect of acid and alkali is not always as straightforward as the Auwers–Skita rule suggests (Wicker, 1956).

A satisfactory mechanism for catalytic hydrogenation must explain not only the observed *cis* addition of hydrogen, but also the fact that olefins are isomerised by hydrogenation catalysts, and that catalytic deuteration of an olefin leads to products containing more and fewer than two atoms of deuterium per molecule (see, for example, Smith and

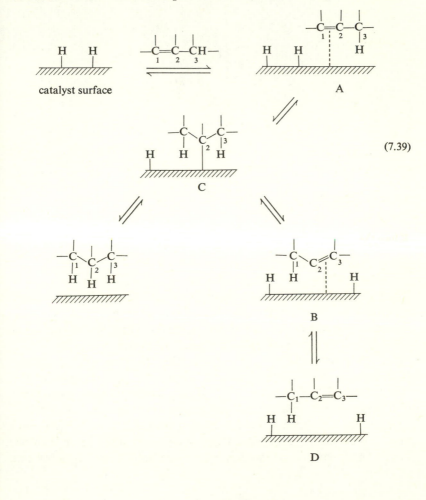

(7.39)

Burwell, 1962), as well as the unexpected formation of *trans* addition products in some hydrogenations. These results can be rationalised on the basis of a mechanism which was first suggested by Horiuti and Polanyi (1934) in which transfer of hydrogen atoms from the catalyst to the adsorbed substrate is supposed to take place in a stepwise manner. The process is thought to involve equilibria between π-bonded forms A and B and a half hydrogenated form C which can either take up another atom of hydrogen or revert to starting material or to an isomeric olefin D (7.39). More recently another mechanism involving the π-allyl intermediate E has been postulated (Gault, Rooney and Kemball, 1962) (7.40). π-Allyl intermediates have also been invoked to explain the

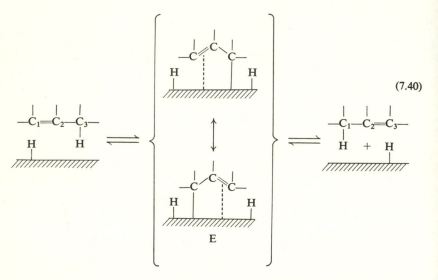

(7.40)

hydrogenolysis of benzyl alcohols and their derivatives. These reactions generally occur with high stereoselectivity, although the steric course depends to a large extent on the catalyst metal. Thus, hydrogenolysis of optically active methyl 3-hydroxy-3-phenylbutyrate over Raney nickel takes place with retention of configuration, whereas with palladium inversion occurs (7.41), (Garbisch, Schreader and Frankel, 1967; Khan, McQuillin and Jardine, 1967; Mitsui, Kudo and Kobayashi, 1969).

Homogeneous hydrogenation. Catalysts for heterogeneous hydrogenation of the types discussed above, although useful, have some disadvantages. They may show lack of selectivity when more than one unsaturated centre is present, they may cause double bond migration and,

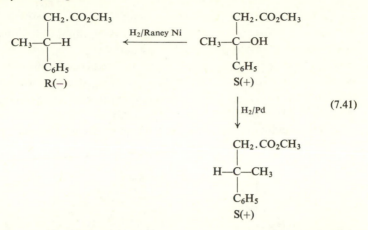

(7.41)

in reactions with deuterium, they usually bring about allylic interchanges with deuterium. This, in conjunction with double-bond migration, results in unspecific labelling with, in many cases, introduction of more than two deuterium atoms. Again, a number of functional groups suffer easy hydrogenolysis over heterogeneous catalysts and this sometimes leads to complications in the reduction of other unsaturated groups in the molecule. The stereochemistry of reduction, despite a number of rules, is difficult to predict since it depends on chemisorption and not on reactions between molecules. Some of these difficulties have been overcome by the recent introduction of soluble catalysts which allow hydrogenation in homogeneous solution.

A number of soluble catalyst systems have been used (cf. Sloan, Matlack and Breslow, 1963), but the most effective found so far are the rhodium and ruthenium complexes tris(triphenylphosphine)chloro-rhodium $(Ph_3P)_3RhCl$, and hydridochlorotris(triphenylphosphine)-ruthenium $(Ph_3P)_3RuClH$.

The rhodium complex is easily made by reaction of rhodium chloride with excess of triphenylphosphine in boiling ethanol (7.42). It is an

$$RhCl_3.3H_2O \xrightarrow[\text{boiling ethanol}]{\text{excess of } Ph_3P} (Ph_3P)_3RhCl \qquad (7.42)$$

extremely efficient catalyst for the homogeneous hydrogenation of non-conjugated olefins and acetylenes at ordinary temperature and pressure in benzene or similar solvents. Functional groups such as oxo, cyano, nitro, chloro, azo are not reduced under these conditions. Mono- and disubstituted double bonds are reduced much more rapidly than tri- or tetrasubstituted ones (Hussey and Takeuchi, 1969), permitting the partial hydrogenation of compounds containing different kinds of double bonds (Osborn, Jardine, Young and Wilkinson, 1966; Biellmann, 1968). Thus, in the reduction of linalool (XXIII) addition of hydrogen occurred selectively at the vinyl group, giving the dihydro compound in 90 per cent yield; carvone (XXIV) was similarly converted into carvotanacetone (7.43). The selectivity of the catalyst is shown further by the remarkable reduction of ω-nitrostyrene to phenylnitroethane (7.44).

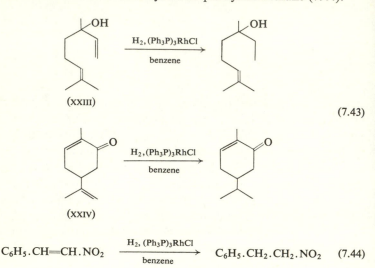

(7.43)

$$C_6H_5.CH{=}CH.NO_2 \xrightarrow[\text{benzene}]{H_2,\ (Ph_3P)_3RhCl} C_6H_5.CH_2.CH_2.NO_2 \quad (7.44)$$

Hydrogenations take place by *cis* addition to the double bond. This was shown by reaction of deuterium with maleic acid to form *meso*-dideuterosuccinic acid, while fumaric acid gave the ($\pm$)-compound (7.45). Likewise, interrupted reduction of hex-2-yne gave a mixture of n-hexane and *cis*-hex-2-ene.

An important practical advantage of this catalyst is that deuterium is introduced without scrambling; that is, only two deuterium atoms are added, at the site of the original double bond. Thus methyl oleate is converted into methyl 9,10-dideuterostearate in quantitative yield, and $\varDelta^{1,4}$-androstadien-3,17-dione (XXV) gave the dideutero compound (XXVI) by *cis* addition to the α-face of the steroid nucleus (7.46).

Another very valuable feature of this catalyst is that it does not bring about hydrogenolysis, thus allowing the selective hydrogenation of olefinic bonds without hydrogenolysis of other susceptible groups in

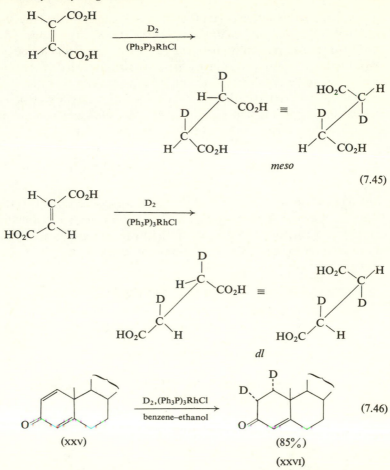

(7.45)

(7.46)

the molecule. Thus, benzyl cinnamate is converted smoothly into the dihydro compound without attack on the benzyl ester group (7.47),

$$PhCH{=}CH.CO_2CH_2Ph \longrightarrow PhCH_2CH_2CO_2CH_2Ph \qquad (7.47)$$

and allyl phenyl sulphide is reduced to propyl phenyl sulphide in 93 per cent yield (7.48).

$$CH_2{=}CH.CH_2{-}S{-}Ph \longrightarrow C_3H_7{-}S{-}Ph \qquad (7.48)$$

Because of the strong affinity of $[(Ph_3P)_3RhCl]$ for carbon monoxide it decarbonylates aldehydes, and olefinic compounds containing aldehyde groups cannot be hydrogenated with this catalyst under the usual conditions (Jardine and Wilkinson, 1967). Thus cinnamaldehyde is converted into styrene in 65 per cent yield, and benzoyl chloride gives chlorobenzene in 90 per cent yield. Acceptable yields of saturated aldehydes can be obtained under special experimental conditions. The iridium complex $IrH_3(Ph_3P)_3$ in presence of acetic acid can be used for the homogeneous hydrogenation of aldehydes and some olefins at 50°C and atmospheric pressure (Coffey, 1967) (7.49). $(Ph_3P)_3RhCl$

$$C_3H_7CHO \xrightarrow[CH_3CO_2H/50°C]{H_2/IrH_3(Ph_3P)_3} C_3H_7CH_2OH \qquad (7.49)$$

dissociates in solvents (S) to give the solvated species $(Ph_3P)_2Rh(S)Cl$ which in presence of hydrogen is in equilibrium with the dihydrido complex $(Ph_3P)_2Rh(S)ClH_2$ in which the hydrogen atoms are directly attached to the metal (7.50). Reduction is believed to take place by dis-

$$(Ph_3P)_2Rh(S)Cl \underset{}{\overset{H_2}{\rightleftharpoons}} (Ph_3P)_2Rh(S)ClH_2$$

$$\overset{RCH=CHR'}{\underset{}{\rightleftharpoons}} (Ph_3P)_2RhCl(RCH=CHR')H_2 \qquad (7.50)$$

$$\longrightarrow R.CH_2.CH_2.R' + (Ph_3P)_2Rh(S)Cl$$

placement of solvent by the olefin, followed by simultaneous stereo-specific *cis*-transfer of the two hydrogen atoms from the metal to the loosely co-ordinated olefin (7.51). Diffusion of the saturated substrate

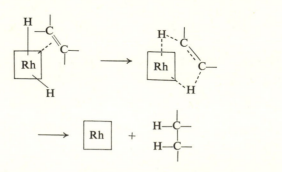

$$(7.51)$$

away from the transfer site leaves the complex again ready to combine with dissolved hydrogen and continue the reduction. However, recent work suggests that in some cases transfer of the hydrogen atoms may not be concerted and an intermediate with a carbon–metal bond may be involved (Heathcock and Poulter, 1969).

The ruthenium complex $(Ph_3P)_3RuClH$, formed *in situ* from $(Ph_3P)_3RuCl_2$ and molecular hydrogen in benzene in presence of a base such as triethylamine, is an even more efficient catalyst which is specific for the hydrogenation of monosubstituted alkenes $RCH{=}CH_2$. Rates of reduction for other types of alkenes are slower by a factor of at least 2×10^3 (Hallman, McGarvey and Wilkinson, 1968). Thus hept-1-ene was rapidly converted into heptane with molecular hydrogen in benzene solution but hept-3-ene was unaffected. Some isomerisation of alkenes is observed with this catalyst but the rate is slow compared with the rate of hydrogenation. Oct-1-ene, for example, gave 8 per cent of *cis*- and *trans*-oct-2-enes after 20 hours. Very similar behaviour is shown by the rhodium complex hydridocarbonyl-tris-(triphenylphosphine)rhodium(I) $(Ph_3P)_3RhH(CO)$. Terminal acetylenes apparently react with the catalyst, but disubstituted acetylenes are converted into *cis* alkenes. Thus stearolic acid gave oleic acid and diphenylacetylene formed *cis*-stilbene (Jardine and McQuillin, 1966) (7.52).

$$CH_3.(CH_2)_7.C{\equiv}C.(CH_2)_7.CO_2H \xrightarrow[\text{benzene}]{H_2,(Ph_3P)_3RuCl_2}$$

$$\underset{H}{\overset{CH_3(CH_2)_7}{\diagdown}}C{=}C\underset{H}{\overset{(CH_2)_7CO_2H}{\diagup}}$$

(7.52)

In contrast to reactions with $(Ph_3P)_2RhClH_2$ hydrogenations with $(Ph_3P)_2RuClH$ and $(Ph_3P)_3RhH(CO)$ are two-step processes which proceed by the reversible formation of a metal–alkyl intermediate. The actual catalyst in the former case is thought to be the square monomer $(Ph_3P)_2RuClH$ with *trans* PPh_3 groups, formed by dissociation in solution (7.53). The high selectivity for reduction of terminal double bonds is attributed to steric hindrance by the bulky PPh_3 groups to the formation of the metal–alkyl intermediate with other types of olefins.

An interesting development in this field has been the use of soluble metal catalysts to effect asymmetric hydrogenations (Abley and McQuillin, 1969). Trichlorotripyridylrhodium, treated with sodium borohydride in dimethylformamide, gives a complex [py$_2$dmf.

$$(Ph_3P)_2Ru\diagdown\begin{matrix}H\\Cl\end{matrix} \xrightarrow{R.CH=CH_2} \left[(Ph_3P)_2Ru\diagdown\begin{matrix}H\\Cl\end{matrix} \longleftarrow \begin{matrix}R\\|\\CH\\||\\CH_2\end{matrix}\right]$$

(7.53)

$$\longrightarrow (Ph_3P)_2Ru\!\!-\!\!\underset{\underset{R}{|}}{\overset{\underset{}{|}}{C}}HCH_3 \xrightarrow{H_2} R.CH_2.CH_3 + (Ph_3P)_2Ru\diagdown\begin{matrix}H\\Cl\end{matrix}$$

$RhCl_2(BH_4)$] which is highly active for the homogeneous hydrogenation of alkenes. By using an asymmetric ligand related to dimethylformamide a catalyst is obtained which can bring about asymmetric hydrogenations. Thus, methyl 3-phenylbut-2-enoate is hydrogenated at a rhodium complex formed in (+)- or (−)-1-phenylethylformamide to give (+)- or (−)-methyl 3-phenylbutanoate in better than 50 per cent optical yield (7.54).

$$PhCMe{=}CH.CO_2CH_3 \longrightarrow Ph\overset{*}{C}HMe.CH_2CO_2CH_3 \qquad (7.54)$$

7.2. Reduction by dissolving metals

General. Chemical methods of reduction are of two main types: those which take place by addition of electrons to the unsaturated compound followed or accompanied by transfer of protons; and those which take place by addition of hydride ion followed in a separate step by protonation.

Reductions which follow the first path are generally effected by a metal, the source of the electrons, and a proton donor which may be water, an alcohol or an acid. They can result either in the addition of hydrogen atoms to a multiple bond, or in fission of a single bond between atoms, usually, in practice, a single bond between carbon and a hetero-atom. In these reactions an electron is transferred from the metal surface (or from the metal in solution) to the organic molecule being reduced giving, in the case of addition to a multiple bond, an anion radical, which in many cases is immediately protonated. The resulting radical subsequently takes up another electron from the metal to form an anion which may be protonated immediately or remain as the anion until work-up. In the absence of a proton source dimerisation or polymerisation of the anion-radical may take place. In some cases a second electron may be added to the anion-radical to form a dianion, or two anions in the case of fission reactions (Birch 1950; House, 1965). These transformations may be represented as shown in (7.55).

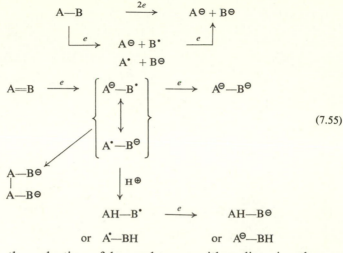

$$(7.55)$$

Thus, in the reduction of benzophenone with sodium in ether or liquid ammonia, the first product is the resonance-stabilised anion radical (XXVII) which, in absence of a proton donor, dimerises to the pinacol. In presence of a proton source, however, protonation leads to the radical (XXVIII) which is subsequently converted into the anion and thence into benzhydrol (7.56). The presence in these anion radicals of an unpaired electron which interacts with the atoms in the conjugated system has been established by measurements of the e.s.r. spectra of various anion radical solutions.

The metals commonly employed in these reductions include the alkali metals, calcium, zinc, magnesium, tin and iron. The alkali metals are often used in solution in liquid ammonia (Birch reduction, see p. 335) or as suspensions in inert solvents such as ether or toluene, frequently with addition of an alcohol or water to act as a proton source. Many reductions are also effected by direct addition of sodium or, particularly, zinc, tin or iron, to a solution of the compound being reduced in a hydroxylic solvent, such as ethanol, acetic acid or an aqueous mineral acid.

Reduction with metal and acid. In the well-known Clemmensen reduction of the carbonyl group of aldehydes and ketones to methyl or methylene, a mixture of the carbonyl compound and amalgamated zinc is boiled with hydrochloric acid, sometimes in presence of a non-miscible solvent which serves to keep the concentration in the aqueous phase low and thus prevent bimolecular condensations at the metal surface (Martin,

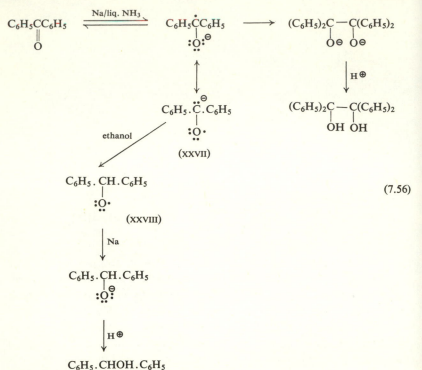

(7.56)

1942). Amalgamation of the zinc raises its hydrogen overvoltage to the point where it survives as a reducing agent in the acid solution and is not consumed in reaction with the acid to give molecular hydrogen. The choice of acid is confined to the hydrogen halides, which appear to be the only strong acids whose anions are not reduced by zinc-amalgam. Thus, stearophenone is converted into n-octadecylbenzene in 88 per cent yield, and 4-bromoindanone gives 4-bromoindane in 70 per cent yield without removal of the bromine atom (7.57). Isolated olefinic

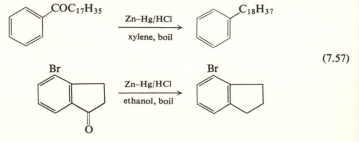

(7.57)

double bonds are not reduced under Clemmensen conditions, but conjugated bonds may be. For example benzalacetone is converted into n-butylbenzene in 50 per cent yield. In 1,3- and 1,4-dicarbonyl compounds and in compounds in which there is a good leaving group adjacent to the carbonyl group, reduction may be accompanied by bond cleavage and rearrangement, as in the examples (7.58) (Buchanan and Woodgate, 1969).

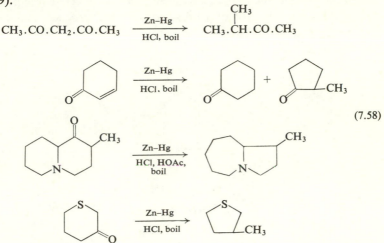

$$(7.58)$$

The mechanism of the Clemmensen reaction is uncertain. Alcohols are not normally reduced under Clemmensen conditions and it would appear that they are not intermediates in the reduction of the carbonyl compounds. The following mechanism (7.59) involving transfer of electrons from the metal surface to the carbonyl carbon atom has been proposed by Nakabayashi (1960). A related mechanism involving initial

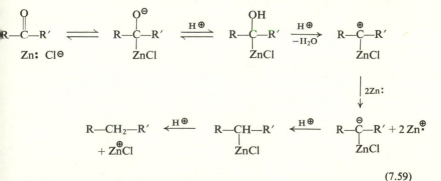

$$(7.59)$$

attack of zinc on the oxygen atom of the carbonyl group has been used by Buchanan and Woodgate (1969) to rationalise the rearrangements observed during Clemmensen reduction of difunctional ketones.

Reductive cleavage of α-substituted ketones, such as α-halo, α-amino-, α-acyloxy- and α-hydroxy-ketones is commonly effected with zinc and acetic acid or dilute mineral acid, rather than under Clemmensen conditions. Metal–amine reducing agents are also effective (see p. 346). The well-known reduction of α-bromoketones with zinc and acetic acid can be represented as in (7.60). α-Ketols are similarly reduced to the ketone (7.61).

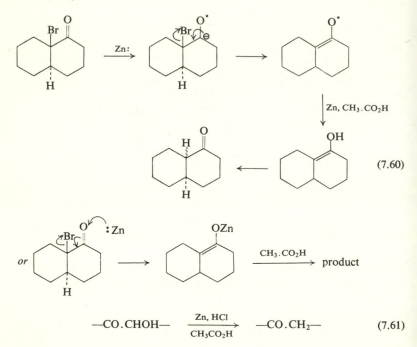

$$-CO.CHOH- \xrightarrow[\text{CH}_3\text{CO}_2\text{H}]{\text{Zn, HCl}} -CO.CH_2- \qquad (7.61)$$

Reductive eliminations of this type proceed most readily if the molecule can adopt a conformation where the bond to the group being displaced is perpendicular to the plane of the carbonyl group. Elimination of the substituent group is then eased by continuous overlap of the developing *p* orbital at the α carbon atom with the π orbital system of the ion radical (7.62). For this reason cyclohexanone derivatives with axial α-substituents are reductively cleaved more readily than their equatorial isomers. For example, the (axial) hydroxyl group of the 20-keto-20*a*-

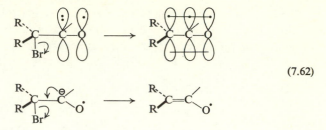

(7.62)

α-hydroxy-20a-β-methyl-D-homo-steroid derivative (XXIX) was cleaved in high yield with zinc and acetic acid, whereas the 20a-β-hydroxy epimer was unaffected (7.63). Vinylogous α-substituted ketones are also

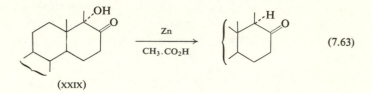

(XXIX)

(7.63)

reductively cleaved. Thus, the (axial) acetoxy compound (XXX) was smoothly converted into cholestenone and the epoxide (XXXI) gave the derivative (XXXII). Again the stereochemistry of the substituent determines the ease of fission (7.64).

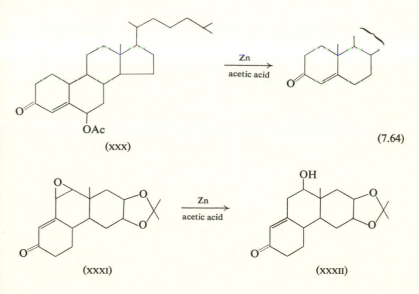

(XXX)

(7.64)

(XXXI)   (XXXII)

Even carbon–carbon bonds may be broken in favourable circumstances. Thus, the cyclopropyl ketone (XXXIII) was converted into the bicyclic compound (XXXIV) with zinc–copper couple in acetic acid (7.65).

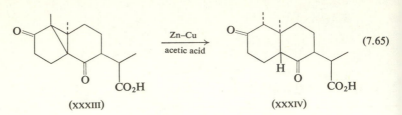

(XXXIII)   (XXXIV)

Although alkyl halides are not normally cleaved by zinc and acid, the benzyl type halides are. For example, 1-chloromethylnaphthalene is readily converted into 1-methylnaphthalene. The more active of two halogen atoms can often be removed selectively, as in (7.66).

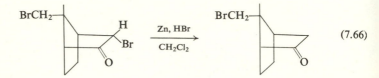

Reduction of carbonyl compounds with metal and an alcohol. Reduction of ketones to secondary alcohols can be effected catalytically (p. 310), with complex hydrides (p. 357) or with sodium (either as the free metal or as a solution in liquid ammonia) and an alcohol. The distinguishing feature of the sodium–alcohol method is that with cyclic ketones it gives rise, in most cases although not always, to the thermodynamically more stable alcohol either exclusively or in preponderant amount. Table 7.2 shows the proportions of more stable *trans* (equatorial) alcohol formed by reduction of 2-methylcyclohexanone with different reducing agents (McQuillin, Ord and Simpson, 1963) (7.67). Similarly 4-t-butylcyclohexanone gives the more stable *trans*-4-t-butylcyclohexanol almost exclusively on reduction with lithium and propanol in liquid ammonia, and numerous other experiments confirm that in the vast majority of cases the more stable alcohol is the main product.

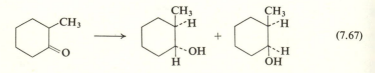

TABLE 7.2 *Proportion of* trans-2-*methylcyclohexanol by reduction of* 2-*methylcyclohexanone with different reagents*

Reagent	Proportion *trans* alcohol (%)
Na–alcohol	99
lithium aluminium hydride	82
sodium borohydride	69
aluminium isopropylate	42
catalyst and hydrogen	7–35

Two main hypotheses have been put forward to account for the high proportion of the more stable epimer formed in these reductions with sodium and an alcohol. In one due to Barton and Robinson (1954) it is supposed that a tetrahedral di-anion is formed initially and adopts the more stable configuration with equatorial oxygen which, on protonation, affords the more stable arrangement of the asymmetric centre (7.68).

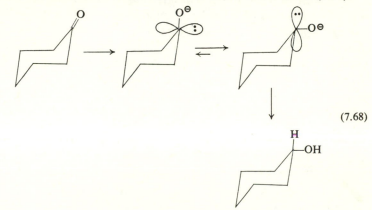

(7.68)

A second suggestion is that the reaction proceeds by transfer of one electron from the metal to the carbonyl group, followed by protonation of the resulting anion radical from the less hindered side and further reduction to form the alkoxide. This alkoxide, once formed, can react with unchanged ketone by a mechanism similar to that of the Meerwein–Pondorff–Verley reduction to give an equilibrium mixture which will favour the more stable of the two alkoxides. Final acidification of the reaction mixture thus affords the more stable alcohol as the main product (7.69).

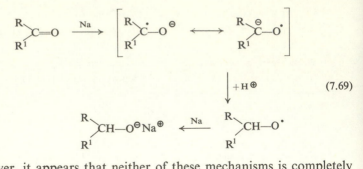

$$(7.69)$$

However, it appears that neither of these mechanisms is completely satisfactory, and modifications have been suggested by Huffman and Charles (1968) and by Taylor (1969). It may be that in the reduction of simple unhindered cyclohexanones protonation of the anion radical (Huffman and Charles) or the dianion (Taylor), contrary to previous opinion, takes place more easily from an axial direction to lead directly to the more stable equatorial alcohol.

In the case of a few strained or sterically hindered ketones reduction with sodium and alcohol has been found to give the less stable product. For example, reduction of norcamphor with lithium and ethanol gave mainly the thermodynamically less stable *endo* alcohol (Huffman and Charles, 1968). This may be because equilibration to the *exo* isomer is prevented by the resistance of the tetrahedral alkoxide ion to reconversion into the strained trigonal ketone.

Aldehydes can be reduced to primary alcohols with sodium and ethanol or aluminium amalgam, but in many cases better yields are obtained by catalytic hydrogenation (p. 310) or with hydride reducing agents (p. 357).

Reduction of ketones with dissolving metals in absence of a proton donor leads to the formation of bimolecular products. The usual reagents are magnesium, magnesium amalgam or aluminium amalgam. Thus, reduction of acetone to pinacol follows the course shown in (7.70).

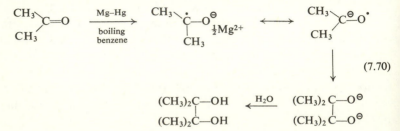

$$(7.70)$$

Bimolecular reduction of this type is often a competing reaction in other reductions with dissolving metals.

Esters are also reduced by sodium and alcohols to form primary alcohols. This, the Bouveault–Blanc reaction, is one of the oldest established methods of reduction used in organic chemistry, but has now been largely replaced by reduction with lithium aluminium hydride or catalytic hydrogenation with a copper chromite catalyst. It follows the same general course as the reduction of aldehydes and ketones. When the reaction is carried out in absence of a proton donor, for example with sodium in xylene or sodium in liquid ammonia, dimerisation, analogous to the formation of pinacols from ketones, takes place, and this is the basis of the well-known and synthetically useful acyloin condensation (McElvain, 1948). Intramolecular reaction gives ring compounds, and the reaction has been extensively used for the synthesis of medium and large carbon rings as well as five- and six-membered rings. Thus, the dicarboxylic ester (xxxv) gives the ten-membered ring ketol (xxxvi), and, in the steroid series, dimethyl marrianolate methyl ether (xxxvii) gives the ketol (xxxviii) which is readily converted into oestrone (7.71).

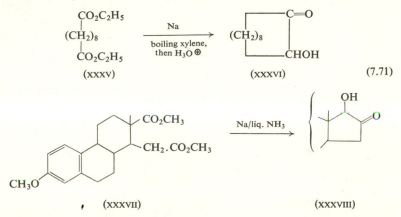

Reduction with metal and ammonia or an amine. Conjugated systems. Isolated carbon–carbon double bonds are not normally reduced by dissolving metal reducing agents because formation of the intermediate electron-addition products requires more energy than ordinary reagents can provide. Reduction is possible when the double bond is conjugated, because then the charges can be stabilised by resonance. By far the best reagent is a solution of an alkali metal in liquid ammonia, with or without addition of an alcohol—the so called Birch reduction conditions. Under these conditions conjugated dienes, $\alpha\beta$-unsaturated ketones, styrene derivatives and even benzene rings can be reduced to dihydro derivatives (Birch, 1950; Birch and Smith, 1958).

Birch reductions are usually effected with solutions of lithium, sodium or potassium in liquid ammonia. In some reactions an alcohol is added to act as a proton donor and as a buffer against the accumulation of the strongly basic amide ion; in other cases acidification with an alcohol or with ammonium chloride is effected at the end of the reaction. Solutions of alkali metals in liquid ammonia contain solvated metal cations and electrons (7.72), and part of the usefulness of these reagents arises

$$M \xrightarrow{\text{liquid NH}_3} M^{\oplus}(NH_3) \cdots e^{\ominus}(NH_3) \tag{7.72}$$

from the small steric requirement of the electrons, which sometimes allows reactions which are difficult to achieve with other reducing agents, and in many cases leads to different stereochemical results. Wasteful reaction of the metal with ammonia to form hydrogen and amide ion is slow in the absence of catalysts; part of the superiority of lithium over sodium and potassium is due to the fact that its reaction with ammonia is catalysed to a lesser extent by colloidal iron present as an impurity in ordinary commercial liquid ammonia. Using distilled ammonia there is little to chose among the three metals (see Dryden, Webber, Burtner and Cella, 1961). Reactions are usually carried out at the boiling point of ammonia ($-33°$), and since the solubility of many organic compounds in liquid ammonia is low at this temperature, co-solvents such as ether, tetrahydrofuran or dimethoxyethane are often added to aid solubility. A common technique is to add the metal to a stirred solution of the compound in liquid ammonia and the co-solvent, in presence of an alcohol where appropriate, but variations of this procedure are often used (Wilds and Nelson, 1953; Djerassi, 1963).

Conjugated dienes are readily reduced to the 1,4-dihydro derivatives with metal–ammonia reagents in the absence of added proton donors. Thus, isoprene is reduced to 2-methylbut-2-ene by sodium in ammonia through the anion radical. The protons required to complete the reduction are supplied by the ammonia (7.73).

Isolated olefinic bonds are not reduced by metal–ammonia reagents alone. In the presence of an alcohol terminal olefinic bonds are reduced. Thus hex-1-ene is converted into hexane by sodium and methanol in liquid ammonia, but hex-2-ene is unaffected.

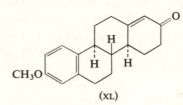

$$(7.73)$$

As was pointed out above (p. 332), reduction of cyclic ketones with metal–ammonia–alcohol reagents leads to the alcohol with the thermodynamically more stable configuration. With $\alpha\beta$-unsaturated ketones the saturated ketone or the saturated alcohol can be obtained depending

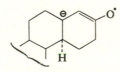

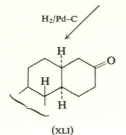

$$(7.74)$$

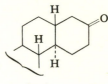

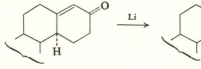

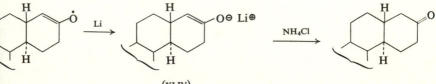

on the conditions. Reduction in absence of a proton donor followed by acidification with ethanol or ammonium chloride leads to the saturated ketone. The same transformation can, of course, be effected by catalytic hydrogenation, but metal–ammonia reduction is more stereoselective and in many cases leads to a product with a different stereochemistry from that obtained by hydrogenation. Thus, the $\alpha\beta$-unsaturated ketone (XL) on reduction with hydrogen and palladium affords the *cis* fused ketone (XLI), while with lithium in liquid ammonia the *trans* product (XLII) is obtained (7.74). The initially formed anion radical (XLIII) is a sufficiently strong base to abstract a proton from the ammonia to form, after addition of another electron the enolate anion (XLIV). This anion is insufficiently basic to abstract a proton from ammonia and, in the absence of a stronger acid it retains its negative charge and resists addition of another electron which would correspond to further reduction. Acidification during isolation leads then to the saturated ketone. In the presence of 'acids' sufficiently strong to protonate the enolate anion, however, further reduction to the saturated alcohol occurs (7.75). The presence of enolate anions of the type of (XLIV) in the metal–

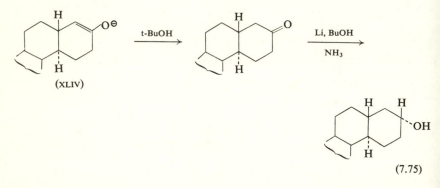

ammonia reduction product before acidification is shown by the ready formation of α-methyl ketones by treatment with methyl iodide (7.76).

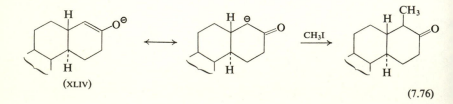

Reduction of cyclic $\alpha\beta$-unsaturated ketones in which there are substituents on the β- and γ-carbon atoms could apparently give rise to two epimeric products (7.77), but it is found in practice that in most cases

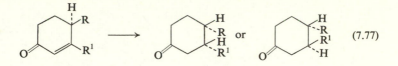

only one isomer is formed, and that is generally the more stable of the two. Barton and Robinson (1954) have suggested that this is because there is a rapid equilibration of a carbanion intermediate leading to the thermodynamically more stable system in which the groups R and R^1 are *trans* to each other.

More recently, however, Stork and Darling (1964) have uncovered a number of cases where the thermodynamically more stable isomer is not the main product, and they have put forward a more comprehensive scheme to account for these reactions. For example, reduction of the octalones ((XLV), R = H or CH_3) with lithium and ethanol in ammonia followed by oxidation with chromic acid gave the *trans*-decalones ((XLVI), R = H or CH_3), even although, because of the steric effect of the substituents, the *cis*-decalones are more stable (7.78). It is suggested

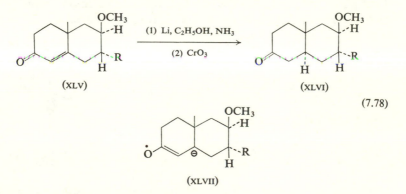

that for stereoelectronic reasons protonation of the β-carbon atom always takes place in a direction which is axial to the cyclohexenone ring, and to the most stable conformation/configuration of the carbanion which allows continuous overlap of the p orbital on the β-carbon atom with the π-orbital system of the enolate radical. Thus, there are three possible structures for the transition state for protonation of the anion radical (XLVII) in the reduction of (XLV) (7.79). Of these the first (A) is

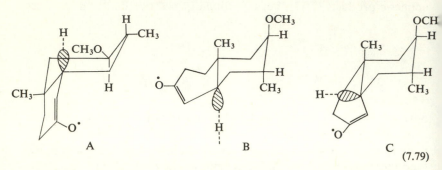

$$(7.79)$$

ruled out because it does not allow continuous overlap of the *p*- and π-orbital systems. Of the other two in both of which orbital overlap is possible, (B) is the more stable for ordinary steric reasons, and protonation takes place through this transition state to give the *trans*-decalone system.

Another example of the formation of the less stable isomer on metal–ammonia reduction of an αβ-unsaturated ketone is shown in the conversion of 3,4-diphenylcyclopent-2-enone into *cis*-3,4-diphenylcyclopentanone. The two possible transition states here are (XLVIII) and (XLIX) and formation of the *cis*-diphenylcyclopentanone reflects the strong steric hindrance to close approach of the proton carrier to the anion (XLIX) caused by the α-phenyl group (7.80).

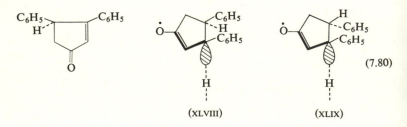

$$(7.80)$$

After protonation of the β-carbon atom of an αβ-unsaturated carbonyl system, the final product is obtained by addition of another electron followed by protonation of the α-carbon atom during work-up (7.81). The stereochemistry of this process is not easy to predict. This protonation is kinetically controlled and usually takes place from the less hindered side of the enolate system, although other factors may influence

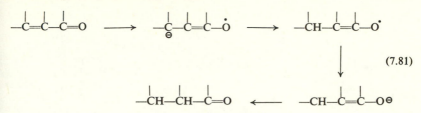

$$(7.81)$$

the direction of addition (Zimmerman, 1961), to give a ketone which is not necessarily the thermodynamically more stable isomer. In many cases, however, the more stable isomer can be obtained by subsequent equilibration and it is sometimes formed directly by equilibration during isolation. For example, reduction of 2-methyl-3-phenylindone (L) with lithium in liquid ammonia, followed by protonation with ammonium chloride, affords a mixture in which *cis*-2-methyl-3-phenylindanone predominates, by addition of a proton to the least hindered side of the enolate anion. Protonation with ethanol, however, leads to the more stable *trans* isomer as a result of equilibration of the initial product by ethoxide ion (7.82).

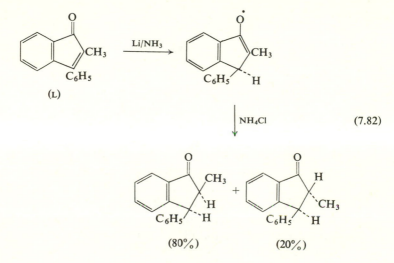

$$(7.82)$$

In the reduction of styrene derivatives also, in which the double bond forms part of a ring, the isomer produced by protonation of the highly basic β-anion corresponds to the most stable conformation of the anion. Thus, reduction of the steroid intermediate (LI) with lithium in ammonia gave the more stable $8\beta,9\alpha$-isomer in major amount (7.83).

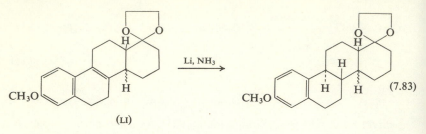

(LI)   (7.83)

Aromatic compounds. One of the most useful synthetic applications of metal–ammonia–alcohol reducing agents is in the reduction of benzene rings to 1,4-dihydro derivatives. The reagents are powerful enough to reduce benzene rings, but specific enough to add only two hydrogen atoms. Benzene itself is reduced with lithium and ethanol in liquid ammonia to 1,4-dihydrobenzene by way of the anion radical (LII) (Birch, 1950; Birch and Nasipuri, 1959) (7.84). The presence

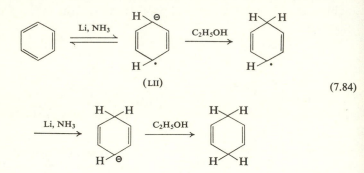

(LII)   (7.84)

of an alcohol as a proton donor is necessary in these reactions, for the initial anion radical is an insufficiently strong base to abstract a proton from the ammonia. The alcohol also acts to prevent the accumulation of the strongly basic amide ion which might bring about isomerisation of the 1,4-dihydro compound to the conjugated 1,2-dihydro isomer which would be further reduced to tetrahydrobenzene.

Naphthalene is converted into the 1,4,5,8-tetrahydro compound with sodium and alcohol in liquid ammonia (7.85). Alkylbenzenes are reduced to 2-alkyl-1,4-dihydrobenzenes. Particularly useful synthetically is the

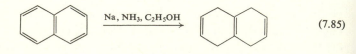

(7.85)

reduction of anisoles and anilines to dihydro compounds which are readily hydrolysed to cyclohexenones. Under mild conditions the $\beta\gamma$-unsaturated ketones are obtained, but these are readily isomerised to the conjugated $\alpha\beta$-unsaturated compounds (7.86). With substituted benzenes addition of

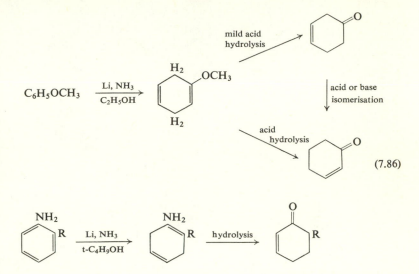

(7.86)

the two hydrogen atoms could take place in more than one way. A general rule is that hydrogen addition takes place at positions which are *para* to each other avoiding, if possible, carbon atoms to which electron-donating substituents are attached. The directive effect of —OCH$_3$ and —NMe$_2$ is stronger than that of alkyl groups so that 5-methoxytetralin gives (LIII) (7.87). On the other hand, reduction of benzene rings substituted by the electron-withdrawing carboxyl group (the only electron-withdrawing group which is not reduced before the benzene ring) gives derivatives of 1,4-dihydrobenzoic acid. These orientation effects are in line with molecular orbital calculations of the sites of highest electron density in the anion radicals (Zimmerman, 1961).

Selective reduction of a benzene ring in presence of another reducible group is possible if the other group can be protected by reversible conversion into a saturated derivative, or into a derivative containing only an isolated carbon–carbon double bond or into an anion by salt formation. Ketones, for example, may be converted into ketals or enol ethers, as shown in (7.88). But this is ineffective if the carbonyl group is conjugated with the benzene ring.

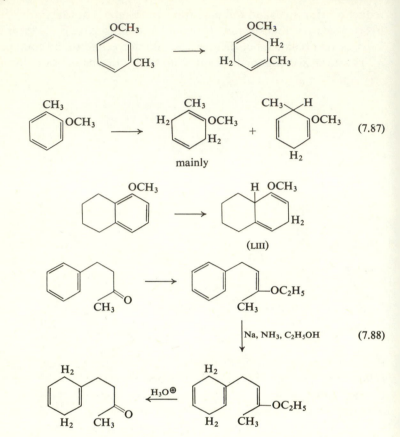

(7.87)

(LIII)

(7.88)

Conversely, reduction of benzene rings takes place only slowly in the absence of a proton donor and selective reduction of an $\alpha\beta$-unsaturated carbonyl system can be effected if no alcohol is added (7.89).

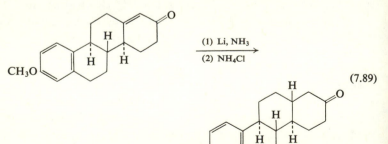

(7.89)

A spectacular demonstration of the high stereoselectivity shown by the metal–ammonia–alcohol reagents is found in the conversion of the compound (LIV) into the ketones (LV) and (LVI) by simultaneous reduction of all the unsaturated centres. Although thirty-two racemates of ketone (LV) and sixty-four racemates of ketone (LVI) are possible, in fact the mixture of the two ketones with the stereochemistry shown was obtained in 25 per cent yield (Johnson *et al.*, 1953) (7.90).

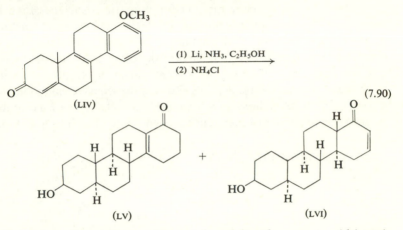

Reduction of acetylenes with metal and liquid ammonia. Although olefinic double bonds are not normally reduced by metal–ammonia reducing agents, the partial reduction of acetylenic bonds is very conveniently effected by these reagents. The procedure is highly selective for the partial reduction of the triple bond, and none of the saturated product is formed. Furthermore, the reduction is completely stereospecific and the only product from a disubstituted acetylene is the corresponding *trans* olefin. This method thus complements the formation of *cis* olefins by catalytic hydrogenation of acetylenes discussed on p. 307. The reaction presumably takes place by stepwise addition of two electrons to the triple bond to form a dianion, the more stable *trans* configuration of which, with maximum separation of the two negatively charged sp^2 orbitals, is protonated to form the *trans* olefin (7.91).

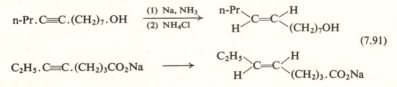

Terminal acetylenic bonds may be reduced or not as required. Under ordinary conditions, reaction of the acetylene with the metal solution leads to the formation of the metallic acetylide which resists further reduction because of the negative charge on the ethynyl carbon atom. In presence of ammonium sulphate, however, the free ethynyl group is liberated and on reduction yields the corresponding olefin. Thus hepta-1,6-diyne is converted into hepta-1,6-diene in high yield under these conditions. This is a better method for preparing 1-alkenes than the alternative catalytic hydrogenation of 1-alkynes, which sometimes leads to small amounts of the saturated hydrocarbons which may be difficult to separate from the olefin. On the other hand, reduction of a terminal acetylene can be suppressed by converting it into its sodium salt by reaction with sodamide, thus allowing the selective reduction of an internal triple bond in the same molecule. Undeca-1,7-diyne, for example, is directly converted into *trans*-undeca-7-en-1-yne in high yield (7.92).

$$CH_3(CH_2)_2C\equiv C(CH_2)_4.C\equiv CH \xrightarrow[\text{liq. NH}_3]{\text{NaNH}_2} R-C\equiv C^{\ominus} \ Na^{\oplus}$$

$$\Big\downarrow \begin{array}{l} \text{(1) Na, NH}_3 \\ \text{(2) H}^{\oplus} \end{array} \qquad (7.92)$$

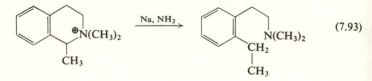

Reductive fission. Metal–amine reducing agents and other dissolving metal systems can bring about a variety of reductive fission reactions, some of which are useful in synthesis. For example, quaternary ammonium salts are cleaved as in the well known Emde reaction (7.93).

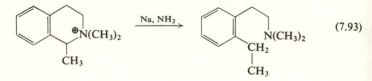

Most of these fission reactions probably proceed by direct addition of two electrons from the metal to the bond which is broken. The anions produced may be protonated by an acid in the reaction medium, or may survive until work-up. In some cases it may be desirable to maintain the fission products as anions to prevent further reduction.

Reaction is facilitated when the anions are stabilised by resonance or by an electronegative atom. As expected, therefore, bonds between electronegative atoms or between an electronegative atom and an unsaturated system which can stabilise a negative charge by resonance, are particularly easily cleaved. Thus allyl and benzyl halides, ethers and esters, and sometimes even the alcohols themselves, are readily cleaved by metal–amine systems and other dissolving metal reagents, as well as by catalytic hydrogenation (see p. 300). Fission reactions of this type, particularly with solutions of an alkali metal in liquid ammonia, have been widely used in structural studies, and also for the reductive removal of unsaturated groups used as protecting agents for amino, imino, hydroxyl and thiol groups. They are of particular value in peptide synthesis, and have an obvious general advantage over hydrolytic or catalytic methods for compounds which are labile or contain sulphur. Benzyl, toluene-*p*-sulphonyl and benzyloxycarbonyl groups are all efficiently replaced by hydrogen, and cystinyl-peptides are cleaved to cysteinyl derivatives (7.94). For example in the synthesis of the naturally

$$RS.CH_2Ph \longrightarrow RSH + CH_3Ph$$

$$RNH.SO_2.C_6H_4CH_3\text{-}p \longrightarrow RNH_2 + HS.C_6H_4CH_3\text{-}p$$

$$RNH.CO_2CH_2Ph \longrightarrow RNH_2 + CO_2 + CH_3Ph \tag{7.94}$$

$$R.SS.R. \longrightarrow 2RSH$$

$$R.O.CH_2Ph \longrightarrow ROH + CH_3Ph$$

$$\begin{array}{c} CH_2.S.CH_2.C_6H_5 \\ | \\ CH_2.CO.NH.CH.CO.NH.CH_2.CO_2H \\ | \\ CH_2 \\ | \\ C_6H_5CH_2.O.CO.NH.CH.CO_2H \end{array}$$

$$\Big\downarrow Na, NH_3 \tag{7.95}$$

$$\begin{array}{c} CH_2SH \\ | \\ CH_2.CO.NH.CH.CO.NH.CH_2.CO_2H \\ | \\ CH_2 \\ | \\ H_2N.CH.CO_2H \end{array}$$

occurring peptide, glutathione, *N*-benzyloxycarbonyl-*γ*-glutamyl-*S*-benzylcysteinylglycine was prepared as the key intermediate in which the benzyloxycarbonyl group was used for protection of the amino group and the benzyl group for protection of the thiol. The protective groups were finally removed to yield glutathione by cleavage with sodium in liquid ammonia (7.95).

Reductive fission has been of great assistance in the elucidation of the structures of a number of naturally occurring allyl and benzyl alcohols, ethers and esters. Thus, the structure of the alcohol lanceol was neatly confirmed by reduction with sodium and alcohol in liquid ammonia to the known sesquiterpene hydrocarbon bisabolene (7.96).

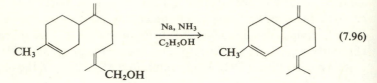

(7.96)

The antibiotic mycelianamide similarly gave methylgeraniolene and a derivative of *p*-hydroxybenzoic acid (7.97). In these examples the

$$
\underset{CH_3}{\overset{CH_3}{\diagdown}}C{=}CH.CH_2.CH_2.\overset{\overset{\displaystyle CH_3}{\displaystyle |}}{C}{=}CH.CH_2.O.C_6H_4R
$$

$$\downarrow$$

$$
\underset{CH_3}{\overset{CH_3}{\diagdown}}C{=}CH.CH_2.CH_2.\overset{\overset{\displaystyle CH_3}{\displaystyle |}}{C}{=}CH.CH_3 + HOC_6H_4R
$$

(7.97)

allylic groups are cleaved specifically without reduction of the olefinic double bonds.

The reductive fission of allyl or benzyl alcohols can be prevented, if necessary, by converting them into the corresponding alkoxide ions by reaction with sodamide or other strong base, as in (7.98).

$$
Ph.CH{=}CH{-}CMe_2OH \xrightarrow[\text{(2) Na–liq. NH}_3]{\text{(1) NaNH}_2} Ph.CH_2.CH_2.CMe_2OH
$$

(7.98)

In contrast to benzyl alcohols, benzylamines are not readily reduced with metal–ammonia reagents, because nitrogen is less electronegative than oxygen, but quaternary salts are easily cleaved (see p. 346).

Solutions of sodium in liquid ammonia also cleave aryl ethers. Diaryl ethers react particularly readily. Thus methyl *m*-cresyl ether is converted into *m*-cresol by reaction with sodium and liquid ammonia in ether and acidification of the sodium salt, and phenyl *p*-tolyl ether similarly gives a mixture of *p*-cresol and phenol in the ratio 3:1.

Organic halides can be converted into the halogen-free compounds with dissolving metal reagents, but since the highly electropositive metals such as sodium induce Wurtz coupling metal–amine systems are unsuitable and best results are obtained with magnesium or zinc in a protonic solvent (7.99). The mechanism of these reactions is uncertain

$$\text{n-}C_{15}H_{31}CH_2I \xrightarrow[\text{CH}_3\text{.COOH}]{\text{Zn, HCl}} \text{n-}C_{15}H_{31}CH_3 \qquad (7.99)$$

but it is thought that they may proceed by way of short-lived organo-metallic intermediates. This is supported by the observation that if a substituent which can be lost as a stable anion (e.g. —OH, —OR, —OCOR, halogen) is present on the adjacent carbon atom elimination rather than simple reductive cleavage is observed (7.100).

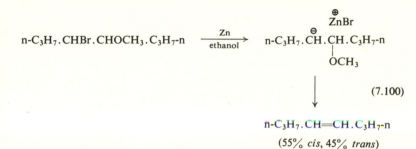

$$\text{(55\% }\textit{cis},\text{ 45\% }\textit{trans)}$$

Reductive cleavage of alkyl halides may or may not occur with retention of configuration, but cleavage of vinyl chlorides with sodium in liquid ammonia gives olefins with the same configuration as the halide, reflecting the great stability of vinyl carbanions compared with alkyl carbanions (7.101).

$$\begin{array}{c}
C_2H_5 \\ \diagdown \\ H \diagup \end{array}\!\!C\!\!=\!\!C\!\!\begin{array}{c}\diagup C_2H_5 \\ \diagdown Cl\end{array} \xrightarrow{\text{Na, NH}_3} \begin{array}{c}C_2H_5 \\ \diagdown \\ H \diagup \end{array}\!\!C\!\!=\!\!C\!\!\begin{array}{c}\diagup C_2H_5 \\ \diagdown H\end{array} \qquad (7.101)$$

It has recently been found that alkyl, vinyl and aromatic halides can be conveniently reduced in high yield by reaction with magnesium in presence of a secondary alcohol (Bryce-Smith, Wakefield and Blues,

1963). β-Bromostyrene, for example is converted into styrene in 72 per cent yield with magnesium and isopropanol at 80°C.

Other useful reagents for reduction of carbon–halogen bonds are trialkyltin hydrides and chromous salts. Reduction with trialkyltin hydrides is a general method and can be applied to both alkyl and aryl halides (Kuivila, 1968). It proceeds by a free-radical chain process. From the synthetic viewpoint, stepwise reduction of geminal poly-halides is a useful application (7.102). Reduction of simple alkyl or

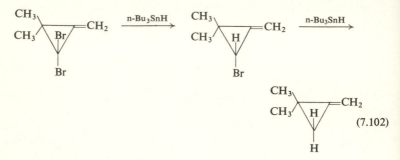

(7.102)

aralkyl halides with chromous salts usually effects replacement of halogen by hydrogen (Hanson and Premuzic, 1968). Reaction is thought to proceed by two one-electron transfer steps as in (7.103). Compounds

$$\ce{>C-Br ->[Cr^{2+}] >C\cdot ->[Cr^{2+}] >C-Cr \longrightarrow >C-H} \qquad (7.103)$$

carrying a vicinal substituent which can be eliminated as an anion (e.g. OH, OR, Cl) usually give olefins (7.104). Elimination can be

$$\ce{CH2Br.CHBr.CH2Cl ->[CrSO4][aq. dimethylformamide] CH2=CH.CH2Cl} \qquad (7.104)$$

circumvented by effecting the reduction with chromous acetate in presence of a hydrogen atom transfer agent such as butanethiol, thus providing, for example, an excellent route from a bromohydrin to the corresponding alcohol as in the route to 11-β-hydroxy-steroids (7.105).

Lithium in alkylamines. Solutions of lithium in aliphatic amines of low molecular weight are powerful reducing agents, but are less selective than metal–ammonia reagents. The amines have the advantage that they

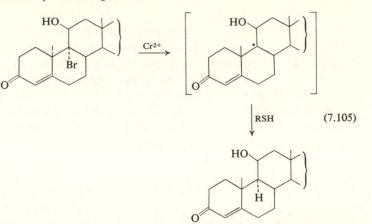

(7.105)

are better solvents for organic substances than ammonia is, and have higher boiling points which makes for easier handling. The higher working temperatures facilitate the initial stages of the reduction, but they also favour conjugation (and therefore further reduction) of the initial product, and unless precautions are taken reduction of an aromatic compound leads to the tetrahydro or even the hexahydro derivative (7.106). Recent work by Benkeser and his co-workers (1963; 1964) has

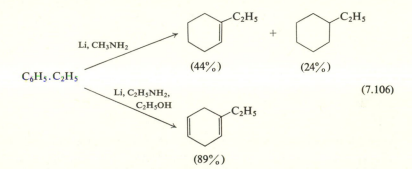

(7.106)

shown that under suitable conditions yields of 1,4-dihydro compounds comparable to those formed under Birch conditions can be obtained.

Solutions of lithium in alkylamines have been used successfully for reductive cleavage of allylic ethers and esters. Reduction of steroid epoxides leads to the axial alcohols, and in this reaction they are more powerful and specific reagents than lithium aluminium hydride (Hallsworth and Henbest, 1957).

7.3. Reduction by hydride-transfer reagents. Reactions which proceed by transfer of hydride ions are widespread in organic chemistry (Deno, Peterson and Saines, 1960), and they are important also in biological systems. Reductions involving the reduced forms of coenzymes I and II, for example, are known to proceed by transfer of a hydride ion from a 1,4-dihydropyridine system to the substrate. In the laboratory the most useful reagents of this type in synthesis are aluminium isopropoxide and the various metal hydride reducing agents.

Aluminium alkoxides. The reduction of carbonyl compounds to alcohols with aluminium isopropoxide has long been known under the name of the Meerwein–Pondorff–Verley reduction (Wilds, 1944). The reaction is easily effected by heating the components together in solution in isopropanol. An equilibrium is set up and the product is obtained by using an excess of the reagent or by distilling off the acetone as it is formed. The reaction is thought to proceed by transfer of hydride ion from the isopropoxide to the carbonyl compound through a six-membered cyclic transition state (7.107). Aldehydes are reduced to primary alcohols and ketones give secondary alcohols, often in high yield. The

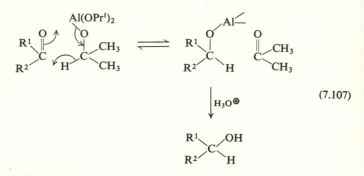

(7.107)

reaction owes its usefulness to the fact that olefinic double bonds and many other unsaturated groups are unaffected, thus allowing selective reduction of carbonyl groups. Cinnamaldehyde, for example, is converted into cinnamyl alcohol, *o*-nitrobenzaldehyde gives *o*-nitrobenzyl alcohol and phenacyl bromide gives styrene bromohydrin.

Reductions of a similar type can be brought about by other metallic alkoxides, but aluminium alkoxide is particularly effective because it is soluble in both alcohols and hydrocarbons and, being a weak base, it shows little tendency to bring about wasteful condensation reactions of the carbonyl compounds.

Reduction of cyclohexanones by the Meerwein–Pondorff–Verley method usually produces a higher proportion of the axial alcohol than other chemical methods of reduction (see Table 7.2, p. 333). Prolonged reaction may lead to the establishment of equilibrium through continuous reduction and re-oxidation, with gradual enrichment of the more stable equatorial alcohol.

Lithium aluminium hydride and sodium borohydride. A number of metal hydrides have been employed as reducing agents in organic chemistry, but the most commonly used are lithium aluminium hydride and sodium borohydride, both of which are commercially available. Another useful reagent, diborane, is prepared *in situ* as required (see p. 207).

The anions of the two complex hydrides can be regarded as derived from lithium or sodium hydride and either aluminium hydride or borane.

$$LiH + AlH_3 \longrightarrow Li^{\oplus} AlH_4^{\ominus}$$

$$NaH + BH_3 \longrightarrow Na^{\oplus} BH_4^{\ominus}$$

These anions are nucleophilic reagents and as such they normally attack polarised multiple bonds such as C=O or C≡N by transfer of hydride ion to the more positive atom. They do not usually react with isolated carbon–carbon double or triple bonds. Thus reduction of acetone with sodium borohydride proceeds as in (7.108). In aprotic

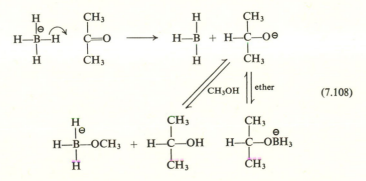

(7.108)

solvents the alkoxyborohydride is formed, but in hydroxylic solvents reaction with the solvent may occur to give the alcohol directly.

With both reagents all four hydrogen atoms may be used for reduction, being transferred in a stepwise process as illustrated for reduction of a ketone (7.109). For reductions with lithium aluminium hydride (but

TABLE 7.3   *Common functional groups reduced with lithium aluminium hydride*

Functional group	Product
$>\!C\!=\!O$	$>\!CH\!-\!OH$
$-CO_2R$	$-CH_2OH + ROH$
$-CO_2H$	$-CH_2OH$
$-CONHR$	$-CH_2NHR$
$-CONR_2$	$-CH_2NR_2$
	or
	$\left[\begin{array}{c}-CH\!-\!NR_2\\ \mid\\ OH\end{array}\right] \longrightarrow -CHO + R_2NH$
$-C\!\equiv\!N$	$-CH_2NH_2$
	or $[-CH\!=\!NH] \xrightarrow{H_2O} -CHO$
$>\!C\!=\!NOH$	$>\!CH\!-\!NH_2$
$-\overset{\mid}{\underset{\mid}{C}}\!-\!NO_2$ (aliphatic)	$-\overset{\mid}{\underset{\mid}{C}}\!-\!NH_2$
$ArNO_2$	$ArNHNHAr$
	or $ArN\!=\!NAr$
$-CH_2OSO_2C_6H_5$	$-CH_3$
or	
$-CH_2Br$	
$>\!CHOSO_2C_6H_5$	$>\!CH_2$
or $>\!CH\!-\!Br$	
$-CH\!-\!C\!\underset{O}{\overset{}{<}}$	$-CH_2\!-\!\overset{\mid}{\underset{OH}{C}}\!-$

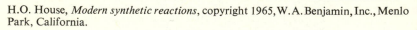

H.O. House, *Modern synthetic reactions*, copyright 1965, W.A. Benjamin, Inc., Menlo Park, California.

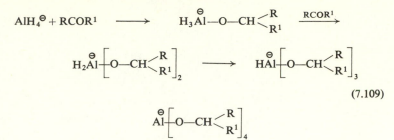

$$(7.109)$$

not with sodium borohydride) each successive transfer of hydride ion takes place more slowly than the one before, and this has been exploited for the preparation of modified reagents which are less reactive and more selective than lithium aluminium hydride itself by replacement of two or three of the hydrogen atoms of the anion by alkoxy groups (see p. 363).

Lithium aluminium hydride is more reactive than sodium borohydride. It reacts readily with water and other compounds containing active hydrogen atoms, and must be used under anhydrous conditions in a non-hydroxylic solvent; ether and tetrahydrofuran are often employed. Sodium borohydride reacts only slowly with water or methanol at room temperature, and reductions with this reagent can be effected in methanol solution at room temperature. Table 7.3, due to House (1965), shows some common functional groups which are reduced with lithium aluminium hydride. Sodium borohydride, being less reactive, is more discriminating in its action than lithium aluminium hydride. It generally does not attack esters or amides, and it is normally possible to reduce aldehydes and ketones selectively with this reagent in presence of a variety of other functional groups. Some typical reductions are illustrated in the equations (7.110).

Some exceptions to the general rule that olefinic double bonds are not attacked by hydride reducing agents have been noted in the reduction of β-aryl-$\alpha\beta$-unsaturated carbonyl compounds with lithium aluminium hydride. Even in these cases selective reduction of the carbonyl group can often be achieved by working at low temperatures and with short reaction times or by using sodium borohydride or aluminium hydride (p. 366) as reducing agent (7.111). This type of reduction of the double bond of allylic alcohols is thought to proceed through a cyclic organo-aluminium compound (LVII), for it is found experimentally that only one of the two hydrogen atoms added to the double bond is derived from the hydride, and acidification with a deuterated solvent leads to the deuterated alcohol shown in (7.112).

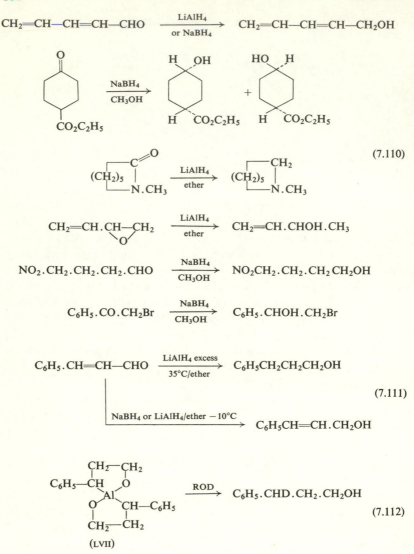

$$CH_2=CH-CH=CH-CHO \xrightarrow[\text{or NaBH}_4]{\text{LiAlH}_4} CH_2=CH-CH=CH-CH_2OH$$

(7.110)

$$CH_2=CH.CH-CH_2 \xrightarrow[\text{ether}]{\text{LiAlH}_4} CH_2=CH.CHOH.CH_3$$

$$NO_2.CH_2.CH_2.CH_2.CHO \xrightarrow[\text{CH}_3\text{OH}]{\text{NaBH}_4} NO_2CH_2.CH_2.CH_2 CH_2OH$$

$$C_6H_5.CO.CH_2Br \xrightarrow[\text{CH}_3\text{OH}]{\text{NaBH}_4} C_6H_5.CHOH.CH_2Br$$

$$C_6H_5.CH=CH-CHO \xrightarrow[\text{35}^\circ\text{C/ether}]{\text{LiAlH}_4 \text{ excess}} C_6H_5CH_2CH_2CH_2OH$$

(7.111)

NaBH$_4$ or LiAlH$_4$/ether -10°C $C_6H_5CH=CH.CH_2OH$

(7.112)

A similar type of aluminium compound is thought to be involved in the reduction of the triple bond of propargylic alcohols to a *trans* olefin with lithium aluminium hydride. Only triple bonds flanked by hydroxyl groups are reduced, as illustrated in the example (7.113).

Hydride reducing agents have probably found their most widespread

$$CH{\equiv}C.(CH_2)_2.C{\equiv}C.CO_2Et \xrightarrow{\text{LiAlH}_4} CH{\equiv}C.(CH_2)_2.CH{=}CH.CH_2OH$$

$$trans \quad (7.113)$$

use in the reduction of carbonyl-containing groups. Aldehydes, ketones, carboxylic acids, esters and lactones can all be reduced smoothly to alcohols under mild conditions (for examples and experimental conditions see Gaylord, 1956; Brown, 1951). Reaction with lithium aluminium hydride is the method of choice for the reduction of carboxylic esters to primary alcohols. Substituted amides are converted into amines or aldehydes, depending on the experimental conditions (see p. 364). When necessary, the carbonyl group of an aldehyde or ketone may be protected against reduction by conversion into the acetal or ketal. For example, acetoacetic ester can be converted into 4-hydroxybutan-2-one by reduction of the corresponding diethyl acetal with lithium aluminium hydride, followed by acid hydrolysis of the acetal group.

Reduction of unsymmetrical open-chain ketones leads, of course, to the racemic carbinol. With asymmetric ketones, however, the two forms of the carbinol may not be formed in equal amount. Thus, in the reduction of the ketone (LVIII) with lithium aluminium hydride, the *threo* form of the alcohol predominates in the product (7.114). The main product

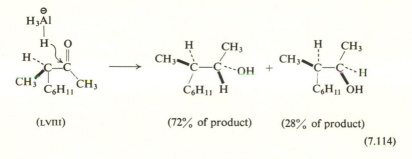

(LVIII)   (72% of product)   (28% of product)

$$(7.114)$$

formed in these reactions can be predicted on the basis of Cram's rule (Cram and Abd Elhafez, 1952) according to which that diastereoisomer predominates in the product which is formed by approach of the reagent from the less hindered side of the carbonyl group when the rotational conformation of the molecule is such that the carbonyl group is flanked by the two least bulky groups on the adjacent asymmetric centre. This may be represented using Newman projection formulae, as (7.115), where S, M and L represent small, medium and large substituents.

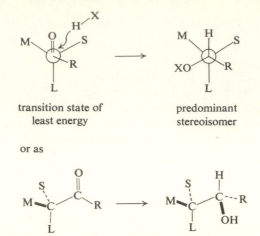

transition state of
least energy

predominant
stereoisomer

(7.115)

or as

Thus, for the reduction of the ketone (LVIII) the predominant *threo*-carbinol arises by attack of the metal hydride anion on the least hindered side of the carbonyl group in the conformation shown.

A somewhat different model for the transition state which is said to predict results more in agreement with experiment has been proposed by Cherest, Felkin and Prudent (1968) but the essential feature of Cram's hypothesis, involving approach of the hydride anion from the less hindered side of the carbonyl group is preserved.

In cases where there is a polar group on the carbon atom adjacent to the carbonyl group Cram's rule may not be followed, because the conformation of the transition state is no longer determined entirely by steric factors. In α-hydroxy- and α-amino-ketones for example, reaction is thought to proceed through a relatively rigid cyclic transition state of the type (LIX) with fixed conformation, and because of this reductions of α-hydroxy- and α-amino-ketones usually proceed with a comparatively high degree of stereoselectivity (7.116). Again, where an adjacent asymmetric carbon atom carries a chlorine substituent, the most reactive conformation of the molecule appears to be the one in which the polar halogen atom and the polar carbonyl group are *trans*, to minimise dipole–dipole repulsion. The predominant product is then formed by approach of the metal hydride anion to the least hindered side of this conformation. Thus, reduction of 4-chloro-octan-3-one with sodium borohydride affords mainly the chlorohydrin shown in (7.117).

With strongly hindered cyclic ketones the main product is again formed by approach of the reagent to the less hindered side of the

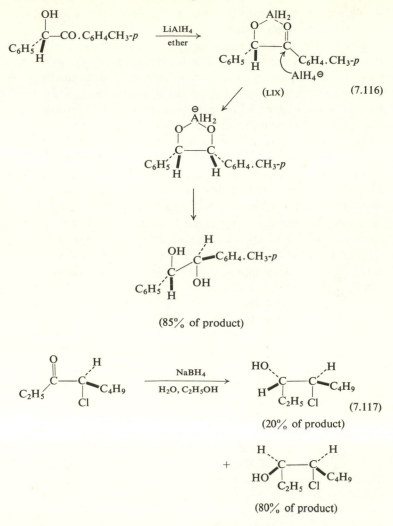

(85% of product)

(20% of product) (7.117)

+

(80% of product)

carbonyl group. Thus, reduction of camphor with lithium aluminium hydride leads mainly to the *exo*-alcohol (isoborneol), whereas nor-camphor, in which approach of the hydride anion is now easier from the side of the methylene bridge, leads mainly to the *endo*-alcohol.

The stereochemical course of the reduction of other less rigid cyclo-hexanones is not so easy to predict. In general, a mixture of products is obtained in which, with comparatively unhindered ketones, the more stable epimer predominates. But the relative proportion of the two isomers varies widely with the substitution of the ring.

It has been suggested (Dauben, Fonken and Noyce, 1956) that the steric course of these reactions is controlled by two usually opposing factors, a kinetic factor related to the degree of steric hindrance of the carbonyl group (steric approach control), and a thermodynamic factor related to the stability of the products (product development control). But it is now believed that the results in all cases can be accounted for by steric factors alone (Kirk, 1969; Cherest and Felkin, 1968). Richer (1965) has shown that, contrary to previous expectation, approach of a hydride ion to the carbonyl group of an unhindered cyclohexanone is easier from the axial side, because equatorial approach is hindered sterically by the axial hydrogen atoms or substituents in the 2- and 6- positions. The result is preferential formation of the more stable equatorial alcohol. But where the approaching group becomes larger, as with Grignard reagents, or where there are bulky axial substituents in the 3- and 5- positions, axial approach of the reagent is less easy and equatorial attack, leading to the axial alcohol, may be favoured. Thus, reduction of 4-t-butyl-2,2-dimethylcyclohexanone gives more equatorial alcohol (resulting from easier axial attack of the hydride on the carbonyl group) than 4-t-butylcyclohexanone because of the greater steric effect of the C-2 axial methyl group compared with an axial hydrogen atom. On the other hand, axial attack on the carbonyl group of 3,3,5-trimethylcyclo-hexanone is hindered by the steric effect of the axial C-3 methyl group, and reduction of this compound leads to a mixture of isomeric alcohols containing approximately equal parts of each.

Improved yields of the more stable alcohol can often be obtained by effecting the reduction of a cyclohexanone with lithium aluminium hydride in presence of aluminium chloride and an excess of the ketone. Thus, 3,3,5-trimethylcyclohexanone is converted entirely into the more stable equatorial alcohol under these conditions. An equilibrium is set up between the aluminium alkoxide and the ketone similar to that found in the Meerwein–Pondorff–Verley reduction, resulting in gradual accumulation of the more stable alkoxide.

Reduction of epoxides and of alkyl halides or sulphonates with lithium aluminium hydride proceeds by S_N2 nucleophilic substitution by hydride ion, with formation of a new C–H bond. Epoxides are thereby reduced to alcohols. With unsymmetrical epoxides reaction takes place at the less highly substituted carbon atom to give the more highly sub-stituted alcohol. Thus, 3,4-epoxybutane yields a mixture in which the

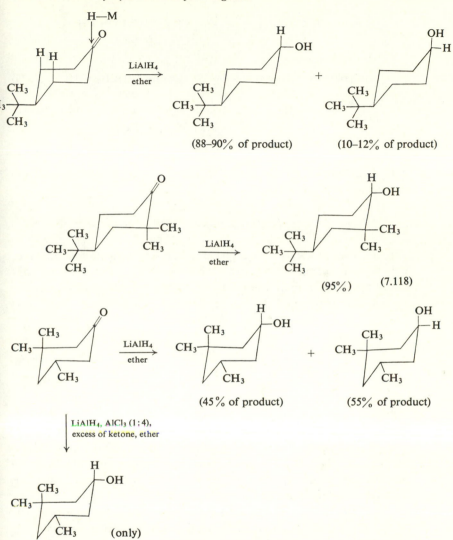

(88–90% of product)   (10–12% of product)

(95%)   (7.118)

(45% of product)   (55% of product)

(only)

secondary alcohol predominates, epoxymethylenecyclohexane is converted into 1-methylcyclohexanol and triphenylethylene oxide gives 1,1,2-triphenylethanol (7.119). Remarkably, however, in presence of 1/3 mole aluminium chloride, the direction of ring opening is reversed, and high yields of the alternative alcohol are obtained, presumably because the reactive species is now the electrophilic aluminium hydride

$$CH_2{=}CH.CH{-}CH_2 \xrightarrow[\text{ether}]{LiAlH_4} CH_2{=}CH.CHOH.CH_3$$
$$\underset{O}{\diagdown\diagup}$$
(mainly)

$$(C_6H_5)_2.\overset{O}{\overset{\diagup\diagdown}{C{-}CH}}.C_6H_5 \xrightarrow[\text{ether}]{LiAlH_4} (C_6H_5)_2COH.CH_2.C_6H_5 \qquad (7.119)$$
only

$$\xrightarrow{LiAlH_4{-}AlCl_3 \ (3:1)} (C_6H_5)_2CH.CHOH.C_6H_5$$
only

formed *in situ* from aluminium chloride and lithium aluminium hydride (Ashby and Prather, 1966).

$$AlCl_3 + 3LiAlH_4 \longrightarrow 4AlH_3 + 3LiCl$$

In accordance with the S_N2 type mechanism, reduction of epoxides with lithium aluminium hydride takes place with inversion of configuration at the carbon atom attacked. Thus 1,2-dimethylcyclohexene epoxide affords *trans*-1,2-dimethylcyclohexanol (7.120).

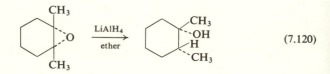

$$(7.120)$$

Primary and secondary alkyl halides are reduced to the hydrocarbons with lithium aluminium hydride. Tertiary halides react only slowly and give mostly olefins. The order of reactivity of the halides is iodides > bromides > chlorides. Aryl and vinyl halides are not attacked, but benzylic and allylic halides are. Aryl iodides and bromides may be reduced to the hydrocarbon with lithium aluminium hydride in boiling tetrahydrofuran (Brown and Krishnamurthy, 1969). Sulphonate esters of primary and secondary alcohols are also readily reduced with lithium aluminium hydride and this reaction has been widely used in synthesis to effect the replacement of an alcoholic hydroxyl group by hydrogen. In recent work it has been found that sodium borohydride in dimethyl sulphoxide or tetramethylene sulphone is an excellent reagent for reductive fission of benzylic and primary, secondary and, in certain cases, tertiary alkyl halides and tosylates in the presence of other reducible groups in the molecule, including carboxylic acid, ester or nitro groups

(Hutchins, Hoke, Keogh and Koharski, 1969; Bell, Vanderslice and Spehars, 1969) (7.121).

$$p\text{-NO}_2\text{C}_6\text{H}_4\text{CH}_2\text{Br} \xrightarrow[\substack{\text{dimethyl} \\ \text{sulphoxide}}]{\text{NaBH}_4} p\text{-NO}_2\text{C}_6\text{H}_4\text{CH}_3 \quad (98\%) \qquad (7.121)$$

Another useful procedure for the formation of a methyl or methylene group is reduction of an acyl toluene-*p*-sulphonylhydrazide, or the toluene-*p*-sulphonylhydrazone of an aldehyde or ketone, with lithium aluminium hydride. Thus, 2-naphthaldehyde is converted into 2-methyl-naphthalene and 1-naphthylacetic acid gives 1-ethylnaphthalene by this procedure.

Lithium hydridoalkoxyaluminates. Lithium aluminium hydride itself is an extremely powerful and versatile reducing agent. More selective reagents can be obtained by modification of lithium aluminium hydride by treatment with alcohols or with aluminium chloride. One such reagent is the sterically bulky lithium hydridotri-t-butoxyaluminate, which is readily prepared by action of the stoichiometric amount of t-butyl alcohol on lithium aluminium hydride.

$$\text{LiAlH}_4 + 3\text{ROH} \longrightarrow \text{LiAlH(OR)}_3 + 3\text{H}_2$$

Analogous reagents are obtained in the same way from other alcohols, and by replacement of only one or two of the hydrogen atoms of the hydride by alkoxy groups, affording a range of reagents of graded activities (Rerick, 1967).

Lithium hydridotri-t-butoxyaluminate is a much milder reducing agent than lithium aluminium hydride itself (Brown, Weissman and Nung Min Yoon, 1966). Thus, although aldehydes and ketones are reduced normally to alcohols, it reacts only slowly with carboxylic esters, and halides, nitriles and nitro groups are not attacked. Aldehydes and ketones can thus be selectively reduced in presence of these groups.

One of the most useful applications of the alkoxy reagents is in the preparation of aldehydes from carboxylic acids by partial reduction of the acid chlorides or dimethylamides. Acid chlorides are readily reduced with lithium aluminium hydride itself or with sodium borohydride to the corresponding alcohols, but with one equivalent of the tri-t-butoxy compound high yields of the aldehyde can be obtained (7.122).

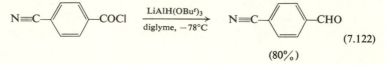

$$(80\%) \qquad (7.122)$$

Although esters, in general, are reduced only slowly, phenyl esters are converted into the aldehyde with LiAlH(OBu)₃. Thus, phenyl cyclohexanecarboxylate gives formylcyclohexane in 60 per cent yield.

Reduction of *N*-substituted amides with excess of lithium aluminium hydride affords the corresponding amine in good yield (7.123). Reaction

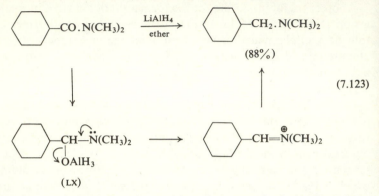

(7.123)

is believed to proceed through an aldehyde-ammonia derivative of the type (LX). By partial reduction of *N,N*-dialkylamides with the less active agents LiAlH(OC₂H₅)₃ or LiAlH₂(OC₂H₅)₂ (the tri-t-butoxy compound is ineffective in this case) reaction stops at the aldehyde-ammonia stage and hydrolysis of the product affords the corresponding

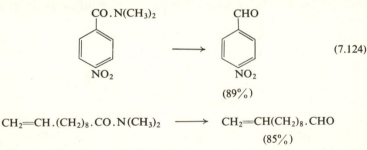

(7.124)

$$CH_2{=}CH.(CH_2)_8.CO.N(CH_3)_2 \longrightarrow CH_2{=}CH(CH_2)_8.CHO$$
(85%)

aldehyde. Yields are usually high, except in the case of αβ-unsaturated amides, and the reaction proceeds readily in presence of other reducible groups (7.124). An alternative procedure is direct reduction with lithium aluminium hydride of the amides derived from ethyleneimine, carbazole, *N*-methylaniline or imidazole. With these amides reaction stops at the aldehyde-ammonia stage because participation of the nitrogen lone pair in the elimination process shown in (LX) is not favoured (7.125).

$$C_3H_7.CO-N\!\!<\!\!\square \xrightarrow{\text{LiAlH}_4} C_3H_7.CHO \quad (88\%) \qquad (7.125)$$

Similarly, reduction of nitriles with lithium aluminium hydride affords a primary amine by way of the imine salt (7.126). With lithium

$$R-C\equiv N \xrightarrow[\text{ether}]{\text{LiAlH}_4} [R-CH=N-\overset{\ominus}{A}lH_3] \xrightarrow[(2)\ H_3O^+]{(1)\ \text{LiAlH}_4} R.CH_2NH_2 \quad (7.126)$$

hydridotriethoxyaluminate, however, reaction stops at the imine stage, and hydrolysis gives the aldehyde. By this procedure trimethylacetonitrile was converted into trimethylacetaldehyde in 75 per cent yield.

Asymmetric reduction of ketones has been achieved with the complex produced from lithium aluminium hydride and the asymmetric alcohol 3-*O*-benzyl-1,2-*O*-cyclohexylidene-α-D-glucofuranose. In presence of ethanol, which gives the mixed alkoxyhydride LiAlH(OEt)(OR)$_2$, carbinols of optical purity as high as 63·5 per cent have been obtained (Landor, Miller and Tatchell, 1967).

Mixed lithium aluminium hydride–aluminium chloride reagents. Further useful modification of the properties of lithium aluminium hydride is achieved by addition of aluminium chloride in various proportions. This serves to release various mixed chloride-hydrides of aluminium as shown in (7.127). The general effect of the addition of

$$3\text{LiAlH}_4 + \text{AlCl}_3 \longrightarrow 3\text{LiCl} + 4\text{AlH}_3$$

$$\text{LiAlH}_4 + \text{AlCl}_3 \longrightarrow \text{LiCl} + 2\text{AlH}_2\text{Cl} \qquad (7.127)$$

$$\text{LiAlH}_4 + 3\text{AlCl}_3 \longrightarrow \text{LiCl} + 4\text{AlHCl}_2$$

aluminium chloride is to lower the reducing power of lithium aluminium hydride and in consequence to produce reagents which are more specific for particular reactions (Eliel, 1961; Rerick, 1968). For example, the carbon–halogen bond is often inert to the mixed hydride reagents. Advantage is taken of this in the reduction of polyfunctional compounds in which retention of halogen is desired, as in the conversion of methyl 3-bromopropionate into 3-bromopropanol; lithium aluminium hydride alone produces n-propanol (7.128).

$$\text{BrCH}_2\text{CH}_2\text{CO}_2\text{CH}_3 \xrightarrow[\text{ether}]{\text{LiAlH}_4-\text{AlCl}_3\ (1:1)} \text{BrCH}_2\text{CH}_2\text{CH}_2\text{OH} \quad (7.128)$$

Similarly, nitro groups are not so easily reduced as with lithium aluminium hydride itself, and *p*-nitrobenzaldehyde can be converted into *p*-nitrobenzyl alcohol in 75 per cent yield. Aldehydes and ketones are reduced to carbinols, and there is no advantage in the use of mixed hydrides in these cases, although it should be noted that the stereochemical result obtained in the reduction of cyclic ketones may not be the same as with lithium aluminium hydride itself. With diaryl ketones and with aryl alkyl ketones however the carbonyl group is reduced to methylene in high yield, and this procedure offers a useful alternative to the Clemmensen or Huang-Minlon methods for reduction of this type of ketone.

Reduction with lithium aluminium hydride–aluminium chloride (3:1) also provides an excellent route from $\alpha\beta$-unsaturated carbonyl compounds to unsaturated alcohols which are difficult to prepare with lithium aluminium hydride alone because of competing reduction of the olefinic double bond. The effective reagent is thought to be aluminium hydride formed *in situ* from lithium aluminium hydride and aluminium chloride (7.129).

$$C_6H_5CH=CH.CO_2C_2H_5 \xrightarrow[\text{ether}]{\text{3LiAlH}_4\text{–AlCl}_3} C_6H_5CH=CH.CH_2OH \quad (7.129)$$
$$(90\%)$$

The most striking difference between lithium aluminium hydride and the mixed hydrides is seen in the reduction of epoxides, where two different modes of ring opening of the epoxide ring are observed (see p. 361). Ethers in general are not attacked by either reagent. Acetals and ketals are reduced with the mixed reagent to ethers (7.130), probably

$$C_6H_5C(OC_2H_5)_2 \underset{\overset{|}{CH_3}}{} \xrightarrow[\text{ether}]{\text{LiAlH}_4\text{–AlCl}_3\ (1:4)} C_6H_5CH.OC_2H_5 \underset{\overset{|}{CH_3}}{} \quad (7.130)$$

through cleavage of the ketal by excess of aluminium chloride to an oxonium salt (LXI) which is then reduced by the aluminium hydride species present (7.131). Esters and lactones may also be reduced to ethers with lithium aluminium hydride and aluminium chloride or

$$C_6H_5.C\underset{\overset{|}{CH_3}}{\overset{\nearrow OC_2H_5}{\underset{\searrow OC_2H_5}{}}} \overset{AlCl_3}{} \longrightarrow$$

$$C_6H_5.C\overset{\oplus}{=}OC_2H_5 \underset{\overset{|}{CH_3}}{} \longrightarrow C_6H_5CH.OC_2H_5 \underset{\overset{|}{CH_3}}{}$$

$$(\text{LXI}) \qquad\qquad\qquad\qquad (7.131)$$

sodium borohydride and boron trifluoride. Thus the lactone (LXII) readily affords the ether (LXIII) (7.132). The actual reducing agent in the

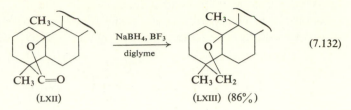

(7.132)

(LXII)   (LXIII) (86%)

latter reaction is diborane B_2H_6, formed by reaction of sodium borohydride and boron trifluoride. It is capable of reducing a number of functional groups but its most useful synthetic application is in the addition to carbon–carbon multiple bonds to form alkylboranes. These reactions are discussed in Chapter 5.

7.4. Other methods; *Wolff–Kishner reduction.* The Wolff–Kishner reduction provides an excellent method for the reduction of the carbonyl group of many aldehydes and ketones to methyl or methylene (Szmant, 1968). As originally described the reaction involved heating the semicarbazone of the carbonyl compound with sodium ethoxide or other base at 200°C in a sealed tube, but it is now more conveniently effected by heating a mixture of the carbonyl compound, hydrazine hydrate and sodium or potassium hydroxide in a high-boiling solvent (diethylene glycol is often used) at 180–200°C for several hours (Huang-Minlon, 1946). Excellent yields of the reduced product are often obtained. The mechanism of the reaction has not been widely studied, but it is believed that the hydrazone initially formed is transformed by reactions similar to those shown in the equations (7.133). A wide variety of aldehydes and

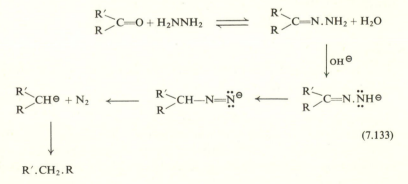

(7.133)

ketones has been reduced by this method. Sterically hindered ketones, such as 11-keto steroids, are resistant, but they too can be reduced by use of anhydrous hydrazine and sodium metal as base.

Reduction of conjugated unsaturated ketones is sometimes accompanied by a shift in the position of the double bond (7.134). In other

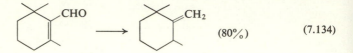

(80%)   (7.134)

cases pyrazoline derivatives may be formed which decompose yielding cyclopropanes isomeric with the expected olefin (7.135).

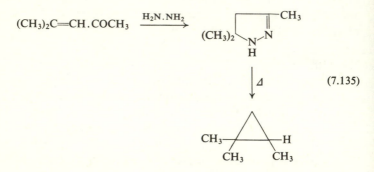

(7.135)

With many α-substituted ketones elimination accompanies reduction. For example the sterol derivative (LXIV) affords the olefin (LXV) (7.136).

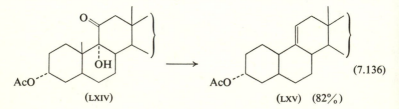

(LXIV)   (LXV)   (82%)   (7.136)

In some cases these eliminations take place under quite mild conditions. Of particular importance is the reductive opening of αβ-epoxyketones which are converted into allylic alcohols. This reaction has been exploited in a synthesis of linalool and other allylic alcohols (7.137).

Raney nickel desulphurisation of thio-ketals and -acetals. Another method for reduction of the carbonyl group of aldehydes and ketones to methyl or methylene which is useful on occasion is desulphurisation

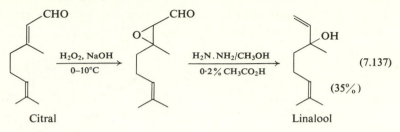

of the corresponding thio-acetals or -ketals with Raney nickel in boiling ethanol (Pettit and van Tamelen, 1962). Hydrogenolysis is effected by the hydrogen absorbed on the nickel during its preparation (7.138).

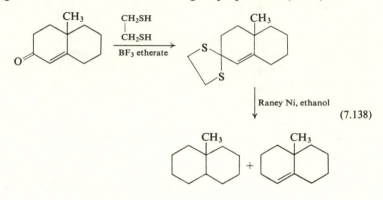

Reaction is effected under fairly mild conditions, but the method suffers from the disadvantage that large amounts of Raney nickel are required, and other unsaturated groups in the compound may also be reduced.

Reductions with di-imide. It has long been known that isolated olefinic bonds can be reduced with hydrazine in presence of oxygen or an oxidising agent. Thus, it was found as early as 1914 that oleic acid is reduced to stearic acid by this method. It has been suggested recently that the actual reducing agent in these reactions is in fact the highly active species di-imide, $HN\!\!=\!\!NH$, formed *in situ* by oxidation of hydrazine, and it has been found that this compound is a highly selective reducing agent which in many cases offers a useful alternative to catalytic hydrogenation for the reduction of carbon–carbon multiple bonds (Miller, 1965).

The reagent is not isolated but is prepared *in situ*, usually by oxidation of hydrazine with oxygen or an oxidising agent (e.g. hydrogen peroxide);

by thermal decomposition of toluene-*p*-sulphonylhydrazide or from azodicarboxylic acid (7.139).

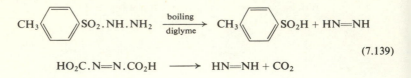

$$(7.139)$$

$$HO_2C.N{=}N.CO_2H \longrightarrow HN{=}NH + CO_2$$

Di-imide is a highly selective reagent. In general, it is found that symmetrical double bonds such as $C{\equiv}C$, $C{=}C$, $N{=}N$, $O{=}O$ are readily reduced, but unsymmetrical, more polar, bonds ($C{\equiv}N$, $N\overset{\displaystyle O}{\underset{\displaystyle O}{\diagup}}$, $C{=}N$, $S{=}O$, $S{-}S$, $C{-}S$) are not. For example, under conditions where oleic acid and azobenzene were readily reduced, methyl cyanide, nitrobenzene and dibenzyl sulphide were unaffected, and the selectivity of the reagent is strikingly demonstrated by the reduction of diallyl sulphide to di-n-propyl sulphide in almost quantitative yield (7.140).

$$(CH_2{=}CH.CH_2)_2S \xrightarrow[\text{boiling glycol}]{\text{tosylhydrazide,}} (CH_3.CH_2CH_2)_2S$$
$$(93\text{--}100\%)$$

$$(7.140)$$

$$C_6H_5.N{=}N.C_6H_5 \xrightarrow[\text{boiling methanol}]{\text{azodicarboxylic acid,}} C_6H_5.NH.NH.C_6H_5$$
$$\text{quantitative}$$

The reactions are highly stereospecific and take place by *cis* addition of hydrogen in all cases. Thus, reduction of fumaric acid with tetra-deuterohydrazine or with potassium azodicarboxylate and deuterium oxide affords ($\pm$)-dideuterosuccinic acid exclusively, while maleic acid gives the *meso* isomer.

Reduction of acetylenes is also a *cis* process. Thus, partial reduction of diphenylacetylene gave, besides starting material and diphenylethane, only *cis*-stilbene; no *trans*-stilbene was detected.

In sterically hindered molecules, addition takes place to the least hindered side of the double bond, but in examples where steric influences are moderate, much less stereochemical discrimination is observed (7.141).

The reactions are regarded as taking place by synchronous transfer of a pair of hydrogen atoms through a cyclic six-membered transition state (7.142). This mechanism explains the high stereospecificity of the

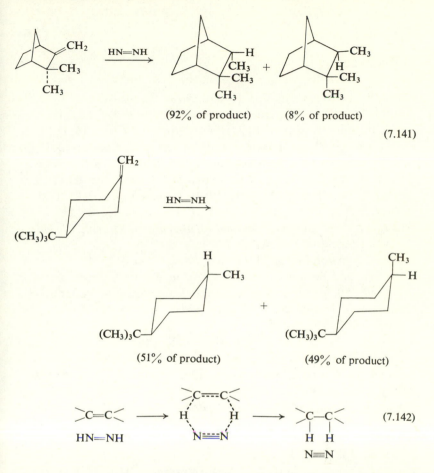

reaction, and couples the driving force of nitrogen formation with the addition reaction. Concerted *cis* transfer of hydrogen is symmetry allowed for the ground-state reaction (Woodward and Hoffmann, 1969).

References

Abley, P. and McQuillin, F. J. (1969). *Chem. Comm.* 477.

Abramovitch, R. A. and Davis, B. A. (1964). *Chem. Rev.* **64**, 149.

Acott, B. and Beckwith, A. L. J. (1964). *Austral. J. Chem.* **14**, 1342.

Akhtar, M. (1962). in *Advances in Photochemistry*, vol. 2, p. 263. Ed. W. A. Noyes, G. S. Hammond and J. N. Pitts (London: Interscience).

Akhtar, A. and Barton, D. H. R. (1961). *J. Am. chem. Soc.* **83**, 2213.

Akhtar, A. and Barton, D. H. R. (1964). *J. Am. chem. Soc.* **86**, 1528.

Akhtar, M. and Pechet, M. M. (1964). *J. Am. chem. Soc.* **86**, 265.

Alder, K. (1948). in *Newer Methods of Preparative Organic Chemistry*, vol. 1 (London: Interscience).

Alder, K. and von Brachel, H. (1962). *Liebigs Ann.* **651**, 141.

Alder, K. and Schumacher, M. (1953). in *Fortschr. Chem. org. Nat. Stoffe*, **10**, 69. Ed. L. Zechmeister (Vienna: Springer Verlag).

Anderson, R. J., Henrick, C. A. and Siddall, J. B. (1970). *J. Am. chem. Soc.* **92**, 735.

Angus, H. J. F. and Bryce-Smith, D. (1960). *J. chem. Soc.* 4791.

Ansell, M. F. *et al.* (1971). *J. chem. Soc.* 1401, 1414, 1423, 1429.

ApSimon, J. W. and Edwards, O. E. (1962). *Canad. J. Chem.* **40**, 896.

Ashby, E. C. and Prather, J. (1966). *J. Am. chem. Soc.* **88**, 729.

Augustine, R. L. (1965). *Catalytic Hydrogenation* (London: Arnold).

Augustine, R. L. (1968). *Reduction* (London: Arnold).

Axelrod, E. H., Milne, G. M. and van Tamelen, E. E. (1970). *J. Am. chem. Soc.* **92**, 2139.

Bacha, J. D. and Kochi, J. K. (1968). *Tetrahedron*, **24**, 2215.

Bachman, W. E. and Struve, W. S. (1942). *Organic Reactions*, vol. 1, p. 38 (New York: Wiley).

Bacon, R. G. R. and Hill, H. A. O. (1965). *Quart. Rev. chem. Soc. Lond.* **19**, 95.

Badger, G. M. (1954). *The Structures and Reactions of Aromatic Compounds* (Cambridge University Press).

Bailey, D. S. and Saunders, W. H. (1970). *J. Am. chem. Soc.* **92**, 6904, 6911.

Bailey, P. S. (1958). *Chem. Rev.* **58**, 925.

Baldwin, J. E., Barton, D. H. R., Dainis, I. and Pereira, J. L. C. (1968). *J. chem. Soc.* (*C*) 2283.

Baldwin, J. E., Hackler, R. E. and Kelly, D. P. (1968). *Chem. Comm.* 537, 538.

Barlow, M. G., Haszeldine, R. N. and Hubbard, R. (1969). *Chem. Comm.* 301.

Bartlett, P. D. (1968). *Science*, **159**, 833.

Barton, D. H. R. (1953). *J. chem. Soc.* 1027, n. 23.

Barton, D. H. R. (1959). *Helv. Chim. Acta*, **42**, 2604.

Barton, D. H. R., Beaton, J. M., Geller, L. E. and Pechet, M. M. (1961). *J. Am. chem. Soc.* **83**, 4076.

Barton, D. H. R., Beckwith, A. L. J. and Goosen, A. (1965). *J. chem. Soc.* 181.

Barton, D. H. R. and Morgan, L. R. (1962). *J. chem. Soc.* 622.

Barton, D. H. R. and Robinson, C. H. (1954). *J. chem. Soc.* 3045.

Barton, D. H. R. and Starratt, A. N. (1965). *J. chem. Soc.* 2445.

Beckwith, A. L. J. and Goodrich, J. E. (1965). *Austral. J. Chem.* **18**, 747.

Bell, H. M., Vanderslice, C. W. and Spehars, A. (1969). *J. org. Chem.* **34**, 3923.

Benkeser, R. A., Agnihotri, R., Burrous, M. L., Kaiser, E. M., Mallan, J. M. and Ryan, P. W. (1964). *J. org. Chem.* **29**, 1313.

Benkeser, R. A., Burrous, M. L., Hazdra, J. J. and Kaiser, E. M. (1963). *J. org. Chem.* **28**, 1094.

Bergelson, L. D., Barsukov, L. I. and Shemyakin, M. M. (1967). *Tetrahedron*, **23**, 2709.

Bergelson, L. D. and Shemyakin, M. M. (1964). *Angew. Chem. internat. Edit.* **3**, 258.

Berson, J. A., Hamlet, Z. and Mueller, W. A. (1962). *J. Am. chem. Soc.* **84**, 297.

Berson, J. A. and Olin, S. S. (1969). *J. Am. chem. Soc.* **91**, 777.

Berson, J. A. and Remanick, A. (1961). *J. Am. chem. Soc.* **83**, 4947.

Berson, J. A., Wall, R. G. and Perlmutter, H. D. (1966). *J. Am. chem. Soc.* **88**, 187.

Bethell, D. (1969). In *Advances in Physical Organic Chemistry*, vol. 7, p. 153. Ed. V. Gold (London: Academic Press).

Biellmann, J. F. (1968). *Bull. Soc. chim. Fr.* 3055.

Birch, A. J. (1950). *Quart. Rev. chem. Soc. Lond.* **4**, 69.

Birch, A. J. and Nasipuri, D. (1959). *Tetrahedron*, **6**, 148.

Birch, A. J. and Smith, H. (1958), *Quart. Rev. chem. Soc. Lond.* **12**, 17.

Birch, A. J. and Walker, K. A. M. (1966). *J. chem. Soc. (C)*, 1894.

Bird, C. L., Frey, H. M. and Stevens, I. D. R. (1967). *Chem. Comm.* 707.

Birkowitz, L. M. and Rylander, P. N. (1958). *J. Am. chem. Soc.* **80**, 6683.

Blackburn, G. M., Ollis, W. D., Plackett, J. D., Smith, C. and Sutherland, I. O. (1968). *Chem. Comm.* 186.

Blackburn, E. V. and Timmons, C. J. (1969). *Quart. Rev. chem. Soc. Lond.* **23**, 482.

Blanchard, E. P. and Simmons, H. E. (1964). *J. Am. chem. Soc.* **86**, 1337, 1347.

Bowers, A., Halsall, T. G., Jones, E. R. H. and Lemin, A. J. (1953). *J. chem. Soc.* 2548.

Brady, S. F., Ilton, M. A. and Johnson, W. S. (1968). *J. Am. chem. Soc.* **90**, 2882.

Breslow, R. and Winnik, M. A. (1969). *J. Am. chem. Soc.* **91**, 3083.

Brimacombe, J. S. (1969). *Angew. Chem. internat. Edit.* **8**, 401.

Brown, C. A. (1969). *J. Am. chem. Soc.* **91**, 5901.

Brown, H. C. (1962). *Hydroboration* (New York: Benjamin).

Brown, H. C. (1969). *Accounts Chem. Res.* **2**, 65.

Brown, H. C. and Bhatt, M. V. (1966). *J. Am. chem. Soc.* **88**, 1440.

Brown, H. C. and Bigley, D. B. (1961). *J. Am. chem. Soc.* **83**, 486.

Brown, H. C., Bowman, D. H., Misumi, S. and Unni, M. K. (1967). *J. Am. chem. Soc.* **89**, 4531.

Brown, H. C. and Brown, C. A. (1962). *J. Am. chem. Soc.* **84**, 2827.

Brown, H. C. and Brown, C. A. (1963). *J. Am. chem. Soc.* **85**, 1003.

Brown, H. C. and Coleman, R. A. (1969). *J. Am. chem. Soc.* **91**, 4606.

Brown, H. C. and Deck, H. R. (1965). *J. Am. chem. Soc.* **87**, 5620.

Brown, H. C. and Dickason, W. C. (1970). *J. Am. chem. Soc.* **92**, 709.

Brown, H. C. and Gallivan, R. M. (1968). *J. Am. chem. Soc.* **90**, 2906.

Brown, H. C. and Garg, C. P. (1961). *J. Am. chem. Soc.* **83**, 2951.

Brown, H. C., Heydkamp, W. R., Breuer, E. and Murphy, W. S. (1964). *J. Am. chem. Soc.* **86**, 3565.

Brown, H. C. and Kabalka, G. W. (1970). *J. Am. chem. Soc.* **92**, 712, 714.

Brown, H. C. and Keblys, K. A. (1964). *J. Am. chem. Soc.* **86**, 1791, 1795.

Brown, H. C., Knights, E. F. and Coleman, R. A. (1969). *J. Am. chem. Soc.* **91**, 2144.

Brown, H. C. and Krishnamurthy, S. (1969). *J. org. Chem.* **34**, 3918.

Brown, H. C. and Murray, K. J. (1961). *J. org. Chem.* **26**, 631.

Brown, H. C., Murray, K. J., Müller, H. and Zweifel, G. (1966). *J. Am. chem. Soc.* **88**, 1443.

Brown, H. C., Nambu, H. and Rogic, M. (1969). *J. Am. chem. Soc.* **91**, 6852.

Brown, H. C. and Negishi, E. (1967). *J. Am. chem. Soc.* **89**(*a*) 5285; (*b*) 5477.

Brown, H. C. and Negishi, E. (1968). *Chem. Comm.* 594.

Brown, H. C. and Nung Min Yoon. (1968). *J. Am. chem. Soc.* **90**, 2686; *Chem. Comm.* 1549.

Brown, H. C. and Pfaffenberger, C. D. (1967). *J. Am. chem. Soc.* **89**, 5475.

Brown, H. C. and Rathke, M. W. (1967). *J. Am. chem. Soc.* **89**, 2737.

Brown, H. C., Rathke, M. W. and Rogic, M. M. (1968). *J. Am. chem. Soc.* **90**, 5038.

Brown, H. C. and Rhodes, S. P. (1969). *J. Am. chem. Soc.* **91**, 2149, 4306.

Brown, H. C. and Rogic, M. M. (1969). *J. Am. chem. Soc.* **91**, 2146, 4304.

Brown, H. C., Rogic, M. M., Rathke, M. W. and Kabalka, G. W. (1967). *J. Am. chem. Soc.* **89**, 5709.

Brown, H. C. and Sharp, R. L. (1966). *J. Am. chem. Soc.* **88**, 5851.

Brown, H. C. and Sharp, R. L. (1968). *J. Am. chem. Soc.* **90**, 2915.

Brown, H. C. and Subba, Rao B. C. (1960). *J. Am. chem. Soc.* **82**, 681.

Brown, H. C. and Unni, M. K. (1968). *J. Am. chem. Soc.* **90**, 2902.

Brown, H. C., Verbrugge, C. and Snyder, C. H. (1961). *J. Am. chem. Soc.* **83**, 1001.

Brown, H. C., Weissman, P. M. and Nung Min Yoon (1966). *J. Am. chem. Soc.* **88**, 1458.

Brown, W. G. (1951). *Organic Reactions*, **6**, 469 (New York: Wiley).

Bryce-Smith, D. (1968). *Pure and appl. Chem.* **16**, 47.

Bryce-Smith, D. Wakefield, B. J. and Blues, E. T. (1963). *Proc. chem. Soc.* 219.

Buchanan, J. G. St. C. and Woodgate, P. D. (1969). *Quart. Rev. chem. Soc. Lond.* **23**, 522.

Butz, L. W. and Rytina, A. W. (1949). *Organic Reactions* **5**, 136 (New York: Wiley).

Cadogan, J. I. G. (1968). *Quart. Rev. chem. Soc. Lond.* **22**, 222.

Cain, E. N., Vukov, R. and Masamune, S. (1969). *Chem. Comm.* 98.

Carlson, R. M. and Helquist, P. M. (1969). *Tetrahedron Lett.* 173.

Carnduff, J. (1966). *Quart. Rev. chem. Soc. Lond.* **20**, 169.

Challand, B. D., Hikino, H., Kornis, G., Lange, G. and de Mayo, P. (1969). *J. org. Chem.* **34**, 794.

Cherest, M., Felkin, H. and Prudent, N. (1968). *Tetrahedron Lett.* 2199.

Cherest, M. and Felkin, H. (1968). *Tetrahedron Lett.* 2205.

Chow, Y. L. and Joseph, T. C. (1969). *Chem. Comm.* 490.

Ciganek, E. (1967). *Tetrahedron Lett.* 3321.

Cimarusti, C. M. and Wolinsky, J. (1968). *J. Am. chem. Soc.* **90**, 113.

Clayton, R. B., Henbest, H. B. and Smith, M. (1957). *J. chem. Soc.* 1982.

Coffey, R. S. (1967). *Chem. Comm.* 923.

Collins, J. C., Hess, W. W. and Frank, F. J. (1968). *Tetrahedron Lett.* 3363.

Colonge, J. and Descotes, G. (1967). In 1,4-*Cycloaddition Reactions.* Ed. J. Hamer (New York: Academic Press).

Colonna, F. P., Fattuta, S., Risaliti, A. and Russo, C. (1970). *J. chem. Soc.* (*C*), 2377.

Conia, J-M. (1963). *Rec. chem. Prog.* **24**, 43.

Conia, J-M. and Leriverend, P. (1960). *C.r. hebd. Séanc. Acad. Sci., Paris,* **250**, 1078.

Cook, A. G. (1968). ed. *Enamines: their Synthesis, Structure and Reactions.* (Marcel Dekker, New York).

Cookson, R. C., Stevens, I. D. R. and Watts, C. T. (1966). *Chem. Comm.* 744.

Cope, A. C. and Trumbull, E. R. (1960). *Organic Reactions* **11**, 317 (New York: Wiley).

Corey, E. J. and Achiwa, K. (1969). *Tetrahedron Lett.* 3257.

Corey, E. J., Bass, J. D., LeMahieu, R. and Mitra, R. B. (1964). *J. Am. chem. Soc.* **86**, 5570.

Corey, E. J. and Cane, D. E. (1969). *J. org. Chem.* **34**, 3053.

Corey, E. J., Carey, F. A. and Winter, R. A. E. (1965). *J. Am. chem. Soc.* **87**, 935.

Corey, E. J. and Chaykovsky, M. (1962). *J. Am. chem. Soc.* **84**, 866.

Corey, E. J. and Chaykovsky, M. (1965). *J. Am. chem. Soc.* **87**, 1345; 1353.

Corey, E. J., Chow, S. W. and Scherrer, R. A. (1957). *J. Am. chem. Soc.* **79**, 5773.

Corey, E. J. and Durst, T. (1968). *J. Am. chem. Soc.* **90**, 5548.

Corey, E. J., Gilman, N. W. and Ganem, B. E. (1968). *J. Am. chem. Soc.* **90**, 5616.

Corey, E. J. and Hamanaka, E. (1967). *J. Am. chem. Soc.* **89**, 2758.

Corey, E. J. and Hertler, W. R. (1959). *J. Am. chem. Soc.* **81**, 5209.

Corey, E. J. and Hertler, W. R. (1960). *J. Am. chem. Soc.* **82**, 1657.

Corey, E. J. and Hortmann, A. G. (1963). *J. Am. chem. Soc.* **85**, 4033.

Corey, E. J. and Jautelat, M. (1967). *J. Am. chem. Soc.* **89**, 3913.

Corey, E. J. and Katzenellenbogen, J. A. (1969). *J. Am. chem. Soc.* **91**, 1851.

Corey, E. J., Katzenellenbogen, J. A., Gilman, N. W., Roman, S. A. and Erickson, B. W. (1968). *J. Am. chem. Soc.* **90**, 5618.

Corey, E. J., Katzenellenbogen, J. A. and Posner, G. H. (1967). *J. Am. chem. Soc.* **89**, 4245.

Corey, E. J. and Kuwajima, I. (1970). *J. Am. chem. Soc.* **92**, 395.

Corey, E. J. and Kwiatkowsky, G. T. (1968). *J. Am. chem. Soc.* **90**, 6816.

Corey, E. J., Mitra, R. B. and Uda, H. (1964). *J. Am. chem. Soc.* **86**, 485.

Corey, E. J. and Posner, G. H. (1967). *J. Am. chem. Soc.* **89**, 3911.

Corey, E. J. and Posner, G. H. (1968). *J. Am. chem. Soc.* **90**, 5615.

Corey, E. J. and Seebach, D. (1965). *Angew. Chem. internat. Edit.* **4**, 1075, 1077.

Corey, E. J. and Semmelhack, M. F. (1967). *J. Am. chem. Soc.* **89**, 2755.

Corey, E. J., Semmelhack, M. F. and Hegedus, L. S. (1968). *J. Am. chem. Soc.* **90**, 2416.

Corey, E. J., Shulman, J. I. and Yamamoto, H. (1970). *Tetrahedron Lett.* 447.

Corey, E. J. and Sneen, R. A. (1956). *J. Am. chem. Soc.* **78**, 6269.

Corey, E. J. and Taylor, W. C. (1964). *J. Am. chem. Soc.* **86**, 3881.

Corey, E. J. and Yamamoto, H. (1970). *J. Am. chem. Soc.* **92**, 226.

Cornforth, J. W., Cornforth, R. H. and Mathew, K. K. (1959). *J. chem. Soc.* 112.

Cornforth, R. H., Cornforth, J. W. and Popják, G. (1962). *Tetrahedron*, **18**, 1351.

Cornforth, J. W., Milborrow, B. V. and Ryback, G. (1965). *Nature, Lond.* **206**, 715.

Cowell, G. W. and Ledwith, A. (1970). *Quart. Rev. chem. Soc. Lond.* **24**, 119.

Cram, D. J. and Abd Elhafez, F. A. (1952). *J. Am. chem. Soc.* **74**, 4828.

Criegee, R. (1957). *Rec. chem. Prog.* **18**, 111.

Criegee, R. (1965). *Oxidation in Organic Chemistry*, p. 297. Ed. K. B. Wiberg. (New York: Academic Press).

Criegee, R. and Gunther, P. (1963). *Chem. Ber.* **96**, 1564.

Dauben, W. G., Fonken, G. J. and Noyce, D. S. (1956). *J. Am. chem. Soc.* **78**, 2579.

De Los, F. de Tar and Relyea, D. I. (1956). *J. Am. chem. Soc.* **78**, 4302.

Deno, N. C., Peterson, H. J. and Saines, G. S. (1960). *Chem. Rev.* **60**, 7.

De Puy, C. H. and King, R. W. (1960). *Chem. Rev.* **60**, 431.

De Puy, C. H., Naylor, C. G. and Beckman, J. A. (1970). *J. org. Chem.* **35**, 2750.

Dilling, W. L. (1966). *Chem. Rev.* **66**, 373.

Dittius, G. (1965). *Methoden der organischen Chemie*, vol VI/3. Ed. E. Müller (Stuttgart: Georg Thieme Verlag).

Djerassi, C. (1951). *Organic Reactions*, **6**, 207 (New York: Wiley).

Djerassi, C. (1963). *Steroid Reactions*, p. 267 (San Francisco: Holden–Day, Inc.).

Doering, W. E. von and Dorfman, E. (1953). *J. Am. chem. Soc.* **75**, 5595.

Dryden, H. L., Webber, G. M., Burtner, R. R. and Cella, J. A. (1961). *J. org. Chem.* **26**, 3237.

Eaborn, C., Pant, B. C., Peeling, E. R. A. and Taylor, S. C. (1969). *J. chem. Soc.* (*C*), 2823.

Eaton, P. E. (1968). *Accounts Chem. Res.* **1**, 50.

Eglinton, G. and McCrae, W. (1963). *Advances in Organic Chemistry. Methods and Results*, vol. 4, p. 225. Ed. R. A. Raphael, E. C. Taylor and H. Wynberg (London: Interscience).

Eliel, E. L. (1961). *Rec. chem. Prog.* **22**, 129.

Emerson, W. S. (1948). *Organic Reactions*, **4**, 174 (New York: Wiley).

Entwistle, I. D. and Johnstone, R. A. W. (1965). *Chem. Comm.* 29.

Epstein, W. W. and Sweat, F. W. (1967). *Chem. Rev.* **67**, 247.

Evans, R. M. (1959). *Quart. Rev. chem. Soc. Lond.* **13**, 61.

Farmer, R. F. and Hamer, J. (1966). *J. org. Chem.* **31**, 2418.

Faulkner, D. J. and Petersen, M. R. (1969). *Tetrahedron Lett.* 3243.

Felkin, H., Swierczewski, G. and Tambuté, A. (1969). *Tetrahedron Lett.* 707.

Fétizon, M. and Golfier, M. (1968). *C.r. hebd. Séanc. Acad. Sci.*, *Paris*, **267**(*C*), 900.

Fieser, L. F. and Fieser, M. (1967). *Reagents for Organic Synthesis*, p. 759 (London: Wiley).

Filler, R. (1963). *Chem. Rev.* **63**, 21.

Fleming, I. and Harley-Mason, J. (1964). *J. chem. Soc.* 2165.

Fleming, I. and Karger, M. H. (1967). *J. chem. Soc.* (*C*), 226.

Fliszar, S. and Carles, J. (1969). *J. Am. chem. Soc.* **81**, 2637.

Foote, C. S. (1968). *Accounts Chem. Res.* **1**, 104.

Foote, C. S. and Wexler, S. (1964). *J. Am. chem. Soc.* **86**, 3879, 3880.

Freeguard, G. F. and Long, L. H. (1964). *Chemy. Ind.* 1582.

Garbisch, E. W., Schreader, L. and Frankel, J. J. (1967). *J. Am. chem. Soc.* **89**, 4233.

Gassman, P. G. and Richmond, G. D. (1966). *J. org. Chem.* **31**, 2355.

Gault, F. G., Rooney, J. J. and Kemball, C. (1962). *J. Catalysis*, **1**, 255.

Gaylord, N. G. (1956). *Reduction with Complex Metal Hydrides.* (New York: Wiley).

Gibson, T. W. and Erman, W. F. (1967). *Tetrahedron Lett.* 905.

Gilchrist, T. L. and Rees, C. W. (1969). *Carbenes, Nitrenes and Arynes* (London: Nelson).

Gillis, B. T. (1967). In *1,4-Cycloaddition Reactions.* Ed. J. Hamer (New York: Academic Press).

Gollnick, K. (1968). *Advan. Photochem.* **6**, 1.

Gollnick, K. and Schlenck, O. (1967). In 1,4-*Cycloaddition Reactions*. Ed. J. Hamer (New York: Academic Press).

Grewe, R. and Hinrich, I. (1964). *Chem. Ber.* **97**, 443.

Grob, C. A. and Kammüller, H. (1957). *Helv. Chim. Acta*, **40**, 2139.

Grob, C. A. and Schiess, P. W. (1967). *Angew. Chem. internat. Edit.* **6**, 1.

Gunstone, F. D. (1960). In *Advances in Organic Chemistry. Methods and Results*. Vol. 1, p. 103. Ed. R. A. Raphael, E. C. Taylor and H. Wynberg (New York: Interscience).

Gutsche, C. D. and Redmore, D. (1968). *Advances in Alicyclic Chemistry. Supplement* 1. *Carbocyclic Ring Expansion Reactions*, p. 111 (London: Academic Press).

Hallman, P. S., McGarvey, B. R. and Wilkinson, G. (1968). *J. chem. Soc.* (*A*), 3143.

Hallsworth, A. S. and Henbest, H. B. (1957). *J. chem. Soc.* 4604.

Hamer, J. and Ahmad, M. (1967). In 1,4-*Cycloaddition Reactions*. Ed. J. Hamer (New York: Academic Press).

Hammond, G. S., Turro, N. J. and Liu, R. S. H. (1963). *J. org. Chem.* **28**, 3297.

Hammond, W. B. and Turro, N. J. (1966). *J. Am. chem. Soc.* **88**, 2880.

Hampton, K. G., Harris, T. M. and Hauser, C. R. (1966). *J. org. Chem.* **31**, 663.

Hanson, J. R. and Premuzic, E. (1968). *Angew. Chem. internat. Edit.* **7**, 247.

Harris, T. M. and Harris, C. M. (1969). *Organic Reactions*, **17**, 155 (London: Wiley).

Hassall, C. H. (1957). *Organic Reactions*, **9**, 73 (London: Wiley).

Hawthorne, M. F. (1961). *J. Am. chem. Soc.* **83**, 2541.

Heathcock, C. H. and Poulter, S. R. (1969). *Tetrahedron Lett.* 2755.

Henbest, H. B. (1965). *Chem. Soc. Spec. Publ.* No. 19, p. 83.

Hennis, H. E. (1963). *J. org. Chem.* **28**, 2570.

Herndon, W. C. and Hall, L. H. (1967). *Tetrahedron Lett.* 3095.

Herz, J. E. and Gonzalez, E. (1969). *Chem. Comm.* 1395.

Heusler, K. and Kalvoda, J. (1964). *Angew. Chem. internat. Edit.* **3**, 525.

Heyns, K. and Paulsen, H. (1963). In *Newer Methods of Preparative Organic Chemistry*, vol. 2, p. 303. Ed. W. Foerst (New York: Academic Press).

Hill, R. K. and Carlson, R. M. (1965). *J. org. Chem.* **30**, 2414.

Hine, J. (1964). *Divalent Carbon* (New York: Ronald Press).

Hines, J. N., Peagram, M. J., Whitham, G. H. and Wright, M. (1968). *Chem. Comm.* 1593.

Hoffmann, H. M. R. (1969). *Angew. Chem. internat. Edit.* **8**, 556.

Hoffmann, R. and Woodward, R. B. (1965). *J. Am. chem. Soc.* **87**, (a) 2046; (b) 4388.

Holmes, H. L. (1948). *Organic Reactions*, **4**, 60 (London: Wiley).

Hooz, J. and Linke, S. (1968). *J. Am. chem. Soc.* **90**, 5936, 6891.

Horiuti, I. and Polanyi, M. (1934). *Trans. Faraday Soc.* **30**, 1164.

Horspool, W. M. (1969). *Quart. Rev. chem. Soc. Lond.* **23**, 223.

House, H. O. (1965). *Modern Synthetic Reactions* (New York: Benjamin).

House, H. O. (1967). *Rec. chem. Prog.* **28**, 99.

House, H. O. and Fischer, W. F. (1968). *J. org. Chem.* **33**, 949.

House, H. O. and Fischer, W. F. (1969). *J. org. Chem.* **34**, 3615.

House, H. O. and Larson, J. K. (1968). *J. org. Chem.* **33**, 61.

House, H. O., Respess, W. L. and Whitesides, G. M. (1966). *J. org. Chem.* **31**, 3128.

House, H. O. and Thompson, H. W. (1963). *J. org. Chem.* **28**, 360.

House, H. O. and Trost, B. M. (1965). *J. org. Chem.* **30**, 2502.

Huang-Minlon, (1946). *J. Am. chem. Soc.* **68**, 2487.

Huffman, J. W. and Charles, J. T. (1968). *J. Am. chem. Soc.* **90**, 6486.

Huisgen, R., Grashey, R. and Sauer, J. (1964). In *The Chemistry of Alkenes*, p. 739. Ed. S. Patai (London: Interscience).

Hussey, A. S. and Takeuchi, Y. (1969). *J. Am. chem. Soc.* **91**, 672.

Hutchins, R. O., Hoke, D., Keogh, J. and Koharski, D. (1969). *Tetrahedron Lett.* 3495.

Inukai, T. and Kojima, T. (1966). *J. org. Chem.* **31**, 2032.

Jardine, F. H. and Wilkinson, G. (1967). *J. chem. Soc.* (*C*), 270.

Jardine, I. and McQuillin, F. J. (1966). *Tetrahedron Lett.* 4871.

Jenner, E. L. (1962). *J. org. Chem.* **27**, 1031.

Johnson, A. W. (1966). *Ylid Chemistry* (London: Academic Press).

Johnson, A. W., Hruby, V. J. and Williams, J. L. (1964). *J. Am. chem. Soc.* **86**, 918.

Johnson, W. S., Bannister, B., Bloom, B. M., Kemp, A. D., Pappo, R., Rogier, E. R. and Szmuszkovicz, J. (1953). *J. Am. chem. Soc.* **75**, 2275.

Johnson, W. S., Tsung-Tee Li, Faulkner, D. J. and Campbell, S. F. (1968). *J. Am. chem. Soc.* **90**, 6225.

Johnson, W. S., Werthemann, L., Bartlett, W. R., Brocksom, T. J., Tsung-Tee Li, Faulkner, D. J. and Petersen, M. R. (1970). *J. Am. chem. Soc.* **92**, 741.

Jones, R. L. and Rees, C. W. (1969). *J. chem. Soc.* (*C*), 2249.

Julia, M., Julia, S. and Tchen, S-Y. (1961). *Bull Soc. chim. Fr.* 1849.

References 381

Kabalka, G. W., Brown, H. C., Suzuki, A., Honma, S., Arase, A. and Itoh, M. (1970). *J. Am. chem. Soc.* **92**, 710.

Kabasakalian, P., Townley, E. R. and Yudis, M. D. (1962). *J. Am. chem. Soc.* **84**, 2718.

Kaufman, G., Cook, F., Schechter, H., Bayless, J. and Friedman, L. (1967). *J. Am. chem. Soc.* **89**, 5736.

Kearns, D. R., Hollins, R. A., Khan, A. U., Chambers, R. W. and Radlick, P. (1967). *J. Am. chem. Soc.* **89**, 5455, 5456.

Khan, A. M., McQuillin, F. J. and Jardine, I. (1967). *J. chem. Soc. (C)*, 136.

Kharasch, M. S. and Tawney, P. O. (1941). *J. Am. chem. Soc.* **63**, 2308.

Kimmer, W. (1955). In E. Müller (ed.) *Methoden der Organischen Chemie (Houben-Weyl)*, vol. 4, part 2 (Stuttgart: Georg Thieme Verlag).

Kingsbury, C. A. and Cram, D. J. (1960). *J. Am. chem. Soc.* **82**, 1810.

Kirk, D. N. (1969). *Tetrahedron Lett.* 1727.

Kirmse, W. (1964). *Carbene Chemistry* (London: Academic Press).

Kloetzel, M. C. (1948). *Organic Reactions*, **4**, 1 (London: Wiley).

Knights, E. F. and Brown, H. C. (1968). *J. Am. chem. Soc.* **90**, 5280, 5281, 5283.

Kobake, Y., Fueno, T. and Furukawa, J. (1970). *J. Am. chem. Soc.* **92**, 6548.

Konaka, R., Terabe, S. and Kuruma, K. (1969). *J. org. Chem.* **34**, 1334.

Kornblum, N., Seltzer, R. and Haberfield, P. (1963). *J. Am. chem. Soc.* **85**, 1148.

Köster, R. (1958). *Liebigs Ann.* **618**, 31.

Köster, R. (1964). *Angew. Chem. internat. Edit.* **3**, 174.

Kroopan, C. G., McKusick, B. C. and Cairns, T. L. (1961). *J. Am. chem. Soc.* **83**, 3428.

Kuivila, H. G. (1968). *Accounts Chem. Res.* **1**, 299.

Kupchan, S. M. and Wormsey, H. C. (1965). *Tetrahedron Lett.* 359.

Landor, S. R., Miller, B. J. and Tatchell, A. R. (1967). *J. chem. Soc. (C)*, 197.

Lee, J. B. and Uff, B. C. (1967). *Quart. Rev. chem. Soc. Lond.* **21**, 429.

Lemieux, R. U. and von Rudloff, E. (1955). *Canad. J. Chem.* **33**, 1701.

Lutz, E. F. and Bailey, G. M. (1964). *J. Am. chem. Soc.* **86**, 3899.

Lwowski, W. (1967). *Angew. Chem. internat. Edit.* **6**, 897.

Maercker, A. (1965). *Organic Reactions*, **14**, 270 (London: Wiley).

Marshall, J. A. (1969). *Rec. chem. Prog.* **30**, 3.

Marshall, J. A. and Anderson, N. H. (1966). *J. org. Chem.* **31**, 667.

Martin, E. L. (1942). *Organic Reactions*, **1**, 155 (London: Wiley).

Martin, J. G. and Hill, R. K. (1961). *Chem. Rev.* **61**, 537.

McElvain, S. M. (1948). *Organic Reactions*, **4**, 256 (London: Wiley).

McKillop, A., Elsom, L. F. and Taylor, E. C. (1968). *J. Am. chem. Soc.* **90**, 2423.

McQuillin, F. J., Ord, W. O. and Simpson, P. L. (1963). *J. chem. Soc.* 5996.

Meinwald, J. and Meinwald, Y. C. (1966). In *Advances in Alicyclic Chemistry*, vol. 1, p. 1. Ed. H. Hart and G. J. Karabatsos (New York: Academic Press).

Meyers, A. I. *et al.* (1969). *J. Am. chem. Soc.* **91**, 763, 764, 765, 2155, 5887; *Tetrahedron Lett.* 1783.

Meyers, A. I. and Smith, E. M. (1970). *J. Am. chem. Soc.* **92**, 1084.

Mihailović, Lj., Čeković, Z. and Stanković, J. (1969). *Chem. Comm.* 981.

Miller, C. E. (1965). *J. chem. Ed.* **42**, 254.

Mitsui, S., Kudo, Y. and Kobayashi, M. (1969). *Tetrahedron*, **25**, 1921.

Moore, W. M., Morgan, D. D. and Stermitz, F. R. (1963). *J. Am. chem. Soc.* **85**, 829.

Moser, W. R. (1969). *J. Am. chem. Soc.* **91**, 1135, 1141.

Müller, E., Kessler, H. and Zeeh, B. (1966). *Fortschr. chem. Forsch.* **7**, 128.

Munch-Petersen, J. (1958). *Acta chem. scand.* **12**, 2007.

Murray, R. W. (1968). *Accounts Chem. Res.* **1**, 313.

Nace, H. R. (1962). *Organic Reactions*, **12**, 57 (London: Wiley).

Nakabayashi, T. (1960). *J. Am. chem. Soc.* **82**, 3900, 3906, 3909.

Nakagawa, K., Konaka, R. and Nakata, T. (1962). *J. org. Chem.* **27**, 1597.

Neckers, D. C. (1967). *Mechanistic Organic Photochemistry* (New York: Reinhold).

Needleman, S. B. and Chang Kuo, M. C. (1962). *Chem. Rev.* **62**, 405.

Nishimuru, J., Kawabata, N. and Furukawa, J. (1969). *Tetrahedron*, **25**, 2647.

Nozake, H., Kato, H. and Noyori, R. (1969). *Tetrahedron*, **25**, 1661.

Nussbaum, A. L. and Robinson, C. H. (1962). *Tetrahedron*, **17**, 35.

Oediger, H., Kabbe, H-J., Möller, F. and Eiter, K. (1966). *Chem. Ber.* **99**, 2012.

Onischenko, A. S. (1964). *Diene Synthesis* (New York: Davey).

Orban, I., Schaffner, K. and Jeger, O. (1963). *J. Am. chem. Soc.* **85**, 3033.

Osborn, J. A., Jardine, F. H., Young, J. F. and Wilkinson, G. (1966). *J. chem. Soc.* (*A*), 1711.

Pappas, J. J. and Keaveney, W. P. (1966). *Tetrahedron Lett.* 4273.

Parham, W. E. and Schweizer, E. E. (1963). *Organic Reactions*, **13**, 55 (London: Wiley).

Parker, A. J. (1962). *Quart. Rev. chem. Soc. Lond.* **16**, 163.

Parker, R. E. and Isaacs, N. S. (1959). *Chem. Rev.* **59**, 737.

Partch, E. (1964). *Tetrahedron Lett.* 3071.

Pettit, G. R. and van Tamelen, E. E. (1962). *Organic Reactions*, **12**, 356 (London: Wiley).

Pfau, M. and Ribière, C. (1970). *Chem. Comm.* 66.

Piatak, D. M., Herbst, G., Wicha, J. and Caspi, E. (1969). *J. org. Chem.* **34**, 112, 116.

Poos, G. I., Arth, G. E., Beyler, R. E. and Sarett, L. H. (1953). *J. Am. chem. Soc.* **75**, 422.

Poulter, C. D., Friedrich, E. C., and Winstein, S. (1969). *J. Am. chem. Soc.* **91**, 6893.

Radlick, P., Klem, R., Spurlock, S., Sims, J. J., van Tamelen, E. E. and Whitesides, T. (1968). *Tetrahedron Lett.* 5117.

Raphael, R. A. (1949). *J. chem. Soc.* Suppl. 44.

Rees, C. W. and Storr, R. C. (1969). *J. chem. Soc.* (*C*), 1474.

Reggel, L., Friedel, R. A. and Wender, I. (1957). *J. org. Chem.* **22**, 891.

Rerick, M. N. (1968). In *Reduction*. Ed. R. L. Augustine (London: Arnold).

Rerick, M. N. and Eliel, E. L. (1962). *J. Am. chem. Soc.* **84**, 2356.

Reucroft, J. and Sammes, P. G. (1971). *Quart. Rev. chem. Soc. Lond.* **25**, 135.

Rhoads, S. J. (1963). In *Molecular Rearrangements*, vol. 1, p. 655. Ed. P. de Mayo (London: Interscience).

Richer, J. C. (1965). *J. org. Chem.* **30**, 324.

Ringold, H. J. and Malhotra, S. K. (1962). *J. Am. chem. Soc.* **84**, 3402.

Robinson, R. (1916). *J. Chem. Soc.* **109**, 1038.

Rona, P. and Crabbé, P. (1968). *J. Am. chem. Soc.* **90**, 4733.

Roth, von W. R. (1966). *Chimia*, **20**, 229.

Russell, G. A. and Mikol, G. J. (1966). *J. Am. chem. Soc.* **88**, 5498.

Russell, G. A. and Ochrymowycz, L. A. (1969). *J. org. Chem.* **34**, 3624.

Rylander, P. N. (1967). *Catalytic Hydrogenation over Platinum Metals* (London: Academic Press).

Rylander, P. N. and Steele, D. R. (1969). *Tetrahedron Lett.* 1579.

Sammes, P. G. (1970). *Quart. Rev. chem. Soc. Lond.* **24**, 37.

Sauer, J. (1966). *Angew. Chem. internat. Edit.* **5**, 211.

Sauer, J. (1967). *Angew. Chem. internat. Edit.* **6**, 16.

Sauer, J. and Kredel, J. (1966). *Tetrahedron Lett.* (a) 731; (b) 6359.

Sauers, R. R. and Ahearn, G. P. (1961). *J. Am. chem. Soc.* **83**, 2759.

Saunders, W. H. (1964). In *The Chemistry of Alkenes*, p. 149. Ed. S. Patai (London: Interscience).

Schaffner, K., Arigoni, D. and Jeger, O. (1960). *Experientia*, **16**, 169.

Schenck, G. O. (1957). *Angew. Chem.* **69**, 579.

Schiller, G. (1955). In E. Müller (ed.) *Methoden der organischen Chemie* (*Houben-Weyl*), vol. 4, part 2 (Stuttgart: Georg Thieme Verlag).

Schlosser, M. and Christmann, K. F. (1966). *Angew. Chem. internat. Edit.* **5**, 126.

Schlosser, M. and Christmann, K. F. (1967). *Liebigs Ann.* **708**, 1.

Schlosser, M. and Christmann, K. F. (1969). *Synthesis*, **1**, 38.

Schmitz, E. and Murawski, D. (1966). *Chem. Ber.* **99**, 1493.

Schönberg, A. (1968). *Preparative Organic Photochemistry* (Berlin: Springer Verlag).

Schulte-Elte, K. H. and Ohloff, G. (1964). *Tetrahedron Lett.* 1143.

Seebach, D. (1969). *Synthesis*, **1**, 17.

Seebach, D., Jones, N. R. and Corey, E. J. (1968). *J. org. Chem.* **33**, 300.

Shapiro, R. H. and Heath, M. J. (1967). *J. Am. chem. Soc.* **89**, 5734.

Sharpless, K. B., Hanzlik, R. P. and van Tamelen, E. E. (1968). *J. Am. chem. Soc.* **90**, 209.

Siddall, J. B., Biskup, M. and Fried, J. H. (1969). *J. Am. chem. Soc.* **91**, 1853.

Simmons, H. E. and Smith, R. D. (1959). *J. Am. chem. Soc.* **81**, 4256.

Siegel, S. and Smith, G. V. (1960). *J. Am. chem. Soc.* **82**, 6082, 6087.

Sloan, M. F., Matlack, A. S. and Breslow, D. S. (1963). *J. Am. chem. Soc.* **85**, 4014.

Smissman, E. E., Suh, J. T., Oxman, M. and Daniels, R. (1962). *J. Am. chem. Soc.* **84**, 1040.

Smith, C. W. and Holm, R. T. (1957). *J. org. Chem.* **22**, 747.

Smith, G. V. and Burwell, R. L. (1962). *J. Am. chem. Soc.* **84**, 925.

Smith, P. A. S. (1963). In *Molecular Rearrangements*, vol. 1, p. 571. Ed. P. de Mayo (London: Wiley–Interscience).

Smolinsky, G. and Feuer, B. I. (1964). *J. Am. chem. Soc.* **86**, 3085.

Snatzke, G. (1961). *Chem. Ber.* **94**, 729.

Sneen, R. A. and Matheny, N. P. (1964). *J. Am. chem. Soc.* **86**, 5503.

Stevens, C. L. and Dykstra, S. J. (1953). *J. Am. chem. Soc.* **75**, 5975.

Stewart, R. (1965). In *Oxidation in Organic Chemistry*, part A, p. 42. Ed. K. B. Wiberg (London: Academic Press).

Stork, G., Brizzolana, A., Landesman, H., Szmuszkovicz, J. and Terrell, R. (1963). *J. Am. chem. Soc.* **85**, 207.

Stork, G. and Darling, S. D. (1964). *J. Am. chem. Soc.* **86**, 1761.

Stork, G. and Dowd, S. R. (1963). *J. Am. chem. Soc.* **85**, 2178.

Stork, G., Grieco, P. A. and Gregson, M. (1969). *Tetrahedron Lett.* 1393.

Stork, G. and Hudrlik, P. F. (1968). *J. Am. chem. Soc.* **90**, 4462, 4464.

Stork, G., Meisels, A. and Davies, J. E. (1963). *J. Am. chem. Soc.* **85**, 3419.

Stork, G., Rosen, P., Golman, N., Coombs, R. V. and Tsuji, J. (1965). *J. Am. chem. Soc.* **87**, 275.

Suzuki, A., Nozawa, S., Itoh, M., Brown, H. C., Negishi, I. and Gupta, S. K. (1969). *Chem. Comm.* 1009.

Swern, D. (1953). In *Organic Reactions*, **7**, 378 (London: Wiley).

Syper, L. (1966). *Tetrahedron Lett.* 4493.

Szmant, H. H. (1968). *Angew. Chem. internat. Edit.* **7**, 120.

Szmuszkovicz, J. (1963). In *Advances in Organic Chemistry. Methods and Results*, vol. 4, p. 1. Eds. Raphael, R. A., Taylor, E. C. and Wynberg, H. (London: Interscience).

Taylor, D. A. H. (1969). *Chem. Comm.* 476.

Taylor, E. C., Hawks, G. H. and McKillop, A. (1968). *J. Am. chem. Soc.* **90**, 2421.

Trippett, S. (1963). *Quart. Rev. chem. Soc. Lond.* **17**, 406.

Tufariello, J. J. and Kissel, W. J. (1966). *Tetrahedron Lett.* 6145.

Tufariello, J. J., Wojtkowski, P. and Lee, L. T. C. (1967). *Chem. Comm.* 505.

Vogel, E., Grimme, W. and Korte, S. (1963). *Tetrahedron Lett.* 3625.

Wagner, P. J. and Hammond, G. S. (1968). *Advances in Photochemistry*, vol. 5, p. 136 (London: Interscience).

Walling, C. (1957). *Free Radicals in Solution* (London: Wiley).

Walling, C. and Padwa, A. (1961). *J. Am. chem. Soc.* **83**, 2207.

Walling, C. and Padwa, A. (1963). *J. Am. chem. Soc.* **85**, 1597.

Wassermann, A. (1965). *Diels–Alder Reactions* (London: Elsevier).

Wawzonek, S., Nelson, M. F. and Thelen, P. J. (1951). *J. Am. chem. Soc.* **73**, 2806.

Wawzonek, S. and Thelen, P. J. (1950). *J. Am. chem. Soc.* **72**, 2118.

Wendler, N. L., Taub, D. and Kuo, H. (1960). *J. Am. chem. Soc.* **82**, 5701.

Wenkert, E., Stenberg, V. I. and Beak, P. (1961). *J. Am. chem. Soc.* **83**, 2320.

Whitesides, G. M., Fischer, W. F., San Filippo, J., Bashe, R. W. and House, H. O. (1969). *J. Am. chem. Soc.* **91**, 4871.

Whitesides, G. M., San Filippo, J., Casey, C. P. and Panek, E. J. (1967). *J. Am. chem. Soc.* **89**, 5302.

Wiberg, K. B. (1965). In *Oxidation in Organic Chemistry*, part A, p. 69. Ed. K. B. Wiberg (London: Academic Press).

Wicker, R. J. (1956). *J. chem. Soc.* 2165.

Wilds, A. L. (1944). *Organic Reactions*, **2**, 178 (London: Wiley).

Wilds, A. L. and Nelson, N. A. (1953). *J. Am. chem. Soc.* **75**, 5366.

Wittig, G. (1962). *Angew. Chem. internat. Edit.* **1**, 415.

Wolff, M. E. (1963). *Chem. Rev.* **63**, 55.

Woodward, R. B. (1967). *Aromaticity.* Chemical Society Special Publication No. 21, p. 242.

Woodward, R. B., Bader, F. E., Bickel, H., Frey, A. J. and Kierstead, R. W. (1958). *Tetrahedron*, **2**, 1.

Woodward, R. B. and Hoffmann, R. (1969). *Angew. Chem. internat. Edit.* **8**, 781.

Woodward, R. B. and Katz, T. J. (1959). *Tetrahedron*, **5**, 70.

Yang, N. C. and Yang, D. H. (1958). *J. Am. chem. Soc.* **80**, 2913.

Yates, P. and Eaton, P. (1960). *J. Am. chem. Soc.* **82**, 4436.

Zimmerman, H. E. (1961). *Tetrahedron*, **16**, 169.

Zimmerman, H. E. (1963). In *Molecular Rearrangements*, vol. 1, p. 345. Ed. P. de Mayo (London: Wiley–Interscience).

Zurflüh, R., Wall, E. N., Siddall, J. B. and Edwards, J. A. (1968). *J. Am. chem. Soc.* **90**, 6224.

Zweifel, G. and Arzoumanian, H. (1967). *J. Am. chem. Soc.* **89**, 5086.

Zweifel, G., Arzoumanian, H. and Whitney, C. C. (1967). *J. Am. chem. Soc.* **89**, 3652.

Zweifel, G. and Brown, H. C. (1963). *Organic Reactions*, **13**, 1 (London: Wiley).

Zweifel, G., Polston, N. L. and Whitney, C. C. (1968). *J. Am. chem. Soc.* **90**, 6243.

Zweifel, G. and Steele, R. B. (1967). *J. Am. chem. Soc.* **89**, (*a*) 2754; (*b*) 5085.

Index

abscissic acid, 127
acetals, reduction to ethers, 366
acetic esters, substituted, from organoboranes and α-bromoesters, 237
acetylacetone, γ-alkylation, 13
acetylenes,
   addition of carbenes, 56
   catalytic hydrogenation, 307–8
   coupling with 1-bromo-acetylenes, 50
   1,3-dienes from, 104–5
   ethylene derivatives from, via vinylalanes and vinylboranes, 100–6
   hydroboration, 218, 221, 222
   photo-addition to αβ-unsaturated ketones, 68
   reduction, 223, 345–6
   terminal: aldehydes from, 222; coupling, 49; vinyl bromides from, 101–2
   αβ-unsaturated acids from, 102
acetylenedicarboxylic acid, as dienophile in Diels–Alder reaction, 116, 117, 131
αβ-acetylenic esters,
   reaction with lithium dialkylcuprates, 97
   reaction with allylic acetates, 98
acid chlorides, reduction to aldehydes, 363
acrylic acid,
   addition to cyclopentadiene, 147, 150
   derivatives, addition to substituted butadienes, 157–8
acrylic esters, ββ-dialkyl, stereoselective synthesis, 97
activating groups, 1, 22
acylation,
   1,3-dicarbonyl compounds, 12
   enamines, 34–5
acylnitrenes, 179
acyloin condensation, 335
Adams' catalyst, 301
alcohols,
   acid-catalysed dehydration, 72
   oxidation: argentic oxide, 251; catalytic, 257–8; Cr^{VI} 244–8; dimethyl sulphoxide, 254–5; N-halo-imides, 258; lead tetra-acetate, 257; manganese dioxide, 249–50; nickel peroxide, 284; Oppenauer, 256; ruthenium tetroxide, 281–2; silver carbonate, 252
   pyrolysis of esters, 75–9
   synthesis from: aldehydes and ketones by reduction, 310, 332–4, 337–41, 352, 357ff; esters, 303, 357; olefins via hydroboration, 218, 219, 224–5, 229, 232–4
   see also allylic alcohols
aldehydes,
   alkylation and acylation, 29–35
   catalytic hydrogenation, 310–12
   conversion into amides and nitriles, 285
   decarbonylation, 324
   oxidation, 251, 290
   photolysis, to cyclobutanols, 202, 203
   reduction: to methyl groups, 327, 363, 367; to primary alcohols, 207–11, 352, 357ff, 363
   synthesis from: acetylenes, 222; alkyl halides, 38–9, 255; aromatic methyl derivatives by oxidation, 243, 244; carboxylic acid derivatives, 210, 363–5; dihydro-1,3-oxazines, 38–9; 1,3-dithianes, 36–7; nitriles, 365; olefins, 232, 233, 279–81; organoboranes and αβ-unsaturated aldehydes, 234–6; ozonides, 275; primary alcohols, 244–50, 252, 254–5, 257–8; Wittig reaction, 90
aldosterone-21-acetate, 189
alkenylation of active methylene compounds, 8
2-alkoxybutadienes, 113, 129
alkoxyl radicals,
   formation, 182–3, 190–4, 197–201
   intramolecular attack on non-activated C–H bonds, 182ff, 190, 194
alkylation,
   aldehydes and ketones, 17ff

alkylation (*continued*)
  π-allylnickel complexes, 45–7
  bisthiocarbanions, 36–7
  carbonyl compounds, role of enolate
    ions, 2
  cyclic ketones, stereochemistry, 27–9
  1,3-dicarbonyl compounds, 4ff: γ-
    alkylation of, 13–14
  enamines, 29ff
  ketones, 17ff
  β-keto-sulphoxides and sulphones, 15
  phenols, 9
  αβ-unsaturated ketones, 24–7, 41–5
  unsymmetrical ketones, selective, 20–7
2-alkylcycloalkanols,
  cis-: from epoxides by reduction with
    diborane, 209; from 2-alkylcyclo-
    alkanones, 211
  trans-: from 2-alkylcycloalkanones,
    211; from 1-alkylcyclohexenes, 218,
    224
alkyl halides,
  coupling with lithium organocuprates,
    47–8
  homologous aldehydes from, 36–7
  olefins from, 71–5
  oxidation to aldehydes, 255
  reduction, 349–50, 360, 362
alkyl sulphonates,
  olefins from, 71–5
  oxidation to aldehydes, 255
allenes,
  formation: from halocyclopropanes,
    56; from propargyl acetates, 98
  in Diels–Alder reaction, 122, 131
allylic acetates, reaction with lithium
    dialkylcuprates, 98
allylic alcohols,
  alkyl derivatives, stereoselective for-
    mation, 87–9, 95–6, 102
  formation: from αβ-epoxyketones, 368;
    from olefins by photo-oxidation,
    291–8; from αβ-unsaturated alde-
    hydes and ketones, 352, 366
  oxidation: to αβ-unsaturated acids and
    esters, 250–1; to αβ-unsaturated car-
    bonyl compounds, 248–50, 256, 258
  photo-oxidation, 293
  reductive cleavage of hydroxyl group,
    348
allylic compounds,
  as dienophiles in the Diels–Alder
    reaction, 119

reductive coupling to 1,5-dienes, 51, 52
reductive fission, 347, 348
allylic hydroperoxides by photo-oxi-
    dation of olefins, 291ff
π-allylnickel complexes, 45–7
allylsulphonium ylids, rearrangement to
    1,5-dienes, 52–3
allyl vinyl ethers, rearrangement to
    γδ-unsaturated aldehydes and ke-
    tones, 112–14
aluminium alkoxides, reduction of alde-
    hydes and ketones, 352
aluminium chloride,
  catalyst in Diels–Alder reaction, 146
  effect on reductions with lithium
    aluminium hydride, 361, 362, 365–6
aluminium hydride, 366
amides,
  formation from aldehydes, 285
  N-substituted, reduction to aldehydes,
    364
amine oxides, pyrolysis to olefins, 75,
    77–80
anthracene,
  addition of oxygen, 125
  as diene in Diels–Alder reaction, 136,
    137
anthraquinone derivatives, formation by
    Diels–Alder reaction, 118, 131
aporphine alkaloids, 65
argentic oxide, 251
aromatic compounds,
  Birch reduction, 341–5
  catalytic hydrogenation, 308–10
  reduction with lithium in alkylamines,
    350–1
aromatic hydrocarbons,
  addition of carbenes, 56–8
  oxidation with chromic acid, 242–3
  see also benzene derivatives *and*
    polycyclic aromatic hydrocarbons
arylation of 1,3-dicarbonyl compounds, 7
aryl ethers, cleavage with sodium–liquid
    ammonia, 349
arynes, dienophiles in Diels–Alder re-
    action, 121
ascaridole, 127
asymmetric synthesis,
  alcohols: by asymmetric reduction of
    ketones, 365; by hydroboration of
    olefins, 218–9
  asymmetric hydrogenation of olefins,
    325–6

in Diels–Alder reaction, 159–60
Auwers–Skita rule, 318
azo compounds, dienophiles in Diels–
   Alder reaction, 124

Baeyer–Villiger oxidation, 287–90
   mechanism, 289
Barton reaction, 181ff
   mechanism, 183, 184
   spatial requirements, 184
   with steroid nitrites, 187–9
benzene derivatives,
   1,4-addition of dienophiles, 136, 137,
      167
   catalytic hydrogenation, 308–10
   formation by Diels–Alder reaction,
      139–40, 142
   oxidation, 242–4
   reduction: alkali metal in liquid
      ammonia 342–5; lithium in alkyl-
      amines, 350–1
benzocyclobutadiene, 121
benzocyclopropene, 145
p-benzoquinone, 118
benzoylacetone, γ-alkylation, 13
benzyloxycarbonyl protecting group,
   hydrogenolysis, 300
   reductive fission, 347
benzyne derivatives, arylation of 1,3-
   dicarbonyl compounds, 7
bicyclo[1,1,0]butane derivatives, synthesis, 66, 67
bicyclo[2,2,1]heptadiene, addition of dienophiles, 167–8
bicyclo[2,2,1]heptene derivatives, synthesis, 138, 139
bicyclo[2,1,1]hexane, 62
bicyclo[2,1,1]hex-2-ene synthesis, 110
bicyclo[2,2,2]octadiene derivatives, synthesis, 139
Birch reduction, 335–46
   acetylenes, 345–6
   aromatic compounds, 342–5
   cyclic ketones, 337
   1,3-dienes, 336
   reductive fission reactions, 346–9
   αβ-unsaturated ketones, 337–41
αω-bisallyl bromides, conversion into
   cyclic 1,5-dienes, 47
bisthio carbanions, alkylation, 36–7
blocking groups in selective alkylation of
   unsymmetrical ketones, 22

9-borabicyclo[3,3,1]nonane,
   addition to propargyl bromide, 216
   B-alkyl- and aryl-derivatives, 213
   carbonylation of B-substituted deri-
      vatives, 233
   preparation, 213
   reaction of B-alkyl derivatives with
      α-bromo-esters, 236, 237
1-bromoacetylenes,
   coupling with terminal acetylenes, 50
   hydroboration, 100
α-bromoesters, reaction with trialkylboranes, 237
bromohydrins, reduction to alcohols, 351
α-bromoketones,
   reaction with trialkylboranes, 236
   reduction to ketones, 330
N-bromosuccinimide, oxidation of alcohols, 258
1,3-butadiene,
   addition of acrylic acid derivatives,
      orientation, 157–8
   2-alkoxy-, in Diels–Alder reaction, 129
   1,4-diacetoxy-, in Diels–Alder reaction,
      129
   reactivity of derivatives in Diels–Alder
      reaction, 128ff
t-butyl chromate, 245

camphor,
   Baeyer–Villiger oxidation, 289
   reduction with lithium aluminium
      hydride, 359
carbanions,
   bisthio, use in synthesis, 36
   formation from: carbonyl compounds,
      1–4; ketones, 17, 20–5; β-keto-
      sulphoxides and -sulphones, 15
   in synthesis, 1ff
   see also enolate anions
carbazole derivatives, 179, 180
carbenes,
   addition to aromatic systems, 56–8
   addition to olefins and acetylenes, 56,
      58, 59
   formation, 54
   insertion reactions, 55
   intramolecular addition, 59
   reactivity, effect of copper salts, 58–9
   structure, 53
carbenoids,
   reactions, 58–9
   structure, 54

carbonylation of organoboranes, 229–34
carboxylic acids,
  derivatives, reduction to aldehydes,
    363, 364
  formation from: aldehydes, 251;
    olefins, 274, 279, 282; primary
    alcohols, 257–8, 281–2, 284
  oxidative decarboxylation, 108–10
  reduction with diborane, 209
carboxylic esters, acyloin condensation,
  335
  pyrolysis to olefins, 75, 76–77
  reduction to: primary alcohols, 303,
    357; ethers, 366
carvone camphor, 67
catalytic deuteration, 306, 322
catalytic hydrogenation, 299ff
  asymmetric, 325–6
  catalysts for, 300–4, 311, 308, 320–1
  functional groups of: acetylenes, 307–
    8; aldehydes and ketones, 310–12;
    aromatic compounds, 308–10;
    nitriles, oximes and nitro com-
    pounds, 312
  general, 299
  homogeneous, 320–6
  mechanism, 319–20, 324
  selective, 303, 321–6
  stereochemistry, 314–18
ceric salts for oxidation of aromatic
  methyl groups, 244
chloral as dienophile in Diels–Alder
  reaction, 122
N-chloroamides,
  conversion into γ-lactones, 197
  conversion into pyrolidine derivatives,
    174
N-chloroamines, cyclisation to pyrrolidine derivatives, 173–8
1-chlorobenzotriazole, 259
chlorohydrins,
  from chloroalkylboranes, 215
  1,4-, by photolysis of hypochlorites,
    190, 191
chromic acid,
  as oxidising agent, 239–41
  oxidation of alcohols, 244–7
  oxidation of aromatic hydrocarbons,
    242–3
  oxidation of olefins, 244
  oxidation of paraffin hydrocarbons, 241
chromium trioxide as oxidising agent,
  239; *see* chromic acid

chromium trioxide–pyridine complex,
  240, 245, 247–8
chromous salts, reduction of organohalides, 350–1
chromyl chloride, 243
cis-principle in Diels–Alder reaction,
  147–9
Claisen rearrangement, allyl vinyl ethers,
  112–13
Clemmensen reduction, 327
  alternatives to, 311, 366
  mechanism, 329
  rearrangements during, 329
conduritol D synthesis, 153
δ-coneceine, 173
copper chromite, 300, 302
copper ion,
  catalytic effect in organic reactions,
    41–2, 49, 50
  effect on reactivity of carbenes, 58–9
coupling reactions,
  allyl derivatives to 1,5-dienes, 51–3
  π-allylnickel complexes, 45–7
  Grignard reagents, 45
  lithium organocuprates, 47–8, 51
  terminal acetylenes, 49
Cram's rule, 357–8
cyclic 1,3-dienes in Diels–Alder reaction,
  137–9
cyclic 1,5-dienes from αω-bisallyl bromides, 47
cyclic ketones,
  catalytic hydrogenation, stereochem-
    istry, 318
  oxidative cleavage of ring, 278, 283
  reduction: lithium aluminium hydride,
    358–60; organoboranes, 211; sodium
    and alcohol, 332–4, 337
  synthesis using organoboranes, 230,
    232
cyclic organoboranes,
  addition to αβ-unsaturated aldehydes
    and ketones, 236
  preparation: from dienes, 220, 221;
    by heating organoboranes, 227
cycloalkenes, dienophiles in Diels–Alder
  reaction, 120
cycloalkynes, dienophiles in the Diels–
  Alder reaction, 121
cyclobutane derivatives, synthesis, 34–5,
  66–8, 165
cyclobutanols by photolysis of aldehydes
  and ketones, 202, 203

cyclobutanone,
  synthesis, 37
  derivatives, 122
cyclobutene,
  derivatives: formation from acetylenes
    and αβ-unsaturated ketones, 68;
    reaction with dienes, 121
  synthesis, 110
cyclodeca-1,6-dienes, synthesis, 107
cis- and *trans*-cyclodec-5-enone synthesis,
  106
1,3-cyclohexadiene derivatives, in Diels–
  Alder reaction, 139
cyclohexanone derivatives
  catalytic hydrogenation, 318
  reduction: aluminium alkoxides, 353;
    lithium aluminium hydride, 358–60;
    sodium and alcohol, 332–4
  synthesis by Diels–Alder reaction, 129
cyclohexenones, preparation by Birch
  reduction of benzene derivatives, 343
cyclo-octyne, 121
cyclopentadiene,
  addition of acrylic acid derivatives,
    147, 150
  diene in Diels–Alder reaction, 138, 147
  dienophile in Diels–Alder reaction, 119
  dimerisation, 119, 138, 165
cyclopentadienone derivatives, in Diels–
  Alder reaction, 140
cyclopropane derivatives, synthesis, 56,
  58, 59–61, 94, 215, 368
cyclopropene,
  derivatives, synthesis, 56
  dienophile in Diels–Alder reaction, 120
cyclopropyl ketones, synthesis, 93–4

Dakin reaction, 290
decalyboranes, fragmentation reactions
  of, 107
dehydrobenzene as dienophile in Diels–
  Alder reaction, 121
desulphurisation with Raney nickel,
  368–9
Dewar benzene synthesis, 109
trans, trans- 1,4-diacetoxybutadiene, 129
ββ-dialkylacrylic esters, stereoselective
  synthesis, 97
dialkylboranes, formation, 212
1,5-diazabicyclo[3,4,0]non-5-ene, 75
diazoalkanes,
  carbenes from, 54
  copper-catalysed decomposition, 59

diazoketones,
  copper-catalysed decomposition, 59
  reaction with trialkylboranes, 238
  Wolff rearrangement, 61–2
diborane,
  addition to: acetylenes, 221; dienes,
    219–20; olefins, 211–19; *see also*
    hydroboration
  contrast with sodium borohydride, 208
  preparation, 207
  reduction of: carbonyl groups, 208;
    carboxylic acids, 209; epoxides, 209;
    unsaturated groups, 207–9
1,3-dicarbonyl compounds,
  acylation, 12
  alkenylation, 8
  alkylation, 4–7, 13–14
  arylation, 7
  enolate anions from, 4
  hydrolysis, 9–10
  preparation from enamines, 34
  vinyl derivatives, 8
2,3-dichloro-5,6-dicyanobenzoquinone,
  259
1,1-dichloro-2,2-difluoroethylene, 165
dicyanoacetylene, 136, 137
Diels–Alder reaction, 115
  asymmetric induction, 159–60
  catalysis, 146–7, 158, 159, 160
  diene, 127, *see* 1,3-dienes
  dienophile, 116, *see* dienophiles
  homo Diels–Alder reaction, 167–8
  mechanism, 154–6, 160–5
  orientation with unsymmetrical com-
    ponents, 156–8
  photosensitised, 165–7
  retro Diels–Alder reaction, 144–5
  stereochemistry: *cis*- principle, 147–9;
    effect of Lewis acids, 146; *endo*
    addition rule, 149–56; orbital sym-
    metry control, 155–6, 164
  transition state, 154, 163, 164
1,3-dienes,
  cyclic ketones from, 230, 232
  in Diels–Alder reaction, 127: acyclic,
    127–31; allenes, 131; benzene deri-
    vatives, 128–9; cyclic dienes, 137–9;
    dienynes, 132; dimethylenecyclo-
    alkanes, 133; enynes, 131; furan
    derivatives, 141–2; heterodienes,
    133; polyenes, 129–30; quinones,
    140; vinylcycloalkenes, 134

1,3-dienes (*continued*)
  homologous diols from, 233
  hydroboration, 219–21
  photocyclisation, 66
  reduction with sodium and liquid
    ammonia, 336
  stereoselective synthesis, 104–6
1,5-dienes, synthesis, 51, 52, 114
dienophiles,
  allenes, 122
  allyl compounds, 119
  arynes, 121
  azo compounds, 124
  cyclo-alkenes and -alkynes, 120–1
  enamines, 120
  ethylene and acetylene derivatives, 116
  heterodienophiles, 122–5
  nitriles, 123
  nitroso compounds, 123
  oxygen, 125–7
  quinones, 118
  αβ-unsaturated carbonyl compounds,
    117, 118
  vinyl compounds, 119, 120
dienynes in Diels–Alder reaction, 132
1,4-dihydrobenzene derivatives, from
  benzene derivatives, 342–4, 350–1
dihydroconessine, 177
5,6-dihydro-1,3(4*H*)-oxazine, 2,4,4,6-
  tetramethyl, use in synthesis of aldehydes, 38–9
5,6-dihydropyran derivatives, formation
  by Diels–Alder reaction, 120, 122,
  133
dihydrosterculic acid, 59
di-imide, 369–71
di-isobutylaluminium hydride, addition
  to acetylenes, 100
di-isopinocampheylborane, 219
1,5-diketones, synthesis, 69
2,3-dimethylbut-2-ylborane,
  addition to olefins, 213, 230
  ketones from olefins, using, 230
  preparation, 213
di(3-methylbut-2-yl)borane,
  addition to: acetylenes, 221, 222;
    dienes, 220, 221; halogeno-olefins,
    215, 216; olefins, 213–14, 215–18
  selective addition to double bonds,
    214
  selective reductions with, 210–11
  stereoselective reduction of ketones,
    211

1,2-dimethylenecycloalkanes as dienes in
  Diels–Alder reaction, 133
4,5-dimethylenecyclohexene, 77
dimethylsulphonium methylide, 93, 94,
  234
dimethyl sulphoxide, 15, 254–5
dimethylsulphoxonium methylide, 93, 94
1,2-diols,
  stereospecific conversion into olefins,
    111
  stereospecific formation from olefins,
    259–64, 269, 270
diols, oxidation with silver carbonate,
  252
diphenylisobenzofuran, 121
diphenylsulphonium isopropylide, 94
disiamylborane, *see* di(3-methylbut-2-yl)-
  borane
1,3-dithianes, alkylation, 36–7

elimination reactions,
  β-eliminations in formation of olefins,
    71–5: stereochemistry, 73–4; struc-
    tural isomers formed, 72–4
  pyrolytic *syn* eliminations, 75–81
enamine reaction, 29ff
enamines,
  alkylation, 29–32, 33–4
  acylation, 34
  dienophiles in Diels–Alder reaction,
    120, 133
  formation, 30
  photo-oxidation, 298
  reaction with αβ-unsaturated carbonyl
    compounds, 33–4
endo-addition rule, 149–56
endoperoxides, 125–7
ene synthesis, 168–71
eneynes in Diels–Alder reaction, 131
enol acetates,
  enolate anions from, 23–4
  epoxidation, 271–2
  ozonolysis, 277
enolate anions,
  alkylation, 4ff
  acylation, 12
  formation from: 1,3-dicarbonyl com-
    pounds, 2–4; enol acetates, 23–4;
    ketones, 17, 20–7
enol ethers,
  dienophiles in Diels–Alder reaction,
    120, 133

epoxidation, 271–2
ozonolysis, 277
epoxidation,
enol acetates and enol ethers, 271–2
olefins, 264–8
αβ-unsaturated carbonyl compounds, 267
epoxides,
formation from: enol acetates and enol ethers, 271–2; olefins with per-acids, 264–8; αβ-unsaturated carbonyl compounds, 267
hydrolysis, 269, 270
reduction: diborane, 209; lithium aluminium hydride, 270, 360–2
rearrangement to carbonyl compounds, 273
αβ-epoxyketones,
from allylic alcohols by photo-oxidation, 293
reduction to allylic alcohol, 368
ergosterol, 126
Étard reaction, 243
ethers,
formation from acetals, ketals, esters and lactones, 366
oxidation to esters and lactones, 282
ethyl azodicarboxylate, dienophile in Diels–Alder reaction, 124, 125
ethylene derivatives,
general syntheses,
from: allyl vinyl ethers, 112–14; carboxylic acids, 107–10; 1,2-diols, 111
by: fragmentation reactions, 106–8
from: toluene-*p*-sulphonylhydrazones, 110
trisubstituted, stereoselective synthesis from: cyclopropyl carbinols, 99–100; vinyl-alanes and -boranes, 100–6
tri- and tetra-substituted, stereoselective synthesis: Cornforth's synthesis, 95; from αβ-acetylenic esters, 97–8; from allylic acetates and lithium dialkylcuprates, 98; from phosphon-bis-*N,N*-dialkyl-amides, 91–3; from propargyl alcohols, 95–6; by Wittig reaction, 87–9

fragmentation reactions in synthesis of olefins, 106–8

free radicals,
formation from: t-alkyl hydroper-oxides, 210; *N*-iodo-amines, 194–5; organic hypohalites, 190; organic nitrites, 183
intramolecular attack on unactivated C–H bonds, 172ff
role in Hofmann–Loeffler–Freytag reaction, 176
role in oxidations with nickel peroxide, 286
furan,
derivatives, as dienes in Diels–Alder reaction, 141–2
stereochemistry of addition of maleic anhydride, 142

Glaser reaction, 49
glutaraldehyde derivatives, formation via the Diels–Alder reaction, 120
Grignard reagents,
1,4-addition to αβ-unsaturated ketones and esters, 41–2
coupling, beneficial effect of thallium salts, 45

N-halo-amides,
γ-lactones from, 194–7
pyrrolidine derivatives from, 174
δ-halo-carbonyl compounds, cyclisation, 49
N-halogenated amines, pyrrolidines from 173–8
heterodienes in Diels–Alder reaction, 133
heterodienophiles, 122–5
hexafluorobut-2-yne, 1,4-addition to benzene derivatives, 136
hexa-1,3,5-trienes, photocyclisation, 63–5
Hofmann–Loeffler–Freytag reaction, 173–8
homoallylic bromides, stereoselective synthesis, 99–100
homo Diels–Alder reaction, 167–8
homogeneous hydrogenation, 320–6
Huang-Minlon reduction, 367
hydridochlorotris(triphenylphosphine)-ruthenium, 321, 325
hydroboration,
acetylenes, 216, 221

hydroboration (*continued*)
  allyl halides, 216
  1-bromo-acetylenes, 100
  crotyl derivatives, 217
  dienes, 219–21
  mechanism, 214
  olefins, 211–19: directive effect of
    substituents, 212, 215–18
  stereochemistry, 218–19
hydrogenolysis,
  allylic and benzylic derivatives, 307,
    309, 346–9
  benzyloxycarbonyl group, 300
  general, 300
  mechanism, 320, 326–7
  α-substituted ketones, 330–2
hydroperoxides, t-alkyl, conversion into
  olefinic alcohols, 201
hydroxyaldehydes,
  from cyclic organoboranes, 236; diols,
    253; lactones, 210
α-hydroxy-(acetoxy-) ketones, 272
hypobromites, organic, photolysis of, 192
hypochlorites, organic, photolysis, 190–1
hypohalites, organic, photolysis, 190–4
hypoiodites,
  organic, formation from alcohols, 192
  photolysis, 192–4

imines, alkylation, 33
N-iodo-amides, photolysis, conversion
  into γ-lactones, 194–6

Jones reagent, 247
juvenile hormone, 97, 108

ketals, reduction to ethers, 366
ketene, reaction with conjugated dienes,
  122
β-keto-esters,
  acylation, 12
  alkylation, 4ff
  hydrolysis and decarboxylation, 9–10
  preparation from enamines, 34
  thallium(I) salts, alkylation and acyla-
    tion of, 12
α-ketols, reduction to ketones, 331
ketones,
  alkylation: cyclic ketones, stereo-
    chemistry, 27–9; direct, 17–29;
    enamines of, 29–34; imines, of, 33;
    selective, of unsymmetrical ketones,
    20–7, 31; αβ-unsaturated ketones,
    24–5
  Baeyer–Villiger oxidation, 287
  catalytic hydrogenation, 310–12
  enolate anions from, 17, 20–7
  formation: alkylation of 1,3-dithianes,
    36–7; from acetylenes, 222; from
    carboxylic acids, 15–16; conjugate
    addition of organo-metallic com-
    pounds to αβ-unsaturated ketones,
    41; by the dihydro-1,3-oxazine
    synthesis, 38, 39; from malonic acid
    derivatives with lead tetra-acetate,
    10; from olefins, 279–81; from
    organoboranes, 225, 229–30, 232,
    234–6, 238; by oxidation of secon-
    dary alcohols, 244–8, 249, 250, 252,
    254–5, 257, 258, 281–2; from ozo-
    nides, 275
  olefins from, 110
  photolysis to cyclobutanols, 202, 203
  oxidation with per-acids, 287–90
  reduction: asymmetric, 365; bimolecu-
    lar reduction, 334; to methylene
    compound, 311, 327, 363, 366, 367;
    to secondary alcohol, 208, 210, 211,
    332–3, 352, 357ff; stereoselective, of
    cyclic ketones, 211, 332, 358
β-keto-sulphoxides and -sulphones,
  alkylation, 15–16
  Michael addition to αβ-unsaturated
    esters, 16–17

lactones,
  formation from: carboxylic acids via
    N-iodo-amides, 194–6; N-chloro-
    amides, 197; diols with silver
    carbonate, 253; ethers with ruth-
    enium tetroxide, 282
  reduction: to ethers, 366; to hydroxy-
    aldehydes, 210; with lithium alu-
    minium hydride, 357
lead tetra-acetate in oxidation of,
  carboxylic acids to olefins, 108, 109
  malonic acid derivatives to ketones, 10
  monohydric alcohols to: tetrahydro-
    furans, 197–201; aldehydes and ke-
    tones, 257
Lindlar's catalyst, 308

liquid ammonia,
  reductions in, 335ff
  reductive fission in, 346ff
lithium aluminium hydride,
  modification of reducing action by:
    alcohols, 363–5; aluminium chlor-
    ide, 361, 362, 365–6
  reduction by, 353: aldehydes and
    ketones, 357; alkyl halides and
    sulphonates, 362; epoxides, 360–2;
    esters and lactones, 357; propargylic
    alcohols, 95, 356; αβ-unsaturated
    carbonyl compounds, 355
lithium dimethylcuprate, conjugate addi-
    tion to αβ-unsaturated ketones, 43
  see also lithium organocuprates
lithium hydridoalkoxyaluminates, re-
    ductions with, 363–5
lithium hydridodi-isobutylmethylalu-
    minate, addition to disubstituted
    acetylenes, 102
lithium organocuprates,
  addition to αβ-acetylenic esters, 97
  conjugate addition to αβ-unsaturated
    ketones, 43
  oxidative coupling, 51
  reaction with organohalides, 47–9
  reaction with vinyl iodides, 96, 100

maleic anhydride,
  as dienophile in Diels–Alder reaction,
    116, 129, 130, 131, 134, 137
  photosensitised addition to benzene
    derivatives, 167
malonic ester,
  acylation, 12
  alkylation, 4ff
  hydrolysis, 9
manganese dioxide, oxidation with, 249–
    50
Meerwein–Pondorff–Verley reduction,
    333, 352
(±)-methyl *trans*-chrysanthemate, 94
methylcopper, 42
2-methylcyclohexanone, reduction to
    alcohol, stereochemistry, 332–3
9-methyl-1-decalone, synthesis, 17, 22
N-methylgranatamine, 175
methylsulphinyl carbanion, 15, 81, 84
Michael reaction,
  with β-ketosulphoxides, 16–17
  alternative, using enamines, 33

Milas reagent, 263
monoalkylboranes, formation, 212

nickel peroxide as oxidising agent, 284–7
nicotine, 173
nitrenes,
  cyclisation, 178–81
  formation from: acyl azides, 179;
    alkyl azides, 179, 180; aromatic
    nitro compounds, 180
nitriles,
  catalytic hydrogenation, 312–13
  dienophiles in Diels–Alder reaction,
    123
  from aldehydes, 285
  reduction to aldehydes, 365
nitrites, organic, photolysis of, 182–9
nitroso-alcohols by photolysis of organic
    nitrites, 182, 183, 185
nitroso compounds, dienophiles in Diels–
    Alder reaction, 123
norascaridole, 126
norcamphor, synthesis by Diels–Alder
    reaction, 139

olefins,
  aldehydes from, via hydroboration,
    232–3, 235–6
  anti-Markownikoff hydration, 224
  catalytic deuteration, 306
  catalytic hydrogenation, 304: migra-
    tion of double bonds, 306; selective,
    306–7, 321–6; stereochemistry, 314,
    315, 316, 317, 318
  cis-disubstituted, by reduction of
    acetylenes, 223, 307
  epoxidation, 264–8
  hydroboration, 211–19
  hydroxylation to 1,2-diols, 259–64,
    269, 270
  ketones from via hydroboration, 229–
    30, 232
  oxidation: chromic acid, 244; iodine
    and silver carboxylates, 263–4;
    per-acids, 264–8; permanganate,
    260–1; osmium tetroxide, 261–3
  oxidative cleavage, 279–80, 281, 282
  ozonolysis, 274–80
  reduction with diborane and organo-
    boranes, 218–23; with di-imide, 369
  reductive dimerisation, 228

olefins (*continued*)
synthesis: β-elimination reactions, 71–5; oxidative decarboxylation of carboxylic acids, 108–10; pyrolytic eliminations, 75–81; stereoselective: from allyl vinyl ethers, 112–14; from 1,2-diols, 111; by fragmentation reactions, 106–8; from phosphonbis-*N,N*-dialkylamides, 90–3; from vinyl-alanes and -boranes, 100–6; by Wittig reaction, 87–9
trans-disubstituted, from acetylenes, 345–46: by Wittig reaction, 81–90, *see also* Wittig reaction
see also ethylene derivatives
orbital symmetry control of Diels–Alder reaction, 155–6, 164
organoboranes, 207ff
1,4-addition to αβ-unsaturated carbonyl compounds, 234–6
carbonylation, 229–34
coupling, 228
cyclic, 220, 221
cyclisation by heat, 227
elimination from, 217
formation, 211–18, 221
fragmentation reactions, 107
isomerisation by heat, 225–6
oxidation, 218, 219, 224, 225
protonolysis, 223
reaction with α-bromo-ketones and -esters, 236–8
reaction with diazoketones, 238
reductions with, 207–10
ring compounds from, 215, 216
synthesis with, 222, 100–6, 107
vinylboranes, 100–1, 104–5, 221
organocopper complexes,
addition to αβ-unsaturated ketones, 41–5
reaction with organohalides, 47–8
role in copper-catalysed decomposition of diazo compounds, 58–9
role in coupling of terminal acetylenes, 49
osmium, catalyst for hydrogenation, 311
osmium tetroxide,
cleavage of olefins, with permanganate, 281
reaction with olefins, 261–3
9-oxabicyclo[2,2,1]hept-2-ene derivatives, synthesis, 141

1,2-oxazine derivatives, formation by Diels–Alder reaction, 123
oxidation,
alcohols: catalytic, 247–8; Cr^{VI} reagents, 244–7, 248; dimethyl sulphoxide, 254–5; *N*-halo-imides, 258; lead tetra-acetate, 197–201, 257; manganese dioxide, 249–50; Oppenauer oxidation, 256; ruthenium tetroxide, 281–2; silver carbonate, 252
aldehydes: argentic oxide, 251; per-acids, 290
allylic and benzylic alcohols, 248, 249–50
aromatic hydrocarbons, 242–3, 244
diols with silver carbonate, 252
ethers, to esters and lactones, 282
ketones, with per-acids, 287–90
olefins: hydroxylation, 259–64, 269; iodine and silver carboxylates (Prévost's reagent), 263–4; osmium tetroxide, 261–3, 281; ozone, 274; per-acids, 264–8; permanganate, 260–1; ruthenium tetroxide, 282; sodium periodate and permanganate, 279–80
organoboranes, 218, 219, 224–5, 227, 228
paraffin hydrocarbons, 241
photo-oxidation, 291–8
with nickel peroxide, 284–7
with ruthenium tetroxide, 281–3
oximes,
catalytic hydrogenation, 312–13
formation from organic nitrites, 182, 188, 189
oxirane derivatives, synthesis, 93–4
oxygen, addition to 1,3-dienes, 125–7
ozonolysis,
enol ethers, 277
mechanism, 277
olefins, 274ff
αβ-unsaturated ketones and acids, 279

paraffin hydrocarbons, oxidation with chromic acid, 241
pentacene,
addition of oxygen, 125
as diene in Diels–Alder reaction, 137
synthesis by Diels–Alder reaction, 133
per-acids, oxidation of,
aldehydes, 290

ketones, 287–90
olefins, 264–8
perhydro-9*b*-boraphenalylhydride, 211
trans-perhydroindanone, 232
phenols, *C*-alkylation, 9
4-phenyl-1,2,4-triazoline-3,5-diene, 259
phosphonate carbanions in Wittig reaction, 85
phosphonbis-*N*,*N*-dialkylamides in synthesis of olefins, 90–3
phosphoranes, *see* Wittig reagents
photocyclisation,
hexa-1,3,5-trienes, 63–5
stilbene derivatives and analogues,
64–5
photocycloaddition reactions,
olefins with enolised 1,5-diketones, 69
olefins and acetylenes with $\alpha\beta$-un-
saturated ketones, 66–8
photo-oxidation,
conjugated dienes, 126
olefins, 291–8: mechanism, 292, 295–7;
stereoselectivity, 294–6
photolysis,
aldehydes and ketones, 202, 203
alkyl and acyl azides, 179
N-halogenated amides, 194–7
N-halogenated amines, 173
organic hypohalites, 190–4
organic nitrites, 182–9
steroid nitrites, 187–9
photosensitised Diels–Alder reactions,
165–7
pinacol reduction, 334
α-pinene, photo-oxidation, 291, 296
platinum metal catalysts,
hydrogenation with, 301–12
hydrogenolysis with, 307, 309
poisoning, 302
preparation, 301
polycyclic aromatic hydrocarbons,
as dienes in Diels–Alder reaction, 136,
137
hydrogenation, 310
oxidation, 243
synthesis by: Diels–Alder reaction,
133, 134, 135; photocyclisation of
stilbenes, 64–5
polyenes,
in Diels–Alder reaction, 129–39
synthesis by Wittig reaction, 90
potassium permanganate, 239, 260–1,
279–80, 281

Prévost reaction, 263–4
propargylic alcohols,
reduction with lithium aluminium
hydride, 95, 356
stereoselective conversion into allylic
alcohols, 95–6
pyridine derivatives, formation by Diels–
Alder reaction, 123
pyrolytic *syn* eliminations in formation of
olefins, 75–81
pyrrolidine derivatives, synthesis, 173,
174–5, 179

quaternary ammonium salts, olefins
from, 71–5
quinones in Diels–Alder reaction, 118,
141

Raney nickel,
catalyst for hydrogenation, 302, 305,
308–10, 312, 313
desulphurisation with, 368–9
reduction, 299ff
acetylenes, 95, 218, 223, 307, 345–6
aldehydes and ketones, 208–9, 211,
310–12, 327–9, 332–5, 357–60
aromatic compounds, 308–10, 341–5,
350–1
1,3-dienes, 336
epoxides, 209, 270, 360–2
nitriles, oximes and nitro compounds,
312
olefins, 218, 223, 304–7, 321–6, 369–70
organo-halides, 349–50, 360, 362
reagents: aluminium alkoxides, 352;
alkali metal–alcohol, 332–5; alkali
metal–alcohol–liquid ammonia,
335–49; diborane, 207–9; di-imide,
369–70; lithium aluminium hydride,
353ff; lithium aluminium hydride–
aluminium chloride, 365–6; lithium
hydridotrialkoxyaluminates, 363–5;
lithium in alkylamines, 350–1;
metal and acid, 327–32; organoboranes, 210–11; sodium borohydride, 353; trialkyltin hydrides,
350
styrene derivatives, 341–2
$\alpha\beta$-unsaturated ketones, 307, 337–41
see also catalytic hydrogenation
reserpine, 153
retro Diels–Alder reaction, 144–5
rhodium, catalyst for hydrogenation, 305,
309, 310

rose-oxides, synthesis, 293
ruthenium, catalyst for hydrogenation, 305, 309, 310
ruthenium tetroxide as oxidising agent, 281–3

α-santalene, 55
sesquicarene, 59
seychellene, 18
shikimic acid, 129
silver carbonate, 252
Simmons–Smith reaction, 59
  stereochemical control by hydroxyl groups, 60–1
small-ring compounds, synthesis, 37, 62, 67–8, 109, 110
sodium borohydride,
  as reducing agent, 353, 355, 356, 362
  contrast with diborane, 208
sodium periodate in cleavage of olefins, 279–81, 282
squalene (all *trans*-), 95, 114
stereochemistry,
  alkylation of cyclic ketones, 27–9
  Birch reduction of αβ-unsaturated ketones, 338–341
  catalytic hydrogenation, 314–18
  Claisen rearrangement of allyl vinyl ethers, 112–14
  conjugate addition of organometallics to αβ-unsaturated ketones, 42
  Diels–Alder reaction, 147–60
  'ene' synthesis, 169–70
  epoxidation of olefins, 268
  hydroboration, 218, 219
  photo-oxidation of olefins, 294–6
  pyrolytic eliminations, 75–6
  olefin-forming β-eliminations, 73
  reduction: acetylenes, 223, 307, 345; cyclic ketones, 211, 332–3, 353, 357–60; with di-imide, 370
  Simmons–Smith reaction, 60–1
  Wittig reaction, 86–9
steroidal alcohols, reaction with lead tetra-acetate, 198–201
steroidal hypoiodites, photolysis, 193
steroidal nitrites, photolysis, 187–9
steroidal ketones, photolysis, 203
steroids, intramolecular attack on un-activated C–H bonds,
  by nitrogen radicals, 178
  by oxy radicals, 187–9, 191, 193, 198–201, 203

stilbene derivatives, photocyclisation, 64–5
sulphoxides, pyrolysis to olefins, 80–1

tetracyanoethylene,
  addition to bicyclo[2,2,1]heptadiene, 167
  as dienophile in Diels–Alder reaction, 116, 117
tetrahydrofuran derivatives, formation from,
  organic hypohalites by photolysis, 190, 191, 192, 193
  monohydric alcohols with lead tetra-acetate, 197–201
1,2,3,4-tetrahydropyridazine derivatives, formation via Diels–Alder reaction, 124
thallium salts,
  catalysts for coupling of Grignard reagents, 45
  in alkylation of 1,3-dicarbonyl compounds, 6
thexylborane, *see* 2,3-dimethylbut-2-ylborane
thiophen-1,1-dioxide Diels–Alder reactions, 143
toluene-*p*-sulphonylhydrazones, conversion into olefins, 110
trialkylboranes,
  1,4-addition to αβ-unsaturated carbonyl compounds, 234–6
  carbonylation, 229–34
  formation, 212
  reaction with α-bromo-ketones and -esters, 236–8
trialkylcarbinols from organoboranes, 229
trialkyltin hydrides, 350
triethyl phosphite, reaction with aromatic nitro compounds, 180, 181
tris(triphenylphosphine)chlororhodium,
  for hydrogenation of olefins, 321–4
  for decarbonylation of aldehydes, 324
tropane, 175
γ-tropolone, 70

unactivated C–H bonds,
  intramolecular attack by nitrogen radicals, 176, 194–5
  intramolecular attack by alkoxyl radicals, 182, 183, 184, 187–9, 193, 198–201, 201–5

intramolecular attack by nitrenes, 178–81
intramolecular attack by carbon radicals, 205
αβ-unsaturated acids,
from αβ-unsaturated aldehydes, 251
ozonolysis of, 279
stereoselective formation from acetylenes, 102
unsaturated alcohols,
from alkyl hydroperoxides, 201
βγ-, thermal decomposition, 169
see also allylic alcohols
αβ-unsaturated aldehydes,
from allylic alcohols, 248, 249–50, 256
epoxidation, 267
reduction with lithium aluminium hydride, 355
selective hydrogenation of olefinic bond, 311
see also αβ-unsaturated carbonyl compounds
αβ-unsaturated carbonyl compounds
1,4-addition of trialkylboranes, 234–6
as dienes in Diels–Alder reaction, 133
as dienophiles in Diels–Alder reaction, 117, 118, 120
reaction with sulphur ylids, 93–4
γδ-unsaturated carbonyl compounds, synthesis from allyl vinyl ethers, 112–13
αβ-unsaturated esters,
from allylic alcohols with manganese dioxide and cyanide ion, 250
synthesis by Wittig reaction, 85
γδ-unsaturated esters, synthesis by Claisen rearrangement, 113–14
unsaturated hydrocarbons, *see* olefins
αβ-unsaturated ketones,
1,4-addition of organometallic compounds, 41–5
α-alkyl-, 25–6
alkylation, 25–6
epoxidation, 267
formation by oxidation of allylic alcohols, 248, 249–50, 256, 259·
photo-addition of olefins and acetylenes, 67–70
oxidative cleavage, 282
ozonolysis, 279
reaction with enamines, 33
reduction with: alkali metal in liquid ammonia, 337–41; lithium aluminium hydride, 355

reductive alkylation, 24–5
selective reduction of carbonyl group, 352–66
selective reduction of olefinic bond, 307, 337–8
βγ-unsaturated ketones, synthesis of αα-dialkyl derivatives, 25–6

vinylalanes,
in stereoselective synthesis of olefins, 100–6
αβ-unsaturated acids and allylic alcohols from, 102
stereoselective synthesis of 1,3-dienes from, 105–6
vinylarenes as dienes in Diels–Alder reaction, 134
vinylboranes, use in stereoselective synthesis of olefins, 100–1, 104–5
vinyl bromides, stereoselective formation from terminal acetylenes, 101–2
vinylcycloalkenes as dienes in Diels–Alder reaction, 134
vinyl derivatives as dienophiles in the Diels–Alder reaction, 119, 120
vinyl halides,
reductive cleavage, 349
stereoselective synthesis, 100
vinyl iodides,
stereoselective formation, from Wittig reagents, 89
from vinyl alanes and vinylboranes, 100
vinylorganoboranes, formation from acetylenes, 221

Wittig reaction, 81–90
mechanism, 82–4
steric control, 87–9
steric course, 86–9
synthesis of olefins, scope, 89–90
using phosphonate carbanions, 85
Wittig reagents, 82–5
Wolff–Kishner reduction, 367
Wolff rearrangement of diazoketones, 61–2

Xanthates, pyrolysis to olefins, 75, 77–9

ylids,
in Wittig reaction, *see* Wittig reagents
sulphur ylids 93–4

zinc as reducing agent, 329, 330